Molecular and Integrative Toxicology

Series editor
Rodney R. Dietert, Department Microbiology & Immunology, Cornell University
College of Veterinary Medicine, Ithaca, New York, USA

Molecular and Integrative Toxicology presents state-of-the-art toxicology in a useful context. Volumes emphasize the presentation of cellular and molecular information aimed toward the protection of human or animal health or the sustainability of environmental systems.

More information about this series at http://www.springer.com/series/8792

Claude L. Hughes • Michael D. Waters
Editors

Translational Toxicology

Defining a New Therapeutic Discipline

Humana Press

Editors
Claude L. Hughes
Therapeutic Science & Strategy Unit
Quintiles, Inc.
Morrisville, NC, USA

Department of Obstetrics and Gynecology
Duke University Medical Center
Durham, NC, USA

Department of Mathematics
North Carolina State University
Raleigh, NC, USA

Michael D. Waters
Michael Waters Consulting
Hillsborough, NC, USA

ISSN 2168-4219 ISSN 2168-4235 (electronic)
Molecular and Integrative Toxicology
ISBN 978-3-319-27447-8 ISBN 978-3-319-27449-2 (eBook)
DOI 10.1007/978-3-319-27449-2

Library of Congress Control Number: 2016934930

Springer Cham Heidelberg New York Dordrecht London

Printed on acid-free paper

Humana Press is a brand of Springer
Springer International Publishing AG Switzerland is part of Springer Science+Business Media (www.springer.com)

Preface

While the scope, principles, and aims of the science of toxicology are broadly accepted, herein we are arguing that an even more expansive perspective is now needed. The modifier "translational" in its current biomedical sense means expediting research and development steps from basic science to human clinical studies with progression into therapeutic trials and iterative cycles of "lessons learned." Juxtaposition of "translational" with "toxicology" is our novel and challenging effort to characterize the research and environmental and public health opportunities that should exist for the development of protective, mitigative, or therapeutic interventions attributable to the human health impact of a wide range of environmental exposures.

As microbiology is the basic science underpinning clinical infectious diseases, we see toxicology and environmental medicine as a corollary. In the realm of microbiology, some infectious diseases are certainly best addressed by avoidance of exposure to the causative (infectious) agent. Since we live in a world in which we are intimately associated with and exposed to innumerable microorganisms, there are other inevitable exposures to infectious agents which sometimes do and sometimes do not cause clinical disease. The entire field of antibiotic research focuses on developing therapeutic interventions that may be used acutely as courses of pharmaceutical therapy to effect a true cure, or antimicrobial drugs may be used in a mitigative matter to enhance human health by providing long-term suppression of clinical disease that would otherwise result from the antecedent exposure to the causative infectious agent. If we now set aside the infectious environmental agents, the other chemical, physical, nutritional, and social exposures also pose major exposure-related health risks. For these exposures, toxicology is arguably the relevant area of basic science, while environmental medicine and public health are the corresponding areas of clinical science. Some exposures to toxicologic agents should be avoided as the best means for preventing consequent associated injury or diseases. In clinical practice settings such as in Poison Control Centers where cases of acute intoxication are managed, clinicians implement interventions aimed at resolving the intoxication in order to save the individual from mortality or harm. However, as is the case for numerous infectious agents, many toxicant exposures are essentially unavoidable. In these nonacute instances, the toxicologic correlate to subacute, subchronic, or chronic infectious diseases has been largely ignored as an area of scientific research for therapeutic development. Given the certainty that gene-environment interactions are central to understanding human health and disease across the life span, we implore our readers to consider the erudite contributions of our multiple authors of the ensuing chapters and think deeply about the ways in which all of you may be able to contribute to making advances in this newly defined field of scientific and medical research to enhance human health in the future.

Durham and Morrisville, NC, USA Claude L. Hughes
Hillsborough, NC, USA Michael D. Waters

Contents

Contributors

David Allen Science and Strategy, Integrated Laboratory Systems, Inc. (ILS), Durham, NC, USA

Rais Ansari Department of Pharmaceutical Sciences, College of Pharmacy, Nova Southeastern University, Fort Lauderdale, FL, USA

Sharnjit Baines Department of Obstetrics & Gynaecology, McMaster University, Hamilton, ON, Canada

Linda Booij Department of Psychology, Concordia University, Montreal, QC, Canada

CHU Sainte-Justine, Montreal, QC, Canada

Leon Chalil Department of Biochemistry and Biomedical Sciences, Obstetrics and Gynecology, and Pediatrics, McMaster University, Hamilton, ON, Canada

Fatima Crispi BCNatal | Barcelona Center for Maternal Fetal and Neonatal Medicine, Hospital Clínic and Hospital Sant Joan de Déu, Universitat de Barcelona, Barcelona, Spain

Edward L. Croom BOV Solutions, INC, Statesville, NC, USA

Elena Demicheva BCNatal | Barcelona Center for Maternal Fetal and Neonatal Medicine, Hospital Clínic and Hospital Sant Joan de Déu, Universitat de Barcelona, Barcelona, Spain

M.A. Dominguez Facultad de Medicina Veterinaria y Zootecnia, Universidad Autónoma de Tamaulipas, Cd. Victoria, Tamaulipas, Mexico

Kelly K. Ferguson Department of Environmental Health Sciences, University of Michigan School of Public Health, Ann Arbor, MI, USA

Warren G. Foster Department of Obstetrics & Gynaecology, McMaster University, Hamilton, ON, Canada

H.C. Furlong Department of Obstetrics & Gynaecology, McMaster University, Hamilton, ON, Canada

M.T. Guerra Department of Morphology, Biosciences Institute, UNESP, Botucatu, São Paulo, Brazil

Cathrine Hoyo Department of Biological Sciences, and Center for Human Health and the Environment, North Carolina State University, Raleigh, NC, USA

Claude L. Hughes Therapeutic Science & Strategy Unit, Quintiles, Inc., Morrisville, NC, USA

Department of Obstetrics and Gynecology, Duke University Medical Center, Durham, NC, USA

Department of Mathematics, North Carolina State University, Raleigh, NC, USA

Kazim Husain Department of Physiology, Pharmacology and Toxicology, Ponce School of Medicine and Health Sciences, Ponce, PR, USA

Darcy Roberto A. Lima Instituto de Neurologia Deolindo Couto, Universidade Federal do Rio de Janeiro, Rio de Janeiro, Brazil

John D. Meeker Department of Environmental Health Sciences, University of Michigan School of Public Health, Ann Arbor, MI, USA

Susan K. Murphy Department of Obstetrics and Gynecology, Duke University Medical Center, Durham, NC, USA

Iyabo Obasanjo African Presidential Center, Boston University, Boston, MA, USA

Nadine Provençal Max Planck Institute of Psychiatry, Munich, Germany

J.C. Sadeu Department of Obstetrics & Gynaecology, McMaster University, Hamilton, ON, Canada

Roseane Maria M. Santos Department of Pharmaceutical Sciences, South University School of Pharmacy, Savannah, GA, USA

Toby Schonfeld Program in Human Research Ethics and Oversight, Office of the Science Advisor, U.S. Environmental Protection Agency, Washington, DC, USA

David A. Skaar Department of Biological Sciences, and Center for Human Health and the Environment, North Carolina State University, Raleigh, NC, USA

Deborah M. Sloboda Department of Biochemistry and Biomedical Sciences, Obstetrics and Gynecology, and Pediatrics, McMaster University, Hamilton, ON, Canada

Farncombe Family Digestive Health Research Institute, Hamilton, ON, Canada

Moshe Szyf Department of Pharmacology and Therapeutics, McGill Faculty of Medicine, Montreal, QC, Canada

Richard E. Tremblay School of Public Health, University College Dublin, Belfield, Dublin, Ireland

Université de Montréal, Montréal, QC, Canada

Michael D. Waters Michael Waters Consulting, Hillsborough, NC, USA

About the Editors

Claude L. Hughes M.D., Ph.D., is the Executive Director in the Therapeutic Science & Strategy Unit within Quintiles Inc. He is board-certified in obstetrics and gynecology and reproductive endocrinology and infertility. He is a Consulting Professor in the Department of Obstetrics and Gynecology, Duke University Medical Center, and an Adjunct Professor in the Department of Mathematics, North Carolina State University, and in the Department of Pathology/Comparative Medicine, Wake Forest University School of Medicine.

Since joining Quintiles in 2001, Dr. Hughes has served as a Medical Advisor on clinical trials or in due diligence assessment teams that evaluated pharmaceuticals, devices, or tests for several medical indications. Therapeutic areas for clinical trials and/or product evaluation have included the following:

1. Reproductive medicine—female sex hormone therapy, male sex hormone therapy, contraception, infertility, in vitro fertilization-embryo transfer, polycystic ovary syndrome, SERMs, overactive bladder/urinary incontinence, endometriosis, preterm labor, ovarian cancer, cervical cancer, microbicides, dysmenorrhea, gynecologic infections, female and male sexual dysfunction, erectile dysfunction, BPH/LUTS, delayed puberty in boys, and primary and secondary hypogonadism in boys.
2. General endocrinology, metabolism, and cardiovascular diseases—osteoporosis prevention and treatment, diabetes mellitus, thyroid hormone replacement, growth hormone therapy in children, hypertension (including use of ambulatory blood pressure monitoring, ABPM), and chronic renal disease.
3. Other medical conditions—GERD, opioid-induced constipation, rheumatoid arthritis, sleep disorders, migraine headaches, neuropathic pain, and psoriasis.
4. Laboratory testing methods—genital cytologic screening and molecular biomarker screening.
5. Drug delivery—new systems for transdermal, buccal, or vaginal administration of drugs.

Prior to joining Quintiles, Dr. Hughes held academic, research, administrative, and clinical practice positions for 15 years in divisions of reproductive endocrinology and infertility in departments of obstetrics and gynecology and clinical and research centers within university-affiliated medical centers. His academic research emphasized investigating the effects of natural and man-made sex hormonelike compounds in the diet in laboratory animal and human studies. Continuing collaborative academic research is supported by a National Science Foundation grant in the Department of Mathematics at North Carolina State University (Dr. Hughes is the Co-principal Investigator).

Since April 2008, Dr. Hughes has served as a Member of the Metabolic Disorders Steering Committee (MDSC) of the Biomarkers Consortium, the public-private biomedical research partnership managed by the Foundation for the National Institutes of Health (FNIH) that includes government, industry, patient advocacy groups, and other nonprofit private sector organizations such as the National Institutes of Health, the Food and Drug Administration, and the Pharmaceutical Research and Manufacturers of America.

Michael D. Waters is a former Government Scientist with more than 35 years of experience in research and research management positions at US EPA and NIH/National Institute of Environmental Health Sciences in Research Triangle Park, NC, USA. He retired as Chief Scientist, Emeritus, from Integrated Laboratory Systems (ILS), Inc., in 2013. He served as Assistant Director of the NIEHS National Center for Toxicogenomics from 2000 to 2007 and was responsible for a major initiative to develop the Chemical Effects in Biological Systems (CEBS) toxicogenomics knowledge base. CEBS is now the data repository of the US National Toxicology Program and is being utilized by the NIH and other government agencies as well as the academic, industrial, and international regulatory scientific communities for the deposition, retrieval, and interpretation of toxicology and toxicogenomics data. At NIEHS, he served on the NIH Bioinformatics and Computational Biology Roadmap Working Group; the US FDA Advisory Committee for Pharmaceutical Science, Pharmacology and Toxicology Subcommittee; the Toxicogenomics and Risk Assessment Committee of the International Programme on Chemical Safety (IPCS); the Advisory Board of the Microarray Gene Expression Data (MGED) Society; and the Scientific Advisory Board of the Rat Genome Database (RGD). For 30 years he directed research in cellular pathology, biochemistry, and genetic toxicology in various capacities at the EPA in RTP, NC. He also served EPA as an Assistant Laboratory Director with programmatic responsibility for international, waste, and multimedia research programs. At EPA he conceived, designed, and helped to develop the EPA Gene-Tox database, now hosted by the National Library of Medicine. He also developed the EPA/International Agency for Research on Cancer (IARC) Genetic Activity Profile (GAP) database, which formed the basis for the use of short-term tests in the evaluation of presumptive human carcinogens by the IARC. His research interests have centered on the evaluation of chemically induced mutations and altered molecular expression in the etiology of cancer and genetic disease. A widely published scientist, he has edited *Mutation Research-Reviews* for nearly 20 years and has held adjunct professorships at both the University of North Carolina at Chapel Hill and Duke University. He has served as President of the Environmental Mutagen Society and the International Association of Environmental Mutagen Societies. The databases he developed and a number of his publications are recognized as important advances that have significantly influenced the fields of genetic toxicology, carcinogenesis, toxicogenomics, and risk assessment.

Part I
Introduction

Chapter 1
The Opportunity to Translate Developmental Toxicology into a Therapeutic Discipline

Claude L. Hughes, Michael D. Waters, David Allen, and Iyabo Obasanjo

Abstract Since presentation of our view of translational developmental toxicology in 2013, numerous investigations from a wide range of approaches have added to the evidence underlying our core premise; namely that as the potential adverse developmental effects from a range of exposures are progressively defined, preventative or mitigative therapies can be conceived, assessed and ethically implemented. The spectrum of studies reported in recent years span a wide range of exposure categories and various developmental outcomes ranging from molecular and cellular to the organismal level up to and including human behavior. Since human exposures to chemicals, physical agents and social factors are inevitable, the human fetus is subject to effects that can have lifelong consequences. In order to apply the translational concept to developmental toxicology, established or potential therapeutic obstetrical, neonatal, childhood and adolescent interventions will be required. Those that undergo testing during developmentally sensitive intervals will likely derive from generally-regarded-as-safe (GRAS) or well-established/repurposed pharmaceutical options. Ultimately if we are to translate environmental health discoveries into safe and effective interventions, we must assert and characterize valid, applicable

C.L. Hughes (✉)
Therapeutic Science & Strategy Unit, Quintiles, Inc.,
5927 S. Miami Blvd., Morrisville, NC 27560, USA

Department of Obstetrics and Gynecology, Duke University Medical Center, Durham, NC, USA

Department of Mathematics, North Carolina State University, Raleigh, NC, USA
e-mail: claude.hughes@quintiles.com

M.D. Waters, Ph.D.
Michael Waters Consulting, 210 N. Wake Street, Hillsborough, NC, USA
e-mail: mdwaters@centurylink.net

D. Allen, Ph.D.
Science and Strategy, Integrated Laboratory Systems, Inc. (ILS), Suite 800,
601 Keystone Park Drive, Research Triangle Park, Durham, NC 27713, USA
e-mail: dallen@ils-inc.com

I. Obasanjo, D.V.M., Ph.D.
Research Fellow, African Presidential Center, Boston University, Boston, MA 02215, USA
e-mail: iobas427@gmail.com

C.L. Hughes, M.D. Waters (eds.), *Translational Toxicology*, Molecular and Integrative Toxicology, DOI 10.1007/978-3-319-27449-2_1

therapies such as GRAS treatments and eventually "ethical pharmaceuticals" for the protective care of these highly vulnerable young people. We can create a safe and efficacious environmental health portfolio of interventional options to improve human health that include both reduction/avoidance of exposure and specific preventative/mitigative/restorative therapeutics. In this chapter we will broadly update new insights that have been gained over the last 2 years regarding the progress of translational developmental toxicology toward becoming a therapeutic discipline.

Keywords Translational toxicology developmental exposures • Obstetrical, neonatal, childhood and adolescent therapies

1.1 Introduction

We have recently asserted that the relatively new term "translational toxicology" should encompass both the existing principles of toxicology and epidemiology but be driven by the aim of developing safe and effective interventions to include prevention, mitigation or reversion of adverse human health effects of exposures. Exposure assessment, basic toxicology and development of certain categories of mathematical models to characterize exposure-outcome relationships are not new areas of research. However, overtly integrating these research processes in order to conceive, assess and validate ethical interventions to prevent, mitigate or reverse adverse effects of environmental exposures is a novel opportunity. The research and development aims within translational toxicology should be the creation of a safe and effective portfolio of interventional options to improve human health that goes beyond merely minimizing exposure(s) so that practitioners have specific preventative/mitigative/restorative therapeutics for human clinical use.

Since presenting our view of translational developmental toxicology in 2013, numerous investigations from a wide range of approaches have added to the evidence underlying our core premise; namely that as the potential harms from developmental effects of a range of exposures are progressively defined, preventative or mitigative therapies can be conceived, assessed and ethically implemented. The spectrum of studies reported in recent years span a wide range of exposure categories and various developmental outcomes ranging from molecular and cellular to the organismal level up to and including human behavior. Since human exposures to chemicals, physical agents and social factors are inevitable, the human fetus is subject to effects that can have lifelong consequences. In order to apply the translational concept to developmental toxicology, established or potential therapeutic obstetrical, neonatal, childhood and adolescent interventions will be required. Those that undergo testing as developmental translational toxicology therapies during developmentally sensitive intervals will likely derive from generally-regarded-as-safe (GRAS) or well-established/repurposed pharmaceutical options. Ultimately if we are to translate environmental health discoveries into safe and effective interventions, we must assert and characterize valid, applicable therapies such as GRAS treatments and eventually "ethical pharmaceuticals" for the protective care of these highly vulnerable young people. We can create a safe and efficacious environmental health portfolio of interventional options to improve human health that include both reduction/avoidance of exposure and specific preventative/mitigative/restorative therapeutics. In this chapter we will broadly update new insights that have been gained over the last 2 years regarding the progress of translational developmental toxicology toward becoming a therapeutic discipline.

In general, contemporary translational research in all areas of medicine relies upon identification, validation and use of various biomarkers for execution of research and development programs. In human health research, the principal outcomes are clinical endpoints (how a patient feels, functions or survives). Short of directly measuring clinical endpoints, surrogate endpoints can be measured, but at a minimum, biomarkers must be explicitly validated as predictors of surrogate endpoints to be reasonably interpreted as indicators of clinical endpoints. The validity of such linkage for many (novel) biomarkers can be challenged, but broad scientific-medical agreement can be reached for many others.

For environmental health research to be translated into actionable preventative medicine and a therapeutic discipline, we must transform the usual logic of "translational" health research into an approach that can plausibly establish in humans a causal relationship between exposure and effect plus development and validation of safe and efficacious preventative/mitigative/restorative therapeutics in humans. We assert that the following clinical study design elements must be included in translational developmental toxicology research and development projects:

(1) Indicator(s) of timing of exposure
(2) Biomarker(s) of current exposure
(3) Molecular/cellular biomarker(s) of systemic effects
(4) Imaging biomarker(s) of target tissue effects
(5) Molecular biomarkers of target tissue effects
(6) Surrogate clinical outcomes
(7) Clinical outcomes
(8) Public health outcomes

Without such a spectrum of demonstrably linked, biologically plausible observations, mitigation strategies are difficult to propose and probably unethical to implement. With such a set of observations, mitigation, compensatory or therapeutic strategies can be conceived, and ethical interventional clinical trials can be designed and conducted.

An abiding challenge is to find some rational basis for narrowing the scientific focus in translational toxicology/translational environmental health. We simply cannot do everything for every possible exposure scenario, so how can we define a manageable scope of research themes with the greatest promise of translating discovery into clinical interventions? The long-standing concept in developmental toxicology is that there are critical lifestages or certain key intervals in the lifespan which are Windows of Susceptibility (of adverse effects due to an exposure) (Kimmel et al. 2011). The same biological reasoning that these critical life stages are particularly susceptible to adverse perturbation should mean that these same key intervals are likely also to be prime targets for intervention which we have called "Windows of Responsivity" (Hughes et al. 2013a, b), to minimize, negate or reverse antecedent or ongoing exposure effects. The preeminent categories of exposed patients that should be addressed via the iterative process of translational toxicology research, advanced modeling and then testing of preventative/mitigative/restorative therapeutic interventions is the exposed gravid woman/mother-infant pairs (the early development window) and the span of the neonatal, childhood and adolescent intervals.

While "Fetal therapy is broadly defined as any intervention administered to or via the mother with a primary indication to improve perinatal or long-term outcomes for the fetus or newborn" (Pauli and Repke 2013), almost reflexively due to concern about unknown consequences of fetal exposures, the "precautionary principle" is generally assumed to be applicable regarding exposures of gravid women, neonates, children and adolescents to exogenous agents of diverse types (Hughes et al. 2013a, b). Beyond immunizations, care of maternal acute and chronic illnesses and specific obstetrical conditions (e.g., pre-eclampsia, gestational diabetes), there are only a few dietary/nutritional or pharmacological treatments that are commonly given during early development where a primary aim is to effect benefits that will obviously or likely primarily manifest later in the lifespan (Hughes et al. 2013a, b). It is noteworthy that over the last 5 years, there have been 25 drugs approved by the US FDA for use in pediatrics/neonatology, but only one of these (a vaccine) is explicitly intended for preventative purpose (Table 1.1). This recent track record with approval of drugs for use in these developmentally sensitive intervals emphasizes that research and implementation in the near-term of protective/mitigative therapeutic interventions are most likely to be derived from nutritional, dietary supplementation or well-established/repurposed pharmaceutical sources. It seems both likely and reasonable that specific novel pharmaceutical interventions will be farther into the future.

Table 1.1 Acute and chronic disease treatment focus for FDA approved drugs for pediatrics/neonatology between 2010-early 2015[a]

Drugs approved in 2015
Bexsero (Meningococcal Group B Vaccine); For the treatment of invasive meningococcal disease caused by serogroup B
Drugs approved in 2014
Alprolix [Coagulation Factor IX (Recombinant), Fc Fusion Protein]; For the treatment of hemophilia B
Arnuity Ellipta (fluticasone furoate inhalation powder); For the treatment of asthma
Qudexy XR (topiramate); For the treatment of partial onset and primary generalized tonic-clonic seizures and Lennox-Gastaut Syndrome
Vimizim (elosulfase alfa); For the treatment of Mucopolysaccharidosis type IVA
Drugs approved in 2013
Actemra (ocilizumab); For the treatment of Polyarticular Juvenile Idiopathic Arthritis
Ilaris (canakinumab); For the treatment of Systemic Juvenile Idiopathic Arthritis
Kineret, anakinra; For the treatment of Cryopyrin-Associated Periodic Syndromes
Kynamro (mipomersen sodium); For the treatment of homozygous familial hypercholesterolemia
Ravicti (glycerol phenylbutyrate); For the treatment of pediatrics and adults with urea cycle disorders
Tivicay (dolutegravir); For the treatment of HIV-1 in adults and children over 12 years of age
Trokendi XR (topiramate); For the treatment of partial onset, tonic-clonic and Lennox-Gastaut Syndrome seizures
Drugs approved in 2012
Fycompa (perampanel); For the treatment of partial-onset seizures with or without secondarily generalized seizures
Oxtellar XR (oxcarbazepine extended release); For the adjunctive therapy of partial seizures in adults and in children 6 years to 17 years of age
Qnasl (beclomethasone dipropionate) nasal aerosol; For the treatment of seasonal and perennial allergic rhinitis
Quillivant XR (methylphenidate hydrochloride); For the treatment of Attention Deficit Hyperactivity Disorder
Sklice (ivermectin) lotion; For the treatment of head lice
Surfaxin (lucinactant); For the treatment of respiratory distress syndrome in premature infants
Drugs approved in 2011
Actemra (tocilizumab); For the treatment of systemic juvenile idiopathic arthritis
Daliresp (roflumilast); For the treatment of chronic obstructive pulmonary disease
Onfi (clobazam); For the adjunctive treatment of seizures associated with Lennox-Gastaut syndrome
Drugs approved in 2010
Cuvposa (glycopyrrolate); For the treatment of chronic severe drooling in pediatrics with neurologic conditions
Dulera (mometasone furoate + formoterol fumarate dihydrate); For the treatment of asthma
Kapvay (clonidine hydrochloride); For the treatment of attention deficit hyperactivity disorder
Veltin (clindamycin phosphate and tretinoin); For the treatment of acne vulgaris

[a]Of these 25 approvals, only one is in fact primarily intended to prevent later disease outcomes, the vaccine for meningococcal Group B vaccine https://www.centerwatch.com/drug-information/fda-approved-drugs/therapeutic-area/15/pediatrics-neonatology

1.2 Key Concepts, Definitions and Needs

1.2.1 The Developmental Origins of Health and Disease (DOHaD) Concept

Long-established principles of teratology acknowledge that exposures of many sorts in early development may adversely affect the morphology and functional normality of exposed individuals. The spectrum of exposures that impacts development spans multiple physical, chemical, nutritional, infectious and social factors. In terms of defining potential molecular mechanisms by which organizational effects occur, the potential complexity is well illustrated by the listing of molecular targets and ligands in the IUPHAR/BPS Guide to Pharmacology 2014 (Table 1.2). Additionally, other mechanisms such as mutagenesis and epigenetic modifications appear to play additional key roles but we will only be able to modestly address epigenetic effects later in this current chapter.

To further highlight how diverse potential mechanisms may be that impact development, even the endogenous proteins molecules involved in transporting exogenous molecules may play a role in mediating development *per se*. In a recent paper, Nigam (2015) profiles some potential roles of the endogenous functions of the drug transporters ATP-binding cassette (ABC) transporter and solute carrier (SLC) transporter families in mammalian development. Since ABC and SLC transporters function in the handling of diverse substrates, including metabolites, antioxidants, signaling molecules, hormones, nutrients and neurotransmitters, these transporters may be part of a larger system of remote communication ('remote sensing and signaling') between cells, organs, body fluid compartments and perhaps even separate organisms. The solute carrier (SLC) and ATP-binding cassette (ABC) transporters are highly expressed in developing neural and other tissues and changes during organ development and thus may play a part in mammalian morphogenesis. Moreover, as the placenta expresses many drug transporters, it is probable that maternal small molecules, of physiological as well as toxicological significance, cross the maternal–fetal barrier in mammals via a set of SLC and ABC transporters. In developing animal organs such as the kidney and liver, the expression of drug transporters and drug-metabolizing enzymes is relatively low at birth and then increases considerably during the postnatal period and through juvenile stages. Low drug transporter expression may be a particularly important issue in premature infants since past studies indicate that there may be a postnatal 'developmental window' of regulation for renal drug elimination (Nigam 2015).

In recent decades there has been more of a focus on understanding how developmental exposures might impact health in more subtle ways across the lifespan. With heightened focus on the developmental effects of hormone-like compounds where likely modes of actions are often well understood, plausible causal pathways between early exposure, changes in system or tissue function and later disease risk has helped recast concepts in this field.

Table 1.2 Molecular targets and ligands: excerpted and modified from the IUPHAR/BPS Guide to Pharmacology[a]

Target class	Number of human molecular targets
7TM receptors	394
G protein-coupled receptors including orphans	388
Orphan G protein-coupled receptors[b]	129
Other 7TM proteins	6
Nuclear hormone receptors	48
Catalytic receptors	240
Ligand-gated ion channels	84
Voltage-gated ion channels	141
Other ion channels	47
Enzymes	1138
Transporters	508
Other protein targets	108
Total number of targets	2708
Chemical class	**Number of ligands**
Synthetic organics	4415
Metabolites	583
Endogenous peptides	730
Other peptides including synthetic peptides	1174
Natural products	217
Antibodies	67
Inorganics	34
Approved drugs	1176
Withdrawn drugs	61
Drugs with INNs	1691
Labelled ligands	580
Total number of ligands	7220

[a]Database content for version 2014.3 released 5th Nov 2014
[b]Orphans are defined as proteins having similarity to receptors but whose endogenous ligands have not yet been conclusively identified

Across the twentieth century several decades of historical studies investigated potential toxicities attributed to hormonal exposures and these insights built the foundation for our more contemporary research regarding hormone mediated developmental effects (McLachlan 1980, 1985; Kincl 1990). The term endocrine-disrupting chemicals (EDCs) has come into use to define a structurally diverse class of synthetic and natural compounds that possess the ability to alter various components of the endocrine system and potentially induce adverse health effects in exposed individuals and populations (Henley and Korach 2006). Research on these compounds has shown that they may act via a variety of both nuclear receptor-mediated and non-receptor-mediated mechanisms to modulate various different components of the endocrine system. Thus some of the either striking or subtle changes in tissue morphology, physiology, and behaviors produced by exposure to exceedingly low doses of such compounds may be endocrine *per se* or maybe due to other very specific molecular or cellular modes of action that do not relate to hormone receptor mediation.

There is a relatively large body of evidence showing that a wide range of EDCs have developmental effects (Vandenberg 2014) and it is plausible that such effects may adversely affect health later in life. A particularly cogent description of this contemporary concept of developmental toxicology is the Developmental Origins of Health and Disease (DOHaD) concept. Barouki et al. (2012) noted that during prenatal and early postnatal life when cell differentiation and specific tissue formation are particularly dynamic, these organizational processes are highly sensitive to environmental factors, such as nutrients, environmental chemicals, drugs, infections and other stressors and in turn alterations in physiological functions can be induced and may lead to the development of non-communicable disease conditions. Their suggestion was that many of the major diseases that have increased substantially in prevalence over the last 40 years may be related in part to developmental factors associated with either nutritional imbalance or exposures to environmental chemicals. The DOHaD concept offers a broad perspective that endorses both policy and public health responses to focus research and disease prevention strategies on early development (*in utero* and during the first years of postnatal life) to reduce potentially adverse consequences for health later in life.

Since hormonal pathway modulation by use of agonists and/or antagonists is a potential strategy to mitigate adverse developmental effects of EDCs (Hughes et al. 2013a, b), we suggest that the experience with endocrine physiology/toxicology is a good template for advancing the general DOHaD concept into an active translational discipline. In essence, a wide range of other developmental exposures such as the *in utero* metabolic milieu, maternal obesity, maternal (gestational) illnesses, maternal substance abuse, maternal/neonatal/childhood exposures to environmental chemicals, pharmaceuticals, over-the-counter drugs, herbal and other bioactive naturally occurring compounds, social stressors, oxidative stress, micronutrient deficiencies, etc. may all be seen as definable exposures whose modes of action can be well-characterized and thus be potentially modified by protective interventions that go beyond active reduction in exposures to some of these factors.

1.2.2 Illnesses During Pregnancy and Drug Use

There is nothing new in the idea that maternal illnesses during pregnancy poses a fetal risk and that use of pharmaceutical drugs to manage those illnesses may often also create a risk for the fetus. Nevertheless, contemporary awareness of the issue of pharmaceutical use during pregnancy continues to be raised and thoughtfully analyzed as investigators and caregivers strive to become more sophisticated about assessment of both benefits and risks for both the mother and the fetus/neonate (Honein et al. 2013). Rates of chronic disease are rising and women are having children at an older age. On one hand, there is a common belief that medication use during pregnancy is unhealthy and unwise, but on the other hand failing to take medication, taking inappropriate doses or not being prescribed a needed treatment can be harmful or fatal for a mother or her unborn child. Today, an American woman will take at least one and on average four medications during her pregnancy. Health care providers, though, often have inadequate information about how a drug works during pregnancy because most medications have not been evaluated in pregnant women (Nolan et al. 2014).

Medications used during pregnancy range from over-the-counter antacids to prescription medications for life-threatening chronic conditions. The safety of use during pregnancy is not always clear because the majority of medications lack sufficient data to appropriately evaluate their teratogenicity until decades after initial marketing. Both acute and chronic medical conditions that are relatively common in reproductive-aged women can require treatment during pregnancy, such as infections, cough and cold symptoms, allergies, depression, asthma, thyroid disorders, diabetes and migraines. However, given that the prevalence of unintended pregnancies in the USA is about 50 %, depending on the timing of pregnancy recognition, women may or may not be aware of their pregnancy when taking medications for these conditions.

Although improving the information available for all medications that might be used in pregnancy would be ideal, priority should be given to those conditions and medications for which more appropriate use is likely to have the greatest public health impact. Some key factors to consider in assessing the potential public health impact include the prevalence of the medical condition among reproductive-aged women, the prevalence of medication use for the maternal condition under consideration, the severity of the maternal medical condition and the anticipated necessity of medication for appropriate treatment of the condition. For example, about 11 % of U.S. reproductive-aged women reported a major depressive episode within the past 12 months; prescription medication treatment to manage the condition in women with a recent major depressive episode was reported by about 40 % of pregnant women and 47 % of nonpregnant reproductive-aged women. There are several medication options available and the fetal risk might vary between these options. Furthermore, stopping antidepressant medications during pregnancy is not a reasonable option for many women given the necessity of this treatment for maternal health. Similarly, an estimated 8–10 % of U.S. reproductive-aged women and 8–9 % of pregnant women have a current diagnosis of asthma. Although 88 % of pregnant women with a current diagnosis of asthma reported asthma symptoms during pregnancy, 41 % reported no use of asthma medication during pregnancy (Honein et al. 2013).

Addressing safer use of medication during pregnancy requires a partnership between public health and clinical medicine with a common goal of improving the health of women and their fetuses/infants. For many relatively common maternal conditions, there is a need to better understand the safety or risk of using specific medication alternatives during pregnancy (Honein et al. 2013).

1.2.3 Hypothyroidism, Hyperthyroidism, and Thyroid Autoimmunity in Pregnancy

As a specific potential source of risk to the fetus, it has long been known that maternal thyroid disease during pregnancy, its treatment, and potential environmental exposures to compounds that impact thyroid function all raise concerns about fetal and neonatal well-being. Thyroid autoimmunity has been associated with increased rate of pregnancy loss, recurrent miscarriage, and preterm delivery. Overt hyperthyroidism and hypothyroidism are responsible for adverse obstetric and neonatal events. Gilbert et al. (2012) showed that children born with normal thyroid function, but who experienced thyroid hormone (TH) insufficiency in utero, display subtle cognitive impairments and abnormalities in brain imaging and noted different patterns of cognitive effects resulting from prenatal versus postnatal TH insufficiency. Several studies suggest that either subclinical hypothyroidism or thyroid autoimmunity increase the risk of complications with one randomized controlled trial showed that pregnant women with subclinical hypothyroidism benefit from treatment in terms of obstetric and neonatal complications, whereas another study demonstrated no benefit in the intelligence quotient of babies born to women with subclinical hypothyroidism (Negro and Stagnaro-Green 2014).

Current guidelines agree that overt hyperthyroidism and hypothyroidism need to be promptly treated and that as potential benefits outweigh potential harm; subclinical hypothyroidism also requires substitutive treatment. The issue of universal thyroid screening at the beginning of pregnancy is still a matter of debate, and aggressive case-finding is supported (Negro and Stagnaro-Green 2014).

1.2.4 Nutritional Guidance Across the Preconception Through Postpartum Period

Since we are predicting that nutrition, dietary supplementation and other GRAS interventions are likely to be the first therapeutic options in translational developmental toxicology, we must first acknowledge that the core diet should be one that promotes maternal wellness, fetal well-being and neonatal/early childhood health. An authoritative current statement is that of the U.S. Academy of Nutrition and Dietetics (Kaiser and Campbell 2014). Their detailed position states that women of childbearing age should adopt a lifestyle optimizing health and reducing risk of birth defects, suboptimal fetal development, and chronic health problems in both mother and child. Components leading to healthy pregnancy outcome include healthy pre-pregnancy weight, appropriate weight gain and physical activity during pregnancy, consumption of a wide variety of foods, appropriate vitamin and mineral supplementation, avoidance of alcohol and other harmful substances, and safe food handling. Nutrition assessment needs to encompass changes in anthropometric, biochemical, and clinical indicators throughout pregnancy. Pregnant women should gain weight according to the 2009 Institute of Medicine Guidelines. Energy needs are no higher than the Estimated Energy Requirement for nonpregnant women until the second trimester; thereafter, the extra energy need per day is 340 kcal and 452 kcal in the second and third trimesters, respectively. Pregnant women can be guided to select a food plan based on age, physical activity, trimester, weight gain, and other considerations. Women are encouraged to participate in at least 150 min of moderate-intensity aerobic activity spread throughout the week or 30 min of moderately intense exercise on most days of the week. When good food choices are made, food consumption to meet extra energy needs and the increased absorption and efficiency of nutrient utilization that occurs in pregnancy are generally adequate to meet most nutrient needs. However, vitamin and mineral supplementation may be important in vulnerable cases including food insecurity; alcohol, tobacco, or other substance dependency; anemia; strict vegetarian (vegan) diet; or poor eating habits. Multiple strategies are needed to support healthy lifestyles for all women, from preconception through the postpartum period.

1.2.5 New FDA Mandated Labeling for Pregnant and Breastfeeding Women

For any therapy intended for use in these developmentally-sensitive intervals which would undergo review and approval by the US Food and Drug Administration (FDA), labeling for use will be subject to new drug labeling rules to clarify risk for pregnant and breastfeeding patients. Some of the recent changes (Cajigal et al. 2015) include:

- The "pregnancy," "labor and delivery," and "nursing mothers" subsections will be replaced with subsections titled "pregnancy," "lactation," and a new subsection titled "females and males of reproductive potential."
- The pregnancy and lactation subsections will include information from available human and animal studies, known or potential maternal or fetal adverse reactions, and dose adjustments needed during pregnancy and the postpartum period.
- The subsection on females and males of reproductive potential will include information on pregnancy testing, birth control, and the drug's possible effect on fertility when needed.

As the field of translational toxicology begins to develop, there will need to be clarity regarding the meaning of "Adverse Event" as used in drug development versus the term "Adverse Effect" as used in environmental health and safety assessments. The term "Adverse Event" as specified by the U.S. Food and Drug Administration in the Code of Federal Regulations part 21 CFR 312.32(a) means any untoward medical occurrence associated with the use of a drug in humans, whether or it is not considered drug-related, while "Adverse Effect" is defined by the US EPA (US EPA 2012) as "a biochemical change, functional impairment, or pathologic lesion that affects the performance of the whole organism, or reduces an organism's ability to respond to an additional environmental challenge."

Therefore as these protective therapies are assessed, developed and utilized, the aim will be to minimize adverse events associated with use of the therapy in order to prevent or mitigate the occurrence of adverse effects in the offspring. This is a distinction with a real difference in meaning.

1.3 Recent Studies: Developmental Exposures and Subsequent Adverse Health Risks

1.3.1 Effects of Maternal Obesity and Maternal Diabetes

Several recent studies regarding the effects of maternal obesity and/or maternal diabetes on the offspring and apparent epigenetic mode of action will be described in Sect. 1.4.2. In more general terms, some recent studies seek to address whether there is a wider range of apparent effects of maternal obesity or diabetes on the fetus including the following questions:

1. Does exposure to maternal obesity impact cardiovascular disease risk in offspring later in life?
2. Does maternal diabetes or obesity only affect metabolic outcomes of offspring or are other systems also affected?

To determine whether maternal obesity during pregnancy is associated with increased mortality from cardiovascular events in adult offspring, birth records of 37,709 people in Aberdeen, Scotland were linked with measures of death and hospital admissions for cardiovascular events in offspring aged 34–61 (Reynolds et al. 2013). Maternal body mass index (BMI) was calculated from height and weight measured at the first antenatal visit. All-cause mortality was increased in offspring of obese mothers (BMI >30) compared with mothers with normal BMI after adjustment for maternal age at delivery, socioeconomic status, sex of offspring, current age, birth weight, gestation at delivery, and gestation at measurement of BMI. Offspring of obese mothers also had an increased risk of hospital admission for a cardiovascular event compared with offspring of mothers with normal BMI. The offspring of overweight mothers also had a higher risk of adverse outcomes. Reynolds et al. (2013) concluded that maternal obesity is associated with an increased risk of premature death in adult offspring and noted that since one in five women in the United Kingdom is obese at the time of initiating prenatal care, strategies to optimize weight before pregnancy are urgently required.

Recent reports raise questions about whether maternal diabetes does or does not impact central nervous system development or indicators of future female fertility in exposed offspring. As recently reviewed by Fraser and Lawlor (2014), while there is evidence from recent human studies suggesting that exposure to maternal diabetes *in utero* adversely affects the risk of offspring for long-term adiposity and adverse cardio metabolic outcomes, there is no strong weight of evidence to suggest that cognitive ability of offspring is or is not affected by development exposure to maternal diabetes.

To investigate the effect of maternal obesity on female gamete development in exposed female offspring, Wu et al. (2015) studied oocytes and blastocysts from obese and lean mice. Oocytes from obese mice gave rise to fetuses that were heavier than controls and had reduced liver and kidney mtDNA content per cell, indicating that maternal obesity before conception had altered the transmission of mitochondria to offspring. Treatment of the obese females with the ER stress inhibitor salubrinal or the chaperone inducer BGP-15 before ovulation increased the amount of the mitochondrial replication factors TFAM and DRP1, and mtDNA content in oocytes as well as completely restored oocyte quality, embryo development and the mtDNA content of fetal tissue to those derived from lean mice.

In summary, there is some scant but suggestive evidence that *in utero* exposure to maternal obesity or diabetes adversely impacts the offspring's risks of obesity and cardiovascular disease and may impact ovarian function later in life. It remains uncertain whether cognitive function may or may not be adversely affected. These are important future research themes that fall within the DOHaD Concept regarding developmental determinants of later chronic disease risks.

1.3.2 Effects and Risks of Drugs During Pregnancy

The inevitability of drug use during pregnancy is most obvious in women who have asthma, are HIV-infected or have psychiatric illnesses. In each of these disease areas, there are large numbers of women who bear children and for whom therapeutic guidelines have been developed and widely implemented.

Asthma has been reported to affect 3.7–8.4 % of pregnant women in the United States (Kwon et al. 2004), making it potentially the most common serious medical problem to complicate pregnancy. Although data have been conflicting, the largest and most recent studies (Demissie et al. 1998; Källén et al. 2000) suggest that maternal asthma increases the risk of perinatal mortality, preeclampsia, preterm birth, and low birth weight infants. More severe asthma is associated with increased risks, while better controlled asthma is associated with decreased risks (Schatz et al. 1995). The management practices for the treatment goal of asthma control are commonly described as a stepwise approach for managing asthma during pregnancy and lactation (NAEPP Working Group 2005). The effort to balance the disease itself with effects/side effects of medications include the following:

- Minimal or no chronic symptoms day or night
- Minimal or no exacerbations
- No limitations on activities; no school/work missed
- Maintain (near) normal pulmonary function
- Minimal use of short acting inhaled beta2-agonist
- Minimal or no adverse effects from medications.

It is noteworthy that the primary responses that drive clinical decision-making are all maternal rather than fetal. In no way does this imply the fetus is ignored but it does demonstrate the essential importance of managing maternal respiratory function for the health of both patients.

In terms of the use of antiretroviral (ARV) drug recommendations for HIV-infected, pregnant women, current recommendations for care (NIH 2014) have been based on the concept that drugs of known benefit to women should not be withheld during pregnancy unless there are known adverse effects to the mother, fetus, or infant and unless these adverse effects outweigh the benefits to the woman. At the present time pregnancy should not preclude the use of optimal drug regimens for the adult woman. Nevertheless it is also advised that the decision to use any ARV drug during pregnancy should be made by a woman after discussing with her health care provider the known and potential benefits and risks to her and her fetus.

The American College of Obstetricians and Gynecologists (ACOG Practice Bulletin 2008) have presented detailed practice recommendations for the care of the estimated more than 500,000 pregnancies in the United States each year involving women who have psychiatric illnesses that either predate or emerge during pregnancy, and the estimated one third of all pregnant women who are exposed to a psychotropic medication at some point during pregnancy.

The following ACOG recommendations and conclusions are based on good and consistent scientific evidence (Level A):

- Lithium exposure in pregnancy may be associated with a small increase in congenital cardiac malformations, with a risk ratio of 1.2–7.7
- Valproate exposure in pregnancy is associated with an increased risk of fetal anomalies, including neural tube defects, fetal valproate syndrome, and long term adverse neurocognitive effects. It should be avoided in pregnancy, if possible, especially during the first trimester.
- Carbamazepine exposure in pregnancy is associated with fetal carbamazepine syndrome. It should be avoided in pregnancy, if possible, especially during the first trimester.
- Maternal benzodiazepine use shortly before delivery is associated with floppy infant syndrome or infantile hypotonia, a condition of decreased muscle tone.

The following ACOG recommendations and conclusions are based on limited or inconsistent scientific evidence (Level B):

- Paroxetine use in pregnant women and women planning pregnancy should be avoided, if possible.
- Fetal echocardiography should be considered for women who are exposed to paroxetine in early pregnancy.
- Prenatal benzodiazepine exposure increased the risk of oral cleft, although the absolute risk increased by 0.01 %.
- Lamotrigine is a potential maintenance therapy option for pregnant women with bipolar disorder because of its protective effects against bipolar depression, general tolerability, and a growing reproductive safety profile relative to alternative mood stabilizers.

In summary, with limited information available on the risks of psychotropic medications, clinical management must incorporate an appraisal of the clinical consequences of offspring exposure, the potential effect of untreated maternal psychiatric illness, and the available alternative therapies. Maternal psychiatric illness, if inadequately treated or untreated, may result in poor compliance with prenatal care, inadequate nutrition, exposure to additional medication or herbal remedies, increased alcohol and tobacco use, deficits in mother-infant bonding, and disruptions within the family environment.

In addition to the risk-benefit assessments that must be made for gestational use of prescription drugs, we must not lose sight of the fact that many over-the-counter medications are commonly used and are typically assumed to be safe for use by anyone including pregnant women, even though this assumption is often unfounded. For example, acetaminophen (paracetamol) use during pregnancy was assessed prospectively in 64 322 live-born children and mothers enrolled in the Danish National Birth Cohort during 1996–2002 (Liew et al. 2014). More than half of all mothers reported acetaminophen use while pregnant. Children whose mothers used acetaminophen during pregnancy were at higher risk for receiving a hospital diagnosis of hyperkinetic disorders (HKDs) or attention-deficit/hyperactivity disorder (ADHD)–like behavioral problems or to use ADHD medications. Because the exposure and outcome are frequent, these results are of public health relevance but further investigations are needed. Multiple other nonprescription products contain one or multiple active medications plus potential combinations of such medications and are readily available for use by the public at large and inevitably fetal exposures occur.

1.3.3 *Effects of Diverse Exposures on Immune and Pulmonary Function*

A modest number of recent studies have shown that environmental exposures may affect immune function and inflammatory pathways that are involved in manifestation of asthma in offspring of exposed women.

Stayner et al. (2014) evaluated maternal exposure to drinking water disinfection by-products (brominated trihalomethanes; BTHM) in relation to micronuclei (MN) frequency in lymphocytes as a marker of genomic damage in maternal and cord blood lymphocytes in 214 mothers and 223 newborns from the Rhea mother–child cohort in Crete, Greece, in 2007–2008. MN frequency in maternal binucleated lymphocytes was found to increase with BTHM concentrations in residential water for exposure during the first and second trimesters and through all routes of BTHM exposure during the first trimester.

Metzger et al. (2013) conducted 2 retrospective case-control analyses of infants born in Washington State from 1987 to 2004 using linked birth certificate, death certificate and hospital discharge records. One assessed morbidity—infants hospitalized due to infectious diseases (IDs) within 1 year of birth (47,404 cases/48,233 controls). The second assessed mortality—infants who died within 1 year due to

IDs (627 cases/2730 controls). Maternal smoking was associated with both hospitalization and mortality due to any ID. In subgroup analyses, maternal smoking was associated with hospitalization due to a broad range of IDs including both respiratory and nonrespiratory IDs.

Hansen et al. (2014) evaluated the association between maternal serum concentrations of persistent organochlorine pollutants (POPs) and the risk of asthma in 965 offspring Denmark after 20 years of follow-up. Concentrations of six polychlorinated biphenyls (PCBs), hexachlorobenzene (HCB), and dichlorodiphenyldichloroethylene (p,p′-DDE) were quantified in 872 maternal sera collected at week 30 of gestation. Maternal serum concentrations of HCB and dioxin-like PCB-118 were positively associated with offspring asthma medication use after 20 years of follow-up.

Whyatt et al. (2014) evaluated associations between asthma diagnosed in inner city children between 5 and 11 years of age and prenatal exposures to butylbenzyl phthalate (BBzP), di-n-butyl phthalate (DnBP), di(2-ethylhexyl) phthalate (DEHP), and diethyl phthalate (DEP). Of 300 children, 154 (51 %) were examined by a physician because of reports of wheeze, other asthma-like symptoms, and/or medication use; 94 were diagnosed with current asthma and 60 without current asthma. The remaining 146 children were classified as nonasthmatic. Compared with levels in nonasthmatics, prenatal metabolites of BBzP and DnBP were associated with a history of asthma-like symptoms and with the diagnosis of current asthma. Risk of current asthma was >70 % higher among children with maternal prenatal BBzP and DnBP metabolite concentrations in the third versus the first tertile.

Cheelo et al. (2015) performed a systematic review and meta-analysis of longitudinal studies that reported an association between paracetamol (acetaminophen) exposure during pregnancy or infancy and the subsequent development of childhood asthma (≥5 years) to determine whether the observed associations are due to confounding by respiratory tract infections. Eleven observational cohort studies met the inclusion criteria. Any paracetamol use during the first trimester was related to increased risk of childhood asthma (five studies) but there was marked between-study heterogeneity and only one of these studies adjusted for maternal respiratory tract infections. The association during early pregnancy exposure was highly variable between studies and exposure during infancy appears to be moderately confounded by respiratory tract infections. There is insufficient evidence to warrant changing guidelines on early life paracetamol exposure at this time on the basis of asthma risk. More studies are needed on the effects of maternal exposure to pollutants and other chemicals on immune function in offspring.

1.3.4 Effects of Diverse Exposures on Reproductive and CNS Development and Function

Results of numerous studies regarding reproductive effects of environmental exposures continue to appear regularly in the current scientific literature. Much of the focus continues to relate to sex steroid hormone action particularly estrogens in the outcomes that often appear superficially contradictory. Some of the past apparent contradictions among the actions of various estrogens are progressively being clarified. Now some two decades after discovery of the ER beta, data suggest that it may be considered as a dominant-negative regulator of ER alpha and that ER beta modulates transcriptional responses to estrogens. The ratio of ER alpha vs. beta within a cell is important in determining the cell sensitivity to estrogens and its biological responses to the hormone such that the multiplicity of receptors largely determines estrogen actions (Böttner et al. 2014). We will not delve into a deep current review of reproductive developmental toxicology herein, however for illustrative purposes we will simply make reference to one recent report each for female and male reproductive effects.

Bellingham et al. (2013) investigated the influence of pre- and/or post-conception exposure to low-level mixtures of environmental chemicals (ECs) on the structure and function of the fetal ovine ovary. Female fetuses were collected at day 110 of gestation, from dams exposed continuously until, and after mating, by grazing in pastures treated with sewage sludge as a fertilizer (TT) or in control fields treated with inorganic fertilizer (CC). In addition, in a cross-over design, fetal ovaries were collected from dams maintained on sludge pastures up to the time of mating but then transferred to control pastures (TC) and, reciprocally, those transferred from control to treated pastures at mating (CT). Their findings indicate that continuous maternal EC exposure before and during gestation, are less deleterious for fetal ovarian development than a change in maternal EC exposure between pre and post-conception. In addition to the period of pregnancy, the pre-conception period appears also as crucial for conditioning long-term effects of EC exposure on ovarian development and primordial follicle reserve and hence future fertility.

Kilcoyne et al. (2014) used adult Leydig cells (ALC) ablation/regeneration in rats to show that dibutyl phthalate (DBP) -induced reduction in intratesticular testosterone in rats reduced ALC stem cell number by ~40 % at birth to adulthood and induced compensated ALC failure (low/normal testosterone and elevated luteinizing hormone). These results suggest that a key component of male reproductive development can fundamentally reprogram adult hormone production through an epigenetic change and may affect lifetime disease risk.

A number of recent studies assess the effects of developmental exposures on central nervous system development and/or function. While those primarily relating to an epigenetic mode of action will be considered in Sect. 1.4.4, some new more general findings will be considered here.

Lead (Pb) exposure inevitably continues to be an important issue in terms of exposures and adverse neurocognitive effects. In terms of routes of exposure, Ettinger et al. (2014) measured lead in 81 maternal blood, plasma, and breast milk samples at 1 month postpartum and in 60 infant blood samples at 3 months of age. Milk-to-plasma (M/P) lead ratios were calculated. Maternal lead levels (mean ± SD) were blood: 7.7 ± 4.0 μg/dL; plasma: 0.1 ± 0.1 μg/L; milk: 0.8 ± 0.7 μg/L. The average M/P lead ratio was 7.7 (range, 0.6–39.8) with 97 % of the ratios being > 1. g/L milk lead ($p < 0.0001$, $R2 = 0.3$). The M/P ratio for lead in humans is substantially higher than previously reported, and transfer of lead from plasma to milk may be higher at lower levels of plasma lead. Breast milk is an important determinant of lead burden among breastfeeding infants.

In a recent study, Boucher et al. (2014) examined the effects of prenatal exposure to polychlorinated biphenyls (PCBs), methylmercury (MeHg), and lead (Pb) on cognitive development in a sample of Inuit infants (n=94) from Arctic Québec. PCBs, mercury (Hg), Pb, and two seafood nutrients, docosahexaenoic acid (DHA) and selenium (Se), were measured in umbilical cord blood. Infants were assessed at 6.5 and 11 months on the Fagan Test of Infant Intelligence (FTII), A-not-B test, and Bayley Scales of Infant Development-2nd Edition (BSID-II). Higher prenatal PCB exposure was associated with decreased FTII novelty preference, indicating impaired visual recognition memory. Prenatal Hg was associated with poorer performance on A-not-B, which depends on working memory and is believed to be a precursor of executive function. Prenatal Pb was related to longer FTII fixation durations, indicating slower speed of information processing. None of these exposures was associated with performance on the BSID-II, a global developmental measure. These findings show that as more subtle testing is performed, more complex profiles of neurocognitive adverse effects can be demonstrated.

To emphasize that environmental exposures are not the only source of concern, Knickmeyer et al. (2014) studied the association between use of selective serotonin reuptake inhibitors (SSRIs) in pregnancy to treat depression (n=33) and the occurrence of Chiari I malformations (CIM) in children exposed to SSRIs *in utero*. These exposed children were compared to 66 children with no history of maternal depression and no SSRI exposure. Another 30 children whose mothers received a diagnosis of depression, but did not receive antidepressants during pregnancy were compared to 60 children with no history of maternal depression and no SSRI exposure. The main outcome was presence/absence of CIM on MRI scans at 1 and/or 2 years of age. The SSRI-exposed children were significantly more likely to be classified as CIM than comparison children with no history of maternal depression and no SSRI exposure. Duration of SSRI exposure, SSRI exposure at conception, and family history of depression increased the risk. This report shows that macroscopic structural defects in the cerebellum and brainstem may be associated with gestational use of this class of psychiatric medication.

1.4 Epigenomic Disruption: The Effects of Early Developmental Exposures

1.4.1 Background

Through DNA methylation, histone modifications, and small regulatory RNAs, epigenomic mechanisms control gene expression during development and across the lifespan. Environmental exposures during developmental intervals impact epigenetic patterns and persistence of these regulatory changes into later life affects the occurrence of a number of diseases.

A number of EDCs have been shown to influence epigenetic programming. For example, Bernal and Jirtle (2010) studied the epigenotoxic effects of bisphenol A (BPA) using the Agouti viable yellow (Avy) mouse model. Dietary BPA exposure was shown to hypomethylate both the Avy and the CabpIAP metastable epialleles. This hypomethylating effect was counteracted with dietary supplementation of methyl donors or the phytoestrogen genistein. As a general concept, epigenotoxicity is now believed to mediate a number of developmental, metabolic, and behavioral disorders in exposed populations. Additionally, evidence suggests that at least some epigenetic changes are also heritable and thus there can be transgenerational inheritance of some phenotypes whether these are associated with increased or decreased risks of later-in-life diseases.

There are now over 150 imprinted genes known in mammals. Two of the earliest discovered were shown to be oppositely imprinted and antagonistic in their regulation of growth; namely the Igf2 gene encodes a paternally expressed ligand that promotes growth, while maternally expressed Igf2r encodes a cell surface receptor that restricts growth by sequestering Igf2 and targeting it for lysosomal degradation. The delta-like 1 gene (Dlk1) encodes a putative ligand that promotes fetal growth and in adults restricts adipose deposition. Conversely, Grb10 encodes an intracellular signaling adaptor protein that, when expressed from the maternal allele, acts to restrict fetal growth and is permissive for adipose deposition in adulthood. Using knockout mice, Madon-Simon et al. (2014) show that these two factors exert their opposite effects on growth and physiology through a common signaling pathway. The major effects are on body size (particularly growth during early life), lean:adipose proportions, glucose regulated metabolism and lipid storage in the liver. Dlk1 and Grb10 appear to define a mammalian growth axis that is separate from the IGF pathway, yet also features an antagonistic imprinted gene pair.

When taken together, recent reports make the point that exposures to both naturally-occurring and xenobiotic EDCs should deservedly be assessed for potential developmental effects via the epigenome as well as other modes of action. Guerrero-Bosagna and Skinner (2014) have recently argued that the epigenetic developmental effects of phytoestrogens/phytochemicals must be assessed. Phytochemicals are one of the largest classes of compounds humans are exposed to throughout life and phytoestrogens act via both sex steroid receptors and have dramatic effects on epigenomic events (Bernal and Jirtle 2010). Another coincidence of at least two modes of action relates to BPA. As cited above, BPA affects the epigenome while Veiga-Lopez et al. (2015) suggest that nitrosative stress may be a biomarker of effect for BPA exposure. These investigators studied samples from human pregnancies during the first trimester and at term, as well as fetal and/or adult samples from prenatally BPA-treated sheep, rats, and mice to assess the impact of BPA on free fatty acid and oxidative stress dynamics. Human mothers exposed to higher BPA during early to mid-pregnancy and their matching term cord samples displayed increased 3-nitrotyrosine (NY), a marker of nitrosative stress. After exposure to a human-relevant dose of BPA, sheep fetuses, adult sheep, prenatally exposed rats, and adult mice all showed increased systemic nitrosative stress. We should expect many apparently simple exposures to act via pleiotropic pathways.

1.4.2 Metabolism and Obesity Effects on Offspring

Azad et al. (2014) investigated the association between early-life antibiotic exposure and subsequent development of overweight and central adiposity. Antibiotic exposure during the first year of life was documented from prescription records. Overweight and central adiposity were determined from anthropometric measurements at ages 9 (n=616) and 12 (n=431). Infants receiving antibiotics in the first year of life were more likely to be overweight later in childhood compared with those who were unexposed. Following adjustment for birth weight, breastfeeding, maternal overweight and other potential confounders, this association persisted in boys but not in girls. Similar gender-specific associations were found for overweight at age 9 and for high central adiposity at age 12. Among boys, antibiotic exposure during the first year of life was associated with an increased risk of overweight and central adiposity in preadolescence, indicating that antibiotic stewardship is particularly important during infancy.

Using data from a large, prospective pregnancy cohort study (n=19 652), with linkage to a national prescription registry, Jensen et al. (2014) evaluated the association between use of hormonal contraceptives before and after conception in relation to offspring overweight or obesity at age 3 years. There was a weak, inverse association between early pregnancy use of a combination oral contraceptive and offspring overweight or obesity at age 3 and a positive, but imprecise, association with use of a progestin-only oral contraceptive in early pregnancy. No association was observed between the use of a hormonal contraceptive before conception and offspring overweight or obesity. Pharmacologic sex hormones in early pregnancy may be inversely or positively associated with offspring overweight or obesity at age 3, depending on the specific formulation used.

Kajantie et al. (2015) studied insulin sensitivity and secretion in 107 Finnish adults born preterm at very low birth weight (VLBW; <1500 g) and 100 controls born at term not small for gestational age (SGA). Compared with controls, VLBW adults had lower calculated insulin sensitivity (Si) and higher insulin secretory response (AIR) than their term-born peers with a similar body size. In young adulthood, this remains compensated by higher insulin secretion leading to higher likelihood of insulin insensitivity occurring.

Ashley-Martin et al. (2014) examined associations between prenatal exposure to several potential endocrine disrupting chemicals (bisphenol A and 11 phthalate metabolites) in first trimester maternal urine samples as well as markers of fetal metabolic dysfunction (leptin and adioponectin) in 1363 cord blood samples pregnancy from ten Canadian sites. Leptin was significantly higher in female than male infants and males showed an inverse, non-linear relationship between BPA and adiponectin. Personal-care items containing phthalates include perfume, eye shadow, moisturizer, nail polish, liquid soap, and hair spray (Rudel and Perovich 2009).

Whisner et al. (2015) determined whether maternal pre-pregnancy BMI (ppBMI), gestational weight gain (GWG) and dietary intake during pregnancy influence fetal fat accretion in utero measured by sonography in 121 pregnant adolescents (ages

13–18 years). After adjusting for infant birth weight, variables significantly associated with fetal abdominal wall thickness (abdominal subcutaneous fat thickness) in late pregnancy included gestational age, maternal race and dietary intake of added sugar.

Emerging evidence suggests that maternal obesity during gestation/lactation "programs" offspring long-term for increased obesity themselves via inflammatory mechanisms linked to two components that are enriched in a "Western diet": saturated fatty acids and branched chain amino acids (BCAAs). Wiley et al. (2014) have shown that in the mouse model, maternal high-fat diet (HFD) can "prime" microglia to elevate levels of proinflammatory cytokines within the hippocampus of adult offspring and BCAAs appear to synergize to result in exacerbated brain and behavioral consequences in offspring.

Dougan et al. (2015) investigated the association between prenatal vitamin intake and obesity among 29,160 mother–daughter dyads in the Nurses' Health Study II. In utero exposure to prenatal vitamins was not associated with body fatness, either in childhood or in adulthood.

Morales et al. (2015) studied the association of maternal circulating 25-hydroxyvitamin D3 (25(OH)D3) concentration in pregnancy with offspring prenatal and postnatal growth and overweight in 2358 mother-infant dyads in Spain. There was no association of maternal 25(OH)D3 concentration with offspring femur length (FL) and a weak inverse association with biparietal diameter (BPD) at 34 weeks. Maternal deficit of 25(OH)D3 was associated with increased risk of fetal overweight and an increased risk of overweight in offspring at age 1 year but not at age 4 years.

van Uitert et al. (2013) studied maternal characteristics and lifestyle factors associated with human embryonic growth trajectories in 87 pregnant women beginning before 8 weeks of gestation by performing weekly three-dimensional ultrasound scans from enrolment up to 13 weeks of gestation. Comparison of embryonic crown-rump length (CRL) measurements to maternal characteristics showed that periconception maternal age is associated with increased, and smoking and alcohol use with decreased embryonic growth trajectories.

van Dijk et al. (2015) reviewed 46 studies investigating the association between obesity and either global, site specific or genome-wide methylation of DNA and the impact of pre- and postnatal interventions on methylation and obesity. There is no consistent evidence for a relationship between global methylation and obesity; however there are multiple obesity-associated differentially methylated sites, mainly in blood cells. Additionally, there are several associations between methylation marks at birth and later life obesity as well as some alterations in methylation at specific sites in weight loss intervention studies.

Jiang et al. (2014) studied the association between serum microRNAs (miRNAs) and macrosomia (birth weight >4000 g) in maternal serum samples collected 1 week before delivery. In comparison to controls, 1 miRNA was significantly upregulated and 10 miRNAs including miR-21 were significantly down-regulated in near term maternal serum samples who delivered macrosomic infants. In future, serum microRNA profiles may be of significant diagnostic utility in monitoring fetal development.

1.4.3 Pulmonary Effects on Offspring

Lung function in infancy predicts pulmonary function throughout life. In utero and early postnatal exposures influence both childhood and adult lung structure and function and may predispose individuals to chronic obstructive lung disease and other disorders. The nutritional and endogenous chemical environment affects development of the lung and can result in altered function in the adult. Studies now suggest that similar adverse impacts may occur in animals and humans after exposure to environmentally relevant doses of certain xenobiotics during critical windows in early life. Potential mechanisms include interference with highly conserved factors in developmental processes such as gene regulation, molecular signaling, and growth factors involved in branching morphogenesis and alveolarization. Miller and Marty (2010) have made key points that assessment of environmental chemical impacts on the lung may require studies that evaluate specific endpoints that are not regularly assessed in standard toxicity tests and that signaling pathways are expected to influence designs of future developmental toxicology studies.

Thacher et al. (2014) followed a birth cohort of 4089 children for 16 years to examine the role of prenatal and postnatal second-hand tobacco smoke (SHS) exposure on occurrence of asthma, rhinitis, and eczema. Exposure to SHS in utero was associated with an overall elevated risk of developing asthma up to 16 years but not for rhinitis or eczema. Exposure to SHS during infancy was associated with an overall elevated risk of asthma, rhinitis, and eczema up to 16 years.

Hart et al. (2015) examined the relationship between maternal vitamin D deficiency at 18 weeks' pregnancy and long-term health outcomes of 901 offspring in Western Australia. Vitamin D deficiency was present in 36 % (323 of 901) of the pregnant women. Maternal vitamin D deficiency during pregnancy was associated with impaired lung development in 6-year-old offspring, neurocognitive difficulties at age 10, increased risk of eating disorders in adolescence, and lower peak bone mass at 20 years.

1.4.4 Neurocognitive Effects on Offspring

Buss et al. (2014) studied 58 mother-child dyads in whom maternal blood samples from early, mid and late pregnancy were measured for concentrations, and a resting state functional magnetic resonance imaging (R-fMRI) scan was performed in newborns within the first 4 weeks after birth. After accounting for the effects of maternal medical gestational complications, pregnancy length, and the newborn's postnatal age at the neuroimaging scan, higher average maternal IL-6 concentrations during pregnancy predict decreased maturity of the newborn functional connectivity of the Default Mode Network (DFM). Maternal IL-6 in early pregnancy was a stronger predictor of the degree of maturity of the newborn DFM network than maternal IL-6 at later gestational stages, suggesting early pregnancy may represent a particularly sensitive period of time for the actions of maternal inflammation on fetal brain development.

Recent studies suggest that exposure to traffic-related air pollutants, including particulate matter (PM), is associated with autism spectrum disorder (autism). Kalkbrenner et al. (2015) identified children with autism by records-based surveillance (n=645 born in North Carolina and n=334 born in the San Francisco Bay Area in California). They were compared with randomly sampled children born in the same counties and years identified from birth records (n=12,434 in North Carolina and n=2232 in California). Exposure to the traffic-related air pollutant particulate matter (PM) less than 10 μm (PM10) at the birth address was assigned from air pollution regulatory monitors. The data showed a relation between traffic-related air pollution and autism in both U.S. states specifically in association with exposure in the third trimester.

Factor-Litvak et al. (2014) studied 328 inner-city mothers and their children by measuring prenatal urinary metabolites of di-n-butyl phthalate (DnBP), butylbenzyl phthalate (BBzP), di-isobutyl phthalate (DiBP), di-2-ethylhexyl phthalate and diethyl phthalate in late pregnancy. The Wechsler Intelligence Scale for Children, 4th edition was administered at child age 7 years to evaluate four areas of cognitive function associated with overall intelligence quotient (IQ). Child full-scale IQ was inversely associated with prenatal urinary metabolite concentrations of DnBP and DiBP. Among children of mothers with the highest versus lowest quartile DnBP and DiBP metabolite concentrations, IQ was 6.7 and 7.6 points lower, respectively.

Rogne et al. (2015) studied the association between fetal growth pattern and cognitive function at 5 and 9 years and regional brain volumes at 15 years in 83 term-born small-for-gestational age (SGA) neonates (13 with fetal growth restriction; 36 with non-fetal growth restriction) and 105 non-SGA neonates. Cognitive function was assessed at 5 and 9 years, and brain volumes were estimated with cerebral magnetic resonance imaging at 15 years. Small-for-gestational-age children had lower intelligence quotient scores at 5 and 9 years and smaller brain volumes at 15 years compared with those in the control group, but these findings were only found in those with fetal growth restriction, indicating a possible relationship to decelerated fetal growth.

1.5 Prenatal, Neonatal and Early Childhood Therapies to Mitigate Adverse Developmental Exposures

Because the therapeutic requirements of the neonate are unique in comparison to older infants and children, the National Institute of Child Health and Human Development and the US Food and Drug Administration (FDA) developed the

Newborn Drug Development Initiative to address the limited study of off-patent drugs in newborns (Ward et al. 2006). A panel of experts met to discuss how to increase the study of drugs for the newborn. The panel identified 4 general categories containing different numbers of criteria as important for ranking drugs for priority investigation: (1) the disease and indication, including elements such as the potential for adverse outcomes, frequency in newborns, and level of evidence for treatment of newborns; (2) drug characteristics, including elements such as duration of dosing, lack of age-appropriate formulation, clinically relevant drug-drug and drug-disease interactions, and drug disposition in newborns; (3) feasibility and methodology for newborn studies, including both analytical considerations and clinical end points; and (4) the ethical basis for study, including elements to address benefit or harm due to exposure to the study drug, study methodology, and benefit of the new treatment relative to established standard therapy. Based on these categories, a list of criteria to warrant study of a drug in newborns was developed.

Since humans born after intrauterine growth restriction (IUGR) are at an increased risk of developing diabetes in adult life and evidence suggests that exercise as adults can reduce the risk of metabolic disease following IUGR, Gatford et al. (2014) have studied the effect of exercise in a rat model of IUGR. Their data in the rat model suggest that exercise in early life might be able to reverse or reprogram the long-term metabolic effects of IUGR.

To determine the effects of lifestyle interventions during pregnancy in obese women on offspring metabolic risk factors and the predictive values of birth weight (BW) and birth abdominal circumference (BAC), Tanvig et al. (2015) studied the offspring of Lifestyle in Pregnancy (LiP) study participants (n = 157) and offspring of normal-weight mothers (external reference group, n = 97). Interventions that included dietary advice, coaching, and exercise during pregnancy had no effect on early childhood metabolic risk factors by these lifestyle interventions.

1.5.1 Potential Future Therapeutic Interventions

Potential therapeutics to mitigate the adverse effects of toxic exposures are likely to come from investigation of compounds that are acknowledged as "GRAS," generally-regarded-as-safe with several different modes-of-action (MOA) or well-established/repurposed pharmaceuticals. Multiple potential modes of action will need to be assessed; safety profiles will need to be developed; and there is minimal expectation that commercial entities will be motivated to make the investments required to develop agents to pre-empt or reduce adverse outcomes later in life due to antecedent environmental exposures. Logical starting points are to consider instances where clinical practice already uses a protective treatment to reduce the risk of acute toxicity and the insights gained from work done in the area of chemoprevention of cancer(s).

1.5.2 Potential Treatment of Pregnant Women to Reduce Fetal Risks due to Smoking

1.5.2.1 Nicotine or N-Acetyl-L-cysteine (NAC) for Maternal Smoke Exposure During Pregnancy

Nicotine-replacement therapy is effective for smoking cessation outside pregnancy and even though its safety has not been thoroughly established, its use is widely recommended during pregnancy. Coleman et al. (2012) found no significant difference in the rate of abstinence from the quit date until delivery between the nicotine-replacement and placebo groups (9.4 % and 7.6 %, respectively; unadjusted odds ratio with nicotine-replacement therapy, 1.26; 95 % confidence interval, 0.82–1.96). Compliance was low; only 7.2 % of women assigned to nicotine-replacement therapy and 2.8 % assigned to placebo used patches for more than 1 month. Rates of adverse pregnancy and birth outcomes were similar in the two groups. Adding a nicotine patch (15 mg per 16 h) to behavioral cessation support for women who smoked during pregnancy did not significantly increase the rate of abstinence from smoking until delivery or the risk of adverse pregnancy or birth outcomes.

Although smoking cessation is the primary goal for the control of cancer and other smoking-related diseases, chemoprevention provides a complementary approach applicable to high risk individuals such as current smokers and ex-smokers (De Flora et al. 2001). On the whole, there is overwhelming evidence that N-Acetyl-L-cysteine (NAC) has the ability to modulate a variety of DNA damage- and cancer-related end-points.

While we do not know of studies in pregnant women, Van Schooten et al. (2002) studied NAC in healthy smoking volunteers in a randomized, double-blind, placebo-controlled, Phase II chemoprevention trial. The subjects were supplemented daily with 2 × 600 mg of oral tablets of NAC (n = 20) or placebo (n = 21) for a period of 6 months, and internal dose markers [plasma and bronchoalveolar lavage (BAL) fluid cotinine, urine mutagenicity], biologically effective dose markers [smoking-related DNA adducts and hemoglobin (Hb) adducts], and biological response markers (micronuclei frequency and antioxidants scavenging capacity) were assessed at both pre- and post-supplementation times (T0 and T1, respectively). Overall, the internal dose markers remained unchanged at T1 as compared with T0 in both NAC and placebo groups. When quantifying the biologically effective dose markers, data showed an inhibitory effect of NAC toward the formation of lipophilic-DNA adducts (5.18 + 0.73 versus 4.08 + 1.03/108 nucleotides; mean + SE; P = 0.05) as well as of 7,8-dihydro-8-oxo-2_-eoxyguanosine adducts in BAL cells (3.9 + 0.6 versus 2.3 + 0.2/105 nucleotides; P = 0.003). There was no effect of NAC on the formation of lipophilic-DNA adducts in peripheral blood lymphocytes. Based on these results, the investigators concluded that NAC has the potential to alter tobacco smoke carcinogenicity in adult humans because it can modulate certain cancer associated biomarkers in specific organs, and by implication could mitigate DNA damage in the fetus.

While women are routinely advised to stop smoking (and most do), many nonsmoking pregnant women are exposed to second hand smoke that can be detected by the biomarker cotinine in urine samples of pregnant women (Swamy et al. 2011). Therefore, this biomarker could be used as a guide for NAC intervention in pregnancy as a possible way to reduce harm to the fetus.

Although the evidence of efficacy is somewhat mixed, NAC is widely used to reduce the risk of acute contrast nephrotoxicity in patients with impaired renal function who undergo computed tomography scanning. To determine whether oral NAC protects against deterioration in renal function in patients with moderate renal insufficiency who undergo elective coronary angiography, Kay et al. (2003) conducted a prospective, randomized, double-blind, placebo-controlled trial in 200 patients aged mean (SD) 68 (6.5) years with stable moderate renal insufficiency (creatinine clearance <60 mL/min [1.00 mL/s]) who were undergoing elective coronary angiography with or without intervention. Participants were randomly assigned to receive oral NAC (600 mg twice per day; n = 102) or matching placebo tablets (n = 98) on the day before and the day of angiography. All patients received low-osmolality contrast agent. NAC treatment significantly increased creatinine clearance from 44.8 mL/min (0.75 mL/s) (95 % CI, 42.7–47.6 mL/min) to 58.9 mL/min (0.98 mL/s) (95 % CI, 55.6–62.3 mL/min) 2 days after the contrast administration (P.001).The increase was not significant in the control group (from 42.1 to 44.1 mL/min [0.70–0.74 mL/s]; P = .15). The benefit of NAC was consistent among various patient subgroups and persistent for at least 7 days. There were no major treatment-related adverse events. The investigators concluded that NAC protects patients with moderate chronic renal insufficiency from contrast-induced deterioration in renal function after coronary angiographic procedures, with minimal adverse effects and at a low cost.

1.5.2.2 AhR Antagonists to Protect the Ovaries and Other Smoking Interventions

It has been shown that benzo[a]pyrene, a key component of cigarette smoke and an aryl hydrocarbon receptor (AhR) ligand, reduced growth of isolated rat follicles in vitro. However, the mechanism underlying the induced changes in folliculogenesis is unknown. Neal et al. (2010) proposed that the reported adverse effects of benzo[a]pyrene on follicle growth are mediated through AhR activation. The objective was to investigate the effect of benzo[a]pyrene with and without AhR antagonists (resveratrol or 3′,4′-dimethoxyflavone (3,4-DMF)) on follicle growth, oestradiol output, anti-Müllerian hormone (AMH) concentration and cell proliferation in isolated rat follicles cultured in vitro. Benzo[a]pyrene treatment significantly inhibited follicle growth and cell proliferation at concentrations of 1.5 ng/ml and higher ($P < 0.05$), an effect attenuated by co-incubation with benzo[a]pyrene and resveratrol or 3,4-DMF. A significant decrease in oestradiol ($P < 0.05$) and AMH output ($P < 0.001$) by cultured follicles was induced by benzo[a]pyrene treatment, an effect attenuated by co-incubation with 3,4-DMF. The results suggest that the adverse effects of benzo[a]pyrene on follicle growth, steroidogenesis and AMH output are mediated through activation of the AhR. Moreover, AhR antagonists such as resveratrol and 3,4-DMF may have therapeutic benefit in protecting the ovary against the adverse effects of AhR ligands, including benzo[a]pyrene.

Cytisine, a plant-based alkaloid, is a partial agonist at the nicotinic acetylcholine receptor and is used for smoking cessation. Walker et al. (2014) conducted an open-label, noninferiority trial in New Zealand in which 1310 adult daily smokers who were motivated to quit smoking and were randomly assigned in a 1:1 ratio to receive cytisine for 25 days or nicotine-replacement therapy for 8 weeks. The effectiveness of cytisine for continuous abstinence was superior to that of nicotine-replacement therapy (overall 40 % versus 31 %) at 1 week, 2 months, and 6 months. Self-reported adverse events of nausea and vomiting and sleep disorders occurred somewhat more frequently in the cytisine group.

McEvoy et al. (2014) randomized 159 newborns of pregnant smokers to vitamin C treatment (500 mg/d; n=76) or placebo (n=83) and compared them to 76 newborns of pregnant nonsmokers by newborn pulmonary function tests (PFTs). Follow-up assessment including wheezing was assessed through age 1 year, and PFTs were performed at age 1 year. Newborns of women randomized to vitamin C compared with those randomized to placebo had improved pulmonary function and had significantly decreased wheezing through age 1 year. Vitamin C in pregnant smokers may be an inexpensive and simple approach to decrease the effects of smoking in pregnancy on newborn pulmonary function and respiratory morbidities.

1.5.3 Optimal Vitamin D for Maternal and Fetal Well-Being

Mounting evidence suggests that vitamin D deficiency could be linked to several chronic diseases, including cardiovascular disease and cancer. Forrest and Stuhldreher (2011) examined the prevalence of vitamin D deficiency and its correlates to test the hypothesis that vitamin D deficiency was common in the US population, especially in certain minority groups. The National Health and Nutrition Examination Survey (NHANES) 2003–2006 data were analyzed for vitamin D levels in adult participants (N=4495). Vitamin D deficiency was defined as a serum 25-hydroxyvitamin D concentrations ≤20 ng/mL (50 nmol/L). The overall prevalence rate of vitamin D deficiency was 41.6 %, with the highest rate seen in blacks (82.1 %), followed by Hispanics (69.2 %). Vitamin D deficiency was significantly more common among those who had no college education, were obese, with a poor health status, hypertension, low high-density lipoprotein cholesterol level, or not consuming milk daily (all $P<.001$). Multivariate analyses showed that being from a non-white race, not college educated, obese, having low high-density lipoprotein cholesterol, poor health, and no daily milk consumption were all significantly, independently associated with vitamin D deficiency (all $P<.05$). Given that vitamin D deficiency is linked to some of the important risk factors of leading causes of death in the United States, it is important that health professionals are aware of this connection and offer dietary and other intervention strategies to correct vitamin D deficiency, especially in minority groups.

Higher total serum 25-hydroxyvitamin D (25(OH)D) concentrations have been associated with better cognitive function mainly in cross-sectional studies in adults. It is unknown if the associations of different forms of 25(OH)D (25(OH)D3 and 25(OH)D2) are similar. Tolppanen et al. (2012) conducted a prospective cohort study (n = 3171) with serum 25(OH)D3 and 25(OH)D2concentrations measured at mean age of 9.8 years and academic performance at age 13–14 years (total scores in English, mathematics and science) and 15–16 years (performance in General Certificates of Education examinations). Serum 25(OH)D3 concentrations were not associated with any educational outcomes. Higher 25(OH)D2 concentrations were associated with worse performance in English at age 13–14 years (adjusted SD change per doubling in 25(OH)D2 (95 % CI) –0.05 (–0.08 to –0.01)) and with worse academic performance at age 15–16 years (adjusted OR for obtaining ≥5 A*–C grades (95 % CI) 0.91 (0.82–1.00)). The null findings with 25(OH)D3 are in line with two previous cross-sectional studies in children. It is possible that the positive association of 25(OH)D with cognitive function seen in adults does not emerge until later in life or that the results from previous cross-sectional adult studies are due to reverse causality. The unexpected inverse association of 25(OH)D2 with academic performance requires replication in further studies. These investigators concluded that their results did not support suggestions that children should have controlled exposure to sunlight, or vitamin D supplements, in order to increase academic performance.

Previous studies indicate reduced risk of type 1 diabetes after intake of vitamin D supplements during pregnancy or early childhood. Sørensen et al. (2012) studied whether lower maternal serum concentrations of 25-hydroxy-vitamin D (25-OH D) during pregnancy were associated with an increased risk of childhood-onset type 1 diabetes. In this case-control study nested within a cohort of 29,072 women in Norway, 25-OH D levels were measured using a radioimmunoassay on samples from late pregnancy in 109 women delivering a child who developed type 1 diabetes before 15 years of age (case subjects) and from 219 control women. Dividing the levels of maternal 25-OH D into quartiles, there was a trend toward a higher risk of type 1 diabetes with lower levels of vitamin D during pregnancy. The odds of type 1 diabetes was more than twofold higher for the offspring of women with the lowest levels of 25-OH D compared with the offspring of those with levels above the upper quartile. The data provide support for the initiation of a randomized intervention trial to prevent type 1 diabetes in children by enhancing maternal 25-OH D status during pregnancy.

1.5.4 Protective Properties of Diet and Dietary Phytochemicals

Increasing attention is being paid to the possibility of applying cancer chemopreventive agents for individuals at high risk of neoplasia. For this purpose, natural compounds have practical advantages with regard to availability, suitability for oral application, regulatory approval and mechanisms of action (Tsuda et al. 2004). Candidate substances such as phytochemicals present in foods and their derivatives have been identified by a combination of epidemiological and experimental studies. Plant constituents include vitamin derivatives, phenolic and flavonoid agents, organic sulfur compounds, isothiocyanates, curcumins, fatty acids and d-limonene among others. Examples of compounds from animals are unsaturated fatty acids and lactoferrin. Recent studies have indicated that mechanisms underlying chemo-preventive potential may be combinations of anti-oxidant, anti-inflammatory, immune-enhancing, and anti-hormone effects, with modification of drug-metabolizing enzymes, influence on the cell cycle and cell differentiation, induction of apoptosis and suppression of proliferation and angiogenesis playing roles in the initiation and secondary modification stages of neoplastic development.

Murphy et al. (2012) estimated usual intakes of nine individual phytonutrients by Americans consuming recommended levels of fruits and vegetables compared to intakes by adults not meeting these recommendations, and sought to identify contributions of food sources to total phytonutrient intakes. The phytonutrients examined in this study are found predominantly in fruits and vegetables. Food consumption data from the NHANES 2003–2006 and phytonutrient concentration data from US Department of Agriculture databases and the published literature were used to estimate energy-adjusted usual intakes. Mean energy-adjusted phytonutrient intakes were compared between subpopulations who consumed recommended amounts of fruits and vegetables versus those who did not. Percentage contributions of each phytonutrient by food source were estimated for all adults. Energy-adjusted intakes of all phytonutrients other than ellagic acid were considerably higher among both men and women meeting dietary recommendations for fruit and vegetable intakes compared to those not meeting the recommendations; energy-adjusted intakes of ellagic acid were higher only among women meeting versus not meeting the recommendations. For five of the nine phytonutrients (α-carotene, β-cryptoxanthin, lycopene, hesperetin, and ellagic acid), a single food typically accounted for 64 % or more of the total intake of the phytonutrient. Energy-adjusted intakes of carotenoids and flavonoids are higher among men and women whose diets conform to dietary guidance for fruits and vegetables. A limited number of foods provide the majority of these phytonutrients. Findings from this research provide important reference information on the phytonutrient contributions of a diet rich in fruits and vegetables.

A key technical challenge in assessing exposures is to measure dietary intake of various foods in the population at large. Labonte' et al. (2012) assessed the validity and the reproducibility of a newly developed web-based, self-administered food frequency questionnaire (web-FFQ). A total of 74 healthy subjects (34 men and 40 women) from the Quebec City metropolitan area were asked to complete, in random order, the web-FFQ, a validated interviewer-administered FFQ (IA-FFQ) and a 3-day food record (3-day FR). The data demonstrate that the web-based FFQ appears to have reasonable validity and good reproducibility for assessing nutrient intakes at the group and individual levels in a population of healthy adults. Thus, there is the potential of using dietary intake of phytonutrients from web-FFQ in pregnant women to assess exposures to these classes of compounds and potentially fetal and infant outcomes.

Clinical outcomes of dietary trials of large doses of carotenoids have not shown consistent results. Eroglu et al. (2012) have developed a potential explanation based on studies of metabolism of β-carotene, the major dietary source of provitamin A. Central cleavage of β-carotene catalyzed by β-carotene oxygenase 1 yields two molecules of retinaldehyde. Subsequent oxidation produces all-trans-retinoic acid (ATRA), which functions as a ligand for a family of nuclear transcription factors, the retinoic acid receptors (RARs). Eccentric cleavage of β-carotene at non-central double bonds is catalyzed by other enzymes and can also occur non-enzymatically. The products of these reactions are β-apocarotenals and β-apocarotenones, whose biological functions in mammals are unknown. The investigators used reporter gene assays to show that none of the β-apocarotenoids significantly activated RARs. Importantly, however, β-apo-14′-carotenal, β-apo-14′-carotenoic acid, and β-apo-13-carotenone antagonized ATRA-induced transactivation of RARs. Competitive radioligand binding assays demonstrated that these putative RAR antagonists compete directly with retinoic acid for high affinity binding to purified receptors. Molecular modeling studies confirmed that β-apo-13-carotenone can interact directly with the ligand binding site of the retinoid receptors. β-Apo-13-carotenone and the β-apo-14′-carotenoids inhibited ATRA-induced expression of retinoid responsive genes in Hep G2 cells. An LC/MS method was developed and demonstrated that 3–5 nM β-apo-13-carotenone was present in human plasma. These findings suggest that β-apocarotenoids function as naturally occurring retinoid antagonists. The antagonism of retinoid signaling by these metabolites may have implications for the activities of dietary β-carotene as a provitamin A and as a modulator of risk for cardiovascular disease and cancer.

The red wine polyphenol resveratrol has been reported to be a calorie restriction mimetic with potential antiaging and antidiabetogenic properties. Park et al. (2012) found that the metabolic effects of resveratrol result from competitive inhibition of cAMP-degrading phosphodiesterases, leading to elevated cAMP levels. The resulting activation of Epac1, a cAMP effector protein, increases intracellular Ca2+ levels and activates the CamK Kβ-AMPK pathway via phospholipase C and the ryanodine receptor Ca2+–release channel. As a consequence, resveratrol increases NAD+ and the activity of Sirt1. Inhibiting PDE4 with rolipram reproduces all of the metabolic benefits of resveratrol, including prevention of diet-induced obesity and an increase in mitochondrial function, physical stamina, and glucose tolerance in mice. Therefore, administration of PDE4 inhibitors may also protect against and ameliorate the symptoms of metabolic diseases associated with aging.

Additional evidence suggesting that resveratrol may have general health benefit is suggested by research into potential for the compound to have preventative effects in cardiovascular disease (CVD). It is generally believed that grape and wine polyphenol resveratrol confers CV benefits, in part by exerting anti-inflammatory effects. Tomé-Carneiro et al. (2012) investigated the effects of a dietary resveratrol-rich grape supplement on the inflammatory and fibrinolytic status of subjects at high risk of CVD and treated according to current guidelines for primary prevention of CVD. Seventy-five patients undergoing primary prevention of CVD participated in this triple-blinded, randomized, parallel, dose–response, placebo-controlled, 1-year follow-up trial. Patients, allocated in 3 groups, consumed placebo (maltodextrin), a resveratrol-rich grape supplement (resveratrol 8 mg), or a conventional grape supplement lacking resveratrol, for the first 6 months and a double dose for the next 6 months. In contrast to placebo and conventional grape supplement, the resveratrol-rich grape supplement significantly decreased high-sensitivity C-reactive protein (−26 %, $p=0.03$), tumor necrosis factor-α (−19.8 %, $p=0.01$), plasminogen activator inhibitor type 1 (−16.8 %, $p=0.03$), and interleukin-6/interleukin-10 ratio (−24 %, $p=0.04$) and increased anti-inflammatory interleukin-10 (19.8 %, $p=0.00$). Adiponectin (6.5 %, $p=0.07$) and soluble intercellular adhesion molecule-1 (−5.7 %, $p=0.06$) tended to increase and decrease, respectively. No adverse effects were observed in any patient. In conclusion, 1-year consumption of a resveratrol-rich grape supplement improved the inflammatory and fibrinolytic status in patients who were on statins for primary prevention of CVD and at high CVD risk. Their results suggest that a dietary intervention with grape resveratrol could be protective in the primary prevention of CVD and thus might offer health benefits and other settings including our consideration as a potential GRAS agent for reducing adverse effects from developmental exposures.

1.5.4.1 Reduction of Oxidative Stress

Harms-Ringdahl et al. (2012) conducted a study to address whether tomato juice protects against formation of reactive oxygen species (ROS) induced by extensive physical exercise in untrained individuals. As a marker of oxidative stress, serum levels of 8-oxodG were monitored using a modified ELISA. An intervention was performed involving 15 untrained healthy subjects who performed a 20 min physical exercise at 80 % of maximum pulse using an ergometer bicycle. Blood samples were taken before and 1 h after the exercise. The procedure was repeated after 5 weeks with a daily intake of 150 ml tomato juice and followed by a 5 weeks wash-out period and another 5 weeks with a daily intake of tomato juice. The results indicated that a daily intake of tomato juice, equal to 15 mg lycopene per day, for 5 weeks significantly reduced the serum levels of 8-oxodG after an extensive physical exercise. Their data suggest that tomato juice has a potential antioxidant effect and may reduce the elevated level of ROS induced by oxidative stress. Since DNA lipids and proteins are constantly exposed to reactive oxygen species (ROS), the molecular injury due to ROS exposure may be reduced by this dietary intervention.

Dibaba et al. (2014) reviewed seven studies including 32,918 participants that measured the association of dietary magnesium (Mg) intake with serum C-reactive protein (CRP) levels in the general population. Since this meta-analysis and systematic review indicates that dietary Mg intake is significantly and inversely associated with serum CRP levels, the potential beneficial effect of Mg intake on chronic diseases may be, at least in part, explained by inhibiting inflammation.

In the setting of fetal alcohol spectrum disorder (FASD) multiple factors such as socioeconomic status, race, genetics, parity, gravidity, age and smoking seem to contribute to the alcohol-induced defects such as abnormal mental, neural, and physical growth. In terms of protective factors, there is evidence that nutrients and prenatal nutritional interventions for FASD could be beneficial (Young et al. 2014). Among these are vitamin A, docosahexaenoic acid, folic acid, zinc, choline, vitamin E, and selenium. Data from studies in animal models and FASD-related research in humans suggest that prenatal multiple-nutrient supplementation could plausibly reduce the impact of FASD (Young et al. 2014).

Kitajima et al. (2015) assessed the effects of long-term supplementation of alpha-Tocopherol on the neurological development of 259 school-aged extremely low birth weight (ELBW) children by comparing subgroups given no alpha-Tocopherol, less than 6 months of alpha-Tocopherol or more than 6 months of alpha-Tocopherol. At age 8 years, the subgroup given long-term supplementation with alpha-Tocopherol had significantly better scores on performance intelligence quotient (IQ).

1.5.4.2 Antidiabetogens

Green tea was suggested as a therapeutic agent for the treatment of diabetes more than 70 years ago, but the mechanisms behind its antidiabetic effect remains elusive. Ortsäter et al. (2012) addressed this issue by feeding a green tea extract (TEAVIGO™) with a high content of epigallocatechin gallate (EGCG) or the thiazolidinedione PPAR-γ agonist rosiglitazone, as positive control, to db/db mice, an animal model for diabetes. Young (7 week-old) db/db mice were randomized and assigned to receive diets supplemented with or without EGCG or rosiglitazone for 10 weeks. Fasting blood glucose, body weight and food intake was measured along the treatment. Glucose and insulin levels were determined during an oral glucose tolerance test after 10 weeks of treatment. Pancreata were sampled at the end of the study for blinded histomorphometric analysis. Islets were isolated and their mRNA expression analyzed by quantitative RT-PCR. The results show that, in db/db mice, EGCG improves glucose tolerance and increases glucose-stimulated insulin secretion. EGCG supplementation reduces the number of pathologically changed islets of Langerhans, increases the number and the size of islets, and heightens pancreatic endocrine area. These effects occurred in parallel with a reduction in islet endoplasmic reticulum stress markers, possibly linked to the antioxidative capacity of EGCG. Their data show that the green tea extract EGCG markedly preserves islet structure and enhances glucose tolerance in genetically diabetic mice. Dietary supplementation with EGCG could potentially contribute to nutritional strategies for the prevention and treatment of type 2 diabetes.

Ruholamin et al. (2014) compared neonatal outcomes in women with gestational diabetes mellitus (GDM) treated with either metformin or insulin in 109 women with GDM who were not adequately control by dietary measures. Treatments were metformin 500 mg once or twice daily or insulin 0.2 IU/kg/day initially. The dose was titrated to achieve target blood glucose values. Neonatal outcomes such as hypoglycemia, birth weight, Apgar score, umbilical artery pH, and hyperbilirubinemia in the 50 women who remained exclusively on metformin were compared with

50 women who treated with insulin. Pregnancy complications or preterm labor were not different significantly between two groups and neonatal outcomes between insulin and metformin groups were similar indicating that metformin appears to be as safe as insulin in the treatment of GDM.

Marques et al. (2014) retrospectively reviewed the clinical records of 186 pregnancies complicated with gestational diabetes mellitus (GDM) to compare the maternal and neonatal outcomes of 32 women who took metformin during pregnancy to 121 women controlled with diet and 33 insulin-treated women. No differences between the metformin group and either the diet or insulin groups with regard to the rates of abortion, prematurity, preeclampsia, macrosomy, small-for-gestational-age (SGA) or large for-gestational-age (LGA) newborns, cesarean deliveries, neonatal intensive care unit (NICU) admissions, and birth malformations or neonatal injuries. No abortions or perinatal deaths were recorded in the metformin group. Ten out of 32 metformin patients required additional insulin. Metformin was not associated with a higher risk of maternal or neonatal complications when compared to the insulin or diet groups.

Samimi et al. (2014) conducted a randomized, double-blind, placebo-controlled clinical trial in 56 women with gestational diabetes mellitus (GDM) assigned to either 1000 mg omega-3 fatty acid supplements containing 180 mg eicosapentaenoic acid and 120 mg docosahexanoic acid (n=28) or placebo (n=28) for 6 weeks. Omega-3 fatty acid supplementation in GDM women had beneficial effects on insulin resistance and a significant reduction in serum high sensitivity C-reactive protein (hs-CRP), however it did not affect plasma glucose, HOMA-B, QUICKI and lipid profiles.

1.5.5 Prospects for HDAC Inhibitors

The reversible acetylation of histones is an important mechanism of gene regulation. During prostate cancer progression, specific modifications in acetylation patterns on histones are apparent. Targeting the epigenome, including the use of histone deacetylase (HDAC) inhibitors, is a novel strategy for cancer chemoprevention. Recently, drugs classified as HDAC inhibitors have shown promise in cancer clinical trials. Ho et al. (2009) have shown that sulforaphane (SFN), a compound found in cruciferous vegetables, inhibits HDAC activity in human colorectal and prostate cancer cells. Based on the similarity of SFN metabolites and other phytochemicals to known HDAC inhibitors, they found that sulforaphane acted as an HDAC inhibitor in the prostate, causing enhanced histone acetylation, derepression of P21 and Bax, and induction of cell cycle arrest/apoptosis, leading to cancer prevention. The ability of SFN to target aberrant acetylation patterns, in addition to effects on phase 2 enzymes, may make it an effective chemoprevention agent. Their studies suggest that high-risk prostate cancer patients might increase survival through simple dietary choices that incorporate readily accessible foods into their diets.

Development of epigenetic drugs holds attraction for translational toxicology applications; however while it is known that histone deacetylases (HDAC) inhibitors impact epigenetics and methylation, the downstream effects are complex. There are four distinct classes of HDAC: class I (HDAC1, 2, 3 and 8), class IIa (HDAC4, 5, 7 and 9), class IIb (HDAC6 and 10) and class IV (HDAC11), with complex and diverse roles (Ratner 2014). Only three HDAC inhibitor drugs have been approved in the US and all for treatment of T-cell lymphoma. Since these drugs are typically pan-HDAC inhibitors there is risk of off-target effects and toxicity. Whether intended for use during development per se or later in life, it is not clear whether the requisite selectivity of actions can or cannot be attained for development of ethical epigenetic pharmaceuticals that could be safety and effectively used to overcome developmental toxicology insults.

1.6 Potential Limitations or Risks

Even in the clearest cases where avoiding or reducing exposure is the key means to avoid harmful effects on development, the reality is often a disappointing failure to achieve avoidance of exposure. This challenge is readily illustrated by the case of ethanol use by pregnant women. While prenatal care guidelines indicate that pregnant women should abstain from alcohol, many pregnant women still consume alcohol during pregnancy. In an effort to understand this behavior, Anderson et al. (2014) studied women's perceptions of information they received about alcohol use during pregnancy after the introduction of abstinence guidelines. By use of semi-structured telephone interviews, women who had completed a pregnancy reported problems with inconsistencies in the information about alcohol use during pregnancy and in the advice provided. Women expressed a need for a clear, consistent message to be provided to women as early as possible and for that message to come from healthcare professionals or another reputable source. Responsible clinicians and public health agencies cannot cease to reinforce the critical value of avoidance of exposure to alcohol and other developmental toxicants but we are also left with the challenge to discern and implement additional protections for the exposed young.

There are a modest number of standard-of-care therapies that constitute in utero fetal exposures to drugs that are approved by regulatory authorities and these serve as a modest guide regarding assessment of risk-benefit relationships for therapies that are intended for use during developmentally sensitive intervals early in life.

Even though the underlying immunological mechanisms and molecular cause-and-effect relationships are not known, vaccines can have nonspecific beneficial effects through their modulation of responses to infections not specifically targeted by the vaccine. This concept of nonspecific vaccine benefits is a challenge to the narrowly focused notion of vaccines as disease-specific interventions. The WHO has noted that the beneficial nonspecific effects provide potentially highly effective and yet relatively easy and affordable solutions to reduce morbidity and mortality in early life (Aaby et al. 2014).

For pregnant women with HIV infection, treatment with combination antiretroviral therapy (ART) is essential to prevent the transmission of HIV from mother to child. Recommendations for ART are similar between pregnant and nonpregnant individuals, in that patients should receive at least three drugs from two different therapeutic classes. Although special consideration must be given to the safety of ART with regard to the fetus, the Antiretroviral Pregnancy Registry which has collected data on live birth outcomes following in utero exposure to antiretroviral agents during all trimesters of pregnancy since 1989, has shown no increased risk for overall birth defects with in utero antiretroviral exposure compared with the national rate of birth defects. Overall, combination ART is safe and effective for use in pregnancy, with preferred treatment regimens having a strong safety profile. Focusing on adherence to a treatment regimen is important to ensure maternal virologic suppression, which is closely associated with minimization of risk for mother-to-child HIV transmission (Gilbert 2014).

Despite the extensive body of evidence that informs regulatory decisions on pharmaceutical products, significant uncertainties persist, including the underlying variability in human biology, factors associated with the chemistry of a drug, and limitations in the research and clinical trial process itself that might limit the generalizability of results. As a result, regulatory reviewers are consistently required to draw conclusions about a drug's safety and efficacy from imperfect data. Identifying and evaluating sources of uncertainty in a regulatory application is an important part of an FDA new drug application reviewers' work; however, drawing conclusions in the face of uncertainty can be a complex and challenging task (Caruso et al. 2014).

Looking forward to the future advances in expanding therapeutic approvals in both children and pregnant women, Gonzalez et al. (2015) have recently argued that the past 15 years of modest progress in seeking drug approvals for pediatric use can help narrow the knowledge gap in pharmacokinetic, safety, and efficacy that would be needed in order for more obstetrical approvals to progress. They argue for the use of innovative clinical trial designs, minimal risk research methods, increased understanding of developmental pharmacology, multidisciplinary research teams, increased clinical pharmacology expertise and training, collaborative research networks, and critical legislative changes. These two populations, pediatric and obstetric, share similar drug development challenges.

1.7 Conclusions and Future Therapeutic Prospects

Human exposures to chemicals, physical agents and social factors are inevitable. All too often the medical practitioner's view of toxicology and human health is the assumption that the old adage of "Absence of evidence (of effect or harm) is not evidence of absence (of effect or harm)" can be ignored! To overcome that therapeutic nihilism, a simple but demanding challenge faces us. To optimize human health, all concerned stakeholders including regulators, commercial entities, health providers, investigators and the public at large need to know the following:

(a) which exposures (dose, duration, window of exposure, other susceptibility factors, etc.) pose a risk that justifies actions of
(b) public health intervention(s) to reduce exposure or
(c) individual health care therapeutic or mitigating treatments. If we are to translate our environmental health science (discovery) into action (safe and effective interventions), we must assert and show how this continuum of discovery will equip medical practitioners with valid, applicable therapies such as effective GRAS treatments and "ethical pharmaceuticals" for care of their patients.

Concerning the timing of interventions for prevention or mitigation, we do need to consider that, for certain adverse effects/disease predispositions due to environmental exposures, there may need to be sustained, continuing interventions. Nevertheless, key windows for interventions are likely to be the same as those in which adverse effects are elicited. Thus, in utero/neonatal/early childhood, adolescent/ peripubertal and then menopausal transition in women may be particularly responsive since organizational/re-organizational processes are biologically dynamic in those intervals.

We must get beyond the simplistic notion that the only method to reduce risk of adverse health outcomes from developmental exposures is to "avoid exposure". While this is not advocacy of a "go ahead and pollute" policy, we must acknowledge that there are different sorts of exposures. Some exposures are intentional (pharmaceutical drug use), some are elective (ethanol consumption, tobacco smoking, etc.) and some are quasi-elective (dietary choices); however many exposures of concern are truly unintended (air quality, water quality, workplace, some dietary contaminants) and essentially inevitable. Some potential toxicants are already ubiquitously present in the environment such that avoiding exposure is already moot. Body burdens of persistent compounds won't "go away" tomorrow by making a decision about what you eat today. Particulates in the air are unavoidable depending on where you live. From ethical and public health perspectives, this category of unintended exposures merits the greatest attention for development of possible intervention/ mitigation beyond the simplistic approach of "Just go and reduce your exposure." Therefore, we need to ascertain how to translate results from research toxicology and epidemiology studies into interventions that can/should plausibly mitigate adverse health effects of exposures. These interventions will almost certainly need to be GRAS (generally-regarded-as-safe) or will depend upon use of well-established/repurposed pharmaceuticals. It seems implausible to speculate that private industry would at the present time be able to justify committing the resources to develop new chemical entities specifically for developmental translational toxicology applications for prevention or reduction of adverse health effects due to early-in-life environmental health exposures. Nevertheless, for the future, the critical window of the maternal-fetal/neonatal/childhood/adolescent interval is wide open for investigation and development of therapeutic/mitigative interventions to reduce diseases in the later lifespan of exposed offspring.

Conflict of Interests All authors [C.H., M.W., D.A., and I.O.] declare that they have no competing interests. Quintiles, Inc. [C.H.] is a pharmaceutical services company that conducts pharmaceutical and biotechnology research and product development for numerous sponsors but does not own pharmaceutical or biotechnology products. Integrated Laboratory Systems, Inc. [D.A.] is a research and testing services company that does not own pharmaceutical or biotechnology products.

Acknowledgments None.

References

Aaby P, Kollmann TR, Benn CS (2014) Nonspecific effects of neonatal and infant vaccination: public-health, immunological and conceptual challenges. Nat Immunol 15(10):895–899

ACOG Practice Bulletin (2008) Clinical management guidelines for obstetrician-gynecologists use of psychiatric medications during pregnancy and lactation. Obstet Gynecol 111:1001–1020

Anderson AE, Hure AJ, Kay-Lambkin FJ, Loxton DJ (2014) Women's perceptions of information about alcohol use during pregnancy: a qualitative study. BMC Public Health 14(1048), 10 pages

Ashley-Martin J, Dodds L, Arbuckle TE, Ettinger AS, Shapiro GD, Fisher M, Morisset A-S, Taback S, Bouchard MF, Monnier P, Dallaire R, Fraser WD (2014) A birth cohort study to investigate the association between prenatal phthalate and bisphenol A exposures and fetal markers of metabolic dysfunction. Environ Heal 13:84, 14 pages

Azad M, Bridgman S, Becker A, Kozyrskyj A (2014) Infant antibiotic exposure and the development of childhood overweight and central adiposity. Int J Obes 38:1290–1298. doi:10.1038/ijo.2014.119

Barouki et al (2012) Developmental origins of non-communicable disease: implications for research and public health. Environ Health 11:42 pp 1–9. http://www.ehjournal.net/content/11/1/42

Bellingham M et al (2013) Exposure to chemical cocktails before or after conception – the effect of timing on ovarian development. Mol Cell Endocrinol 376:156–172

Bernal AJ, Jirtle RL (2010) Epigenomic disruption: the effects of early developmental exposures. Birth Defects Res A Clin Mol Teratol 88(10):938–944. doi:10.1002/bdra.20685

Böttner M et al (2014) Estrogen receptor beta: tissue distribution and the still largely enigmatic physiological function. J Steroid Biochem Mol Biol 139:245–251

Boucher O et al (2014) Domain-specific effects of prenatal exposure to PCBs, mercury, and lead on infant cognition: results from the environmental contaminants and child development study in nunavik. Environ Health Perspect 122:310–316. doi:10.1289/ehp.1206323

Buss C, Graham A, Rudolph M, Rasmussen J, Entringer S, Wadhwa P, Fair D (2014) Maternal interleukin-6 concentrations during pregnancy and newborn functional brain connectivity. Society for Neuroscience 2014 Annual Meeting, Abstract only. Program Control #: 8123

Cajigal S, Tassinari MS, Best J (2015) FDA updates labels for pregnant and breastfeeding women. Medscape. 9 January 2015

Caruso D, English R, Claiborne A (2014) Characterizing and Communicating Uncertainty in the Assessment of Benefits and Risks of Pharmaceutical Products: workshop summary. Forum on Drug Discovery, Development, and Translation; Board on Health Sciences Policy; Institute of Medicine. National Academies Press, Washington, DC. ISBN 978-0-309-31000-0. 140 pages

Cheelo M, Lodge CJ, Dharmage SC, Simpson JA, Matheson M, Heinrich J, Lowe AJ (2015) Paracetamol exposure in pregnancy and early childhood and development of childhood asthma: a systematic review and meta-analysis. Arch Dis Child 100:81–89. doi:10.1136/archdischild-2012-303043

Coleman T, Cooper S, Thornton JG, Grainge MJ, Watts K et al (2012) A randomized trial of nicotine-replacement therapy patches in pregnancy. N Engl J Med 366:808–818
De Flora S, Izzotti A, D'Agostini F, Balansky R (2001) Mechanisms of N-acetylcysteine in the prevention of DNA damage and cancer, with special reference to smoking-related end-points. Carcinogenesis 22:999–1013
Demissie K, Marcella SW, Breckenridge MB, Rhoads GG (1998) Maternal asthma and transient tachypnea of the newborn. Pediatrics 102(1 Pt 1):84–90
Dibaba DT, Xun P, He K (2014) Dietary magnesium intake is inversely associated with serum C-reactive protein levels: meta-analysis and systematic review. Eur J Clin Nutr 68:510–516. doi:10.1038/ejcn.2014.7
Dougan M, Willett W, Michels K (2015) Prenatal vitamin intake during pregnancy and offspring obesity. Int J Obes 39:69–74. doi:10.1038/ijo.2014.107
Eroglu A, Hruszkewycz D, de la Sena C, Narayanasamy Riedl SK et al (2012) Naturally occurring eccentric cleavage products of provitamin A β-carotene function as antagonists of retinoic acid receptors. J Biol Chem 287:15886–15895. doi:10.1074/jbc.M111.325142jbc.M111.325142
Ettinger A et al (2014) Maternal blood, plasma, and breast milk lead: lactational transfer and contribution to infant exposure. Environ Health Perspect 122:87–92. doi:10.1289/ehp.1307187
Factor-Litvak P, Insel B, Calafat AM, Liu X, Perera F et al (2014) Persistent associations between maternal prenatal exposure to phthalates on child IQ at age 7 years. PLoS One 9(12):e114003. doi:10.1371/journal.pone.0114003
Forrest K, Stuhldreher W (2011) Prevalence and correlates of vitamin D deficiency in US adults. Nutr Res 31:48–54
Fraser A, Lawlor DA (2014) Long-term health outcomes in offspring born to women with diabetes in pregnancy. Curr Diab Rep 14:489. doi:10.1007/s11892-014-0489-x
Gatford KL, Kaur G, Falcão-Tebas F, Wadley GD et al (2014) Exercise as an intervention to improve metabolic outcomes after intrauterine growth restriction. Am J Physiol Endocrinol Metab 306:E999–E1012. doi:10.1152/ajpendo.00456.2013
Gilbert E (2014) HIV antiretrovirals in pregnancy: which are safe? Medscape, 22 October 2014
Gilbert ME et al (2012) Developmental thyroid hormone disruption: prevalence, environmental contaminants and neurodevelopmental consequences. Neurotoxicology 33:842–852
Gonzalez D, Boggess KA, Cohen-Wolkowiez M (2015) Lessons learned in pediatric clinical research to evaluate safe and effective use of drugs in pregnancy. Obstet Gynecol 125:953–958
Guerrero-Bosagna CM, Skinner MK (2014) Environmental epigenetics and phytoestrogen/phytochemical exposures. J Steroid Biochem Mol Biol 139:270–276
Hansen S et al (2014) Maternal concentrations of persistent organochlorine pollutants and the risk of asthma in offspring: results from a prospective cohort with 20 years of follow-up. Environ Health Perspect 122:93–99. doi:10.1289/ehp.1206397
Harms-Ringdahl M, Jenssen D, Haghdoost S (2012) Tomato juice intake suppressed serum concentration of 8-oxodG after extensive physical activity. Nutr J 11:1–5. doi:10.1186/1475-2891-11-29
Hart P, Lucas R, Walsh J, Zosky G et al (2015) Vitamin D in fetal development: findings from a birth cohort study. Pediatrics 135(1):e167–e173. doi:10.1542/peds.2014-1860
Henley D, Korach K (2006) Endocrine-disrupting chemicals use distinct mechanisms of action to modulate endocrine system function. Endocrinology 147:S25–S32
Ho E, Clarke J, Dashwood R (2009) Dietary sulforaphane, a histone deacetylase inhibitor for cancer prevention. J Nutr 139:2393–2396
Honein M, Gilboa S, Broussard C (2013) The need for safer medication use in pregnancy. Expert Rev Clin Pharmacol 6(5):453–455
Hughes C, Waters M, Allen D, Obasanjo I (2013a) Translational toxicology: integrated research strategies. BMC Pharmacol Toxicol 14:51. http://www.biomedcentral.com/2050-6511/14/51
Hughes C, Waters M, Obasanjo I, Allen D (2013b) Translational developmental toxicology: prospects for protective therapeutic obstetrical and neonatal interventions. J Neonatal Biol 2:122. doi:10.4172/2167-0897.1000122

IUPHAR/BPS Guide to PHARMACOLOGY (2014) Current database content for version 2014. 3 released 5 November 2014

Jensen E, Daniels J, Stürmer T, Robinson W et al (2014) Maternal hormonal contraceptive use and offspring overweight or obesity. Int J Obes 38:1275–1281. doi:10.1038/ijo.2014.114

Jiang H, Wen Y, Hu L, Miao T, Zhang M, Dong J (2014) Serum MicroRNAs as Diagnostic Biomarkers for Macrosomia. Reprod Sci 22(6):664–671. 17 December 2014. pii: 1933719114561557.

Kaiser L, Campbell C (2014) Practice paper of the Academy of Nutrition and Dietetics: nutrition and lifestyle for a healthy pregnancy outcome. J Acad Nutr Diet 114(9):1447. July 2014, pp 1–13

Kajantie E, Strang-Karlsson S, Hovi P, Wehkalampi K, Lahti J, Kaseva N, Järvenpää A-L, Räikkönen K, Eriksson JG, Andersson S (2015) Insulin sensitivity and secretory response in adults born preterm: the Helsinki study of very low birth weight adults. J Clin Endocrinol Metab 100:244–250. doi:10.1210/jc.2014-3184

Källén B, Rydhstroem H, Aberg A (2000) Asthma during pregnancy – a population based study. Eur J Epidemiol 16(2):167–171

Kalkbrenner A, Windham G, Serre M, Akita Y et al (2015) Particulate matter exposure, prenatal and postnatal windows of susceptibility, and autism spectrum disorders. Epidemiology 26:30–42

Kay J, Chow W, Chan T, Lo S, Kwok O et al (2003) Acetylcysteine for prevention of acute deterioration of renal function following elective coronary angiography and intervention. J Am Med Assoc 289:553–558

Kilcoyne KR et al (2014) Fetal programming of adult Leydig cell function by androgenic effects on stem/progenitor cells. PNAS Early Edition. www.pnas.org/cgi/doi/10.1073/pnas.1320735111

Kimmel et al (2011) Critical windows of children's development and susceptibility to environmental toxins. Encycl Environ Health. 834–843. doi:10.1016/B978-0-444-52272-6.00012-X

Kincl F (1990) Hormone toxicity in the newborn. Springer, Berlin

Kitajima H, Kanazawa T, Mori R, Hirano S, Ogihara T, Fujimura M (2015) Long-term alpha–Tocopherol supplements may improve mental development in extremely low birth weight infants. Acta Paediatr 104(2):e82–e89. doi:10.1111/apa.12854

Knickmeyer R et al (2014) Rate of chiari I malformation in children of mothers with depression with and without prenatal SSRI exposure. Neuropsychopharmacology 39:2611–2621. doi:10.1038/npp.2014.114

Kwon HL, Belanger K, Bracken MB (2004) Effect of pregnancy and stage of pregnancy on asthma severity: a systemic review. Am J Obstet Gynecol 190:1201–1210

Labonte' M, Cyr A, Baril-Gravel L, Royer M, Lamarche B (2012) Validity and reproducibility of a web-based, self-administered food frequency questionnaire. Eur J Clin Nutr 66:166–173. doi:10.1038/ejcn.2011.163

Liew Z et al (2014) Acetaminophen use during pregnancy, behavioral problems, and hyperkinetic disorders. JAMA Pediatr. E1–E8. doi:10.1001/jamapediatrics.2013.4914

Madon-Simon M, Cowley M, Garfield AS, Moorwood K, Bauer SR, Ward A (2014) Antagonistic roles in fetal development and adult physiology for the oppositely imprinted Grb10 and Dlk1 genes. Madon-Simon et al. BMC Biol 12:771. 22 p. doi:10.1186/s12915-014-0099-8

Marques P, Carvalho MR, Pinto L, Guerra S (2014) Metformin safety in the management of gestational diabetes. Endocr Pract 20(10):1022–1031

McEvoy CT et al (2014) Vitamin C supplementation for pregnant smoking women and pulmonary function in their newborn infants: a randomized clinical trial. JAMA 311(20):2074–2082. doi:10.1001/jama.2014.5217

McLachlan J (1980) Estrogens in the environment. Elsevier, New York

McLachlan J (1985) Estrogens in the environment II: influences on development. Elsevier, New York

Metzger MJ, Halperin AC, Manhart LE, Hawes SE (2013) Association of maternal smoking during pregnancy with infant hospitalization and mortality due to infectious diseases. Pediatr Infect Dis J 32(1):e1–e7. doi:10.1097/INF.0b013e3182704bb5

Miller M, Marty M (2010) Impact of environmental chemicals on lung development. Environ Health Perspect 118:1155–1164. doi:10.1289/ehp.0901856

Morales E, Rodriguez A, Valvi D, Iñiguez C, Esplugues A, Vioque J, Marina L, Jiménez A, Espada M, Dehli C, Fernández-Somoano A, Vrijheid M, Sunyer J (2015) Deficit of vitamin D in pregnancy and growth and overweight in the offspring. Int J Obes 39:61–68. doi:10.1038/ijo.2014.165

Murphy M, Barraj L, Herman D, Bi X, Cheatham R et al (2012) Phytonutrient intake by adults in the United States in relation to fruit and vegetable consumption. J Acad Nutr Diet 112:222–229

NAEPP Working Group Report on Managing Asthma During Pregnancy: Recommendations for Pharmacologic Treatment—Update (2005) U.S. Department of Health and Human Services. National Institutes of Health National Heart, Lung, and Blood Institute. NIH Publication No. 05–5236. March 2005

Neal M, Mulligan Tuttle A, Casper R, Lagunov A, Foster W (2010) Aryl hydrocarbon receptor antagonists attenuate the deleterious effects of benzo[a]pyrene on isolated rat follicle development. Reprod Biomed Online 21:100–108

Negro R, Stagnaro-Green A (2014) Clinical aspects of hyperthyroidism, hypothyroidism, and thyroid screening in pregnancy. Endocr Pract 20(6):597–607

Nigam DK (2015) What do drug transporters really do? Nat Rev Drug Discov 14:28–44

NIH Panel on Treatment of HIV-Infected Pregnant Women and Prevention of Perinatal Transmission (2014) Recommendations for use of antiretroviral drugs in pregnant HIV-1-infected women for maternal health and interventions to reduce perinatal HIV transmission in the United States. Available at http://aidsinfo.nih.gov/contentfiles/lvguidelines/PerinatalGL.pdf. Downloaded from http://aidsinfo.nih.gov/guidelines on 9/4/2014

Nolan M, Berghella V, Wisner K (2014) Pregnant Women Must Be Studied Too. Huffington Post. Posted: 10/21/2014 1:27 pm EDT Updated: 10/21/2014 4:59 pm EDT. http://www.huffingtonpost.com/society-for-womens-health-research/

Ortsäter H, Grankvist N, Wolfram S, Kuehn N, Sjöholm A (2012) Diet supplementation with green tea extract epigallocatechin gallate prevents progression to glucose intolerance in db/db mice. Nutr Metab 9:1–10. doi:10.1186/1743-7075-9-11

Park S, Ahmad F, Philp A, Baar K, Williams T et al (2012) Resveratrol ameliorates aging-related metabolic phenotypes by inhibiting cAMP phosphodiesterases. Cell 148:421–433

Pauli JM, Repke JT (2013) Update: obstetrics. OBG Manage 25:28–31

Ratner M (2014) Small biotech steers HDAC inhibitor to clinic. Nat Biotechnol 32(9):853–854

Reynolds RM et al (2013) Maternal obesity during pregnancy and premature mortality from cardiovascular event in adult offspring: follow-up of 1 323 275 person years. BMJ 347:f4539 pp 1–10. doi:10.1136/bmj.f4539

Rogne T et al (2015) Fetal growth, cognitive function, and brain volumes in childhood and adolescence. Obstet Gynecol 125:673–682. doi:10.1097/AOG.0000000000000694

Rudel RA, Perovich LJ (2009) Endocrine disrupting chemicals in indoor and outdoor air. Atmos Environ 43(1):170–181. doi:10.1016/j.atmosenv.2008.09/025

Ruholamin S, Eshaghian S, Allame Z (2014) Neonatal outcomes in women with gestational diabetes mellitus treated with metformin in compare with insulin: a randomized clinical trial. J Res Med Sci 19:970–975

Samimi M, Jamilian M, Asemi Z, Esmaillzadeh A (2014) Effects of omega-3 fatty acid supplementation on insulin metabolism and lipid profiles in gestational diabetes: Randomized, double-blind, placebo-controlled trial. Clin Nutr. pii: S0261-5614(14)00169-1. doi:10.1016/j.clnu.2014.06.005. [Epub ahead of print]

Schatz M, Zeiger RS, Hoffman CP, Harden K, Forsythe A, Chilingar L et al (1995) Perinatal outcomes in the pregnancies of asthmatic women: a prospective controlled analysis. Am J Respir Crit Care Med 151:1170–1174

Sørensen I, Joner G, Jenum P, Eskild A, Torjesen P et al (2012) Maternal serum levels of 25-hydroxy-vitamin D during pregnancy and risk of type 1 diabetes in the offspring. Diabetes 61:175–178

Stayner L et al (2014) Exposure to brominated trihalomethanes in water during pregnancy and micronuclei frequency in maternal and cord blood lymphocytes. Environ Health Perspect 122:100–106. doi:10.1289/ehp.1206434

Swamy G, Reddick K, Brouwer R, Pollak K, Myers E (2011) Smoking prevalence in early pregnancy: comparison of self-report and anonymous urine cotinine testing. J Matern Fetal Neonatal Med 24:86–90

Tanvig M, Vinter CA, Jørgensen JS, Wehberg S, Ovesen PG, Beck-Nielsen H, Christesen HT, Jensen DM (2015) Effects of lifestyle intervention in pregnancy and anthropometrics at birth on offspring metabolic profile at 2.8 years: results from the lifestyle in pregnancy and offspring (LiPO) study. J Clin Endocrinol Metab 100:175–183. doi:10.1210/jc.2014-2675

Thacher JD, Gruzieva O, Pershagen G, Neuman A, Wickman M, Kull I, Melén E, Bergström A (2014) Pre- and postnatal exposure to parental smoking and allergic disease through adolescence. Pediatrics 134(3):428–34. doi:10.1542/peds.2014-0427

Tolppanen A, Sayers A, Fraser W, Lawlor D (2012) Association of serum 25-hydroxyvitamin D3 and D2 with academic performance in childhood: findings from a prospective birth cohort. J Epidemiol Community Health 66:1137–1142. doi:10.1136/jech-2011-200114

Tomé-Carneiro J, Gonzálvez M, Larrosa M, Yáñez-Gascón M, García-Almagro F et al (2012) One-year consumption of a grape nutraceutical containing resveratrol improves the inflammatory and fibrinolytic status of patients in primary prevention of cardiovascular disease. Am J Cardiol 110:356–363. doi:10.1016/j.amjcard.2012.03.030

Tsuda H, Ohshima Y, Nomoto H, Fujita K, Matsuda E et al (2004) Cancer prevention by natural compounds. Drug Metab Pharmacokinet 19:245–263

US EPA (2012) Integrated Risk Information Systems (IRIS) glossary. http://www.epa.gov/iris/help_gloss.htm, http://ofmpub.epa.gov/sor_internet/registry/termreg/searchandretrieve/glossariesandkeywordlists/search.do?details=&glossaryName=IRIS%20Glossary

van Dijk S, Molloy P, Varinli H, Morrison J, Muhlhausler B, members of EpiSCOPE (2015) Epigenetics and human obesity. Int J Obes 39:85–97. doi:10.1038/ijo.2014.34

van Schooten F, Nia A, De Flora S, D'Agostini F, Izzotti A et al (2002) Effects of oral administration of N-acetyl-L-cysteine: a multi-biomarker study in smokers. Cancer Epidemiol Biomarkers Prev 11:167–175

van Uitert E et al (2013) Periconception maternal characteristics and embryonic growth trajectories. Hum Reprod 28(12):3188–3196

Vandenberg L (2014) Low-dose effects of hormones and endocrine disruptors. Vitam Horm 94:129–165. Elsevier Inc. ISSN 0083–6729. doi:10.1016/B978-0-12-800095-3.00005-5

Veiga-Lopez A, Pennathur S, Kannan K, Patisaul HB, Dolinoy DC, Zeng L, Padmanabhan V (2015) Impact of gestational bisphenol A on oxidative stress and free fatty acids: human association and interspecies animal testing studies. Endocrinol. Published online 20 Jan 2015, 12 p. doi:10.1210/en.2014-1863

Walker N, Howe C, Glover M, McRobbie H, Barnes J et al (2014) Cytisine versus nicotine for smoking cessation. N Engl J Med 371:2353–2362. doi:10.1056/NEJMoa1407764

Ward RM, Benitz WE, Benjamin DK, Blackmon L, Giacoia GP, Hudak M, Laskylo T, Rodriguez W, Selen A (2006) Criteria supporting the study of drugs in the newborn. Clin Ther 28(9):1385–1398

Whisner CM, Young BE, Pressman EK, Queenan RA, Cooper EM, O'Brien KO (2015) Maternal diet but not gestational weight gain predicts central adiposity accretion in utero among pregnant adolescents. Int J Obes (Lond) 39:565–574. doi:10.1038/ijo.2014.202.

Whyatt RM, Perzanowski MS, Just AC, Rundle AG et al (2014) Asthma in inner-city children at 5–11 years of age and prenatal exposure to phthalates: the Columbia Center for Children's Environmental Health Cohort. Environ Health Perspect 122:1141–1146. doi:10.1289/ehp.1307670

Wiley MG, Bolton JL, Simmons LA, Ryan B, Truong S, Bilbo S (2014) Developmental programming of body weight, neuroinflammation, and behavior by western diets. Society for Neuroscience 2014 Annual Meeting, Abstract only. Program#/Poster#: 545.19/PP8

Wu L, Russell D, Wong S, Chen M et al (2015) Mitochondrial dysfunction in oocytes of obese mothers: transmission to offspring and reversal by pharmacological endoplasmic reticulum stress inhibitors. Development 142:681–691. doi:10.1242/dev.114850

Young JK, Giesbrecht HE, Eskin MN, Aliani M, Miyoung SM (2014) Nutrition implications for fetal alcohol spectrum disorder. Adv Nutr 5:675–692

Chapter 2
The Role of Toxicokinetics and Toxicodynamics in Developmental and Translational Toxicology

Edward L. Croom

Abstract Toxicokinetics and toxicodynamics vary depending on dose, developmental stage and exposure timing. The embryo is the developmental stage most susceptible to toxicants. Many birth defects occur only as a result of first trimester exposures when organogenesis is occurring and the formation of new organs and structures (e.g., limbs) can be disrupted. The Absorption, Distribution, Metabolism and Elimination (ADME) of toxicants changes dramatically as the embryo grows from a single-celled organism into a fetus with multiple organ systems. Before birth the mother's behavior and body largely determines the toxicokinetics of chemicals. Immediately after birth the newborn's organs must suddenly function independently of the mother. Birth itself is a major trigger causing widespread changes in gene and protein expression patterns. The average infant doubles in size in under a year's time, the first of roughly four doublings in size after birth. Toddlers' mobility and mouthing behavior put them at greatest risk for accidental poisonings. Older children's growth and development makes their metabolism unpredictable. Adolescents develop adult metabolic capacity but adolescent behavior can make them more likely to abuse some toxicants (e.g., inhalants). ADME changes as pregnancy progresses for adolescents and adults. In general adults are the least susceptible to toxicants but occupation becomes a major risk factor for exposure. Elderly adults become somewhat more susceptible to toxicants as organ function declines with age. Beer's list, unit doses of iron, and the X classification system are examples of translational toxicology designed to protect elderly patients, toddlers and embryos respectively.

Keywords Toxicokinetics • Toxicodynamics • Developmental toxicology • Translational toxicology • Teratogens • ADME

E.L. Croom (✉)
VP Chemistry, BOV Solutions, INC, Statesville, NC 28677, USA

230 Union St N, Concord, NC 28025, USA
e-mail: elcroom@gmail.com

C.L. Hughes, M.D. Waters (eds.), *Translational Toxicology*, Molecular and Integrative Toxicology, DOI 10.1007/978-3-319-27449-2_2

2.1 Introduction

That "the dose makes the poison" is a well-known concept in toxicology. Toxicokinetics describes how a toxicant (i.e., a poison) enters the body and reaches a target tissue. Toxicodynamics describes what happens to that tissue once the toxicant reaches an effective dose. Susceptibility to toxicants changes during development. Doses that may be safe for adults can be dangerous to children and devastating to embryos. Translational toxicology is about how our knowledge of toxicants, anatomy, physiology, behavior, analytical chemistry, engineering and medicine can be combined to both prevent and treat damaging exposures to chemicals.

Toxicokinetics can be divided into four major stages; Absorption, Distribution, Metabolism and Excretion or Elimination (ADME). Behavior and genetics can also have an impact on ADME. A simple example of how behavior can impact the absorption of toxicants can be seen with smog exposures. People who wear masks when going outdoors will absorb less smog particles than those who spend the same amount of time outdoors breathing unfiltered air. Behavior can also alter distribution, metabolism and elimination. There can be significant alterations in the toxicokinetic profiles of individuals based on diet and medications taken (e.g., drinking grapefruit juice, taking amoxicillin). Similarly, mutations can significantly alter the toxicokinetic profiles of individuals. These variations in toxicokinetics can contribute to variations in toxicodynamics.

Toxicodynamics can also be divided into different levels; molecular, cellular, tissue and organ. Molecules, cells, tissues and organs form and disappear at different stages of development so the timing of the exposures will also be examined. For a given toxicant there can be a limited window of susceptibility based on developmental changes. Thalidomide when taken early during pregnancy caused many children to be born without limbs. However, this only occurred after exposures early during pregnancy and the mothers were never at risk of losing their already formed arms and legs (Kim and Scialli 2011).

Translational toxicology involves both prevention and treatment efforts. Prevention efforts include screening, safety labels and features on products, personal protection equipment and legal restrictions on prescriptions and purchases. Treatment efforts can be divided into prophylactic treatments and responsive treatments. Prophylactic treatments are for cases where exposures are expected but are generally unavoidable. Responsive treatments are given after an exposure has occurred. There is always a delay between exposure and responsive treatments. This delay can be worsened if significant time has to be spent identifying the toxicant.

This chapter is primarily structured around the different human developmental stages. Nine stages will be described; embryos, fetuses, infants, toddlers, children, adolescents, adults, pregnant women and adolescents, and elderly adults. For each developmental stage clinically relevant toxicokinetic and toxicodynamic changes will be explored through the use of representative toxicants known to cause disproportionate harm to that developmental stage. The toxicokinetics and toxicodynamics of representative toxicants affecting earlier developmental stages will also be compared with adult responses. S-Warfarin, the most active form of one of the most widely used prescription medications, will be followed throughout development as example of how changes during development impact the toxicokinetics and toxicodynamics of a single compound. For all of the representative toxicants the University of Alberta/Metabolomics Innovation Center DrugBank database (Wishart et al. 2006) and the U.S. Environmental Protection Agency's ACToR database (Judson et al. 2012) were searched as sources. Additional sources were used as needed. Finally, examples of translational toxicology prevention or treatment efforts will be provided for each representative toxicant.

2.2 Embryos

2.2.1 Introduction

The embryonic stage starts with fertilization and carries on until the end of the first 8 weeks of gestation. During this time the embryo changes from a single-celled organism into one that has developed rudimentary organs. These periods of organogenesis make the embryonic stage the most susceptible to developmental toxicants. *Absorption* always occurs first through the mother and then through the embryo. The specific changes in maternal absorption will be described later on in the chapter when pregnancy is reviewed.

At the earliest stages of embryonic development absorption is accomplished through diffusion. Active transport and passive transport are relatively limited within the embryo as many transporters have little or no expression during that stage. The embryo is completely dependent upon the mother and is both attached to and contained by the uterus. Maternal behavior plays a major role in determining what toxicants are absorbed (e.g., drinking alcohol during pregnancy).

Distribution within the embryo is limited to diffusion at first but becomes more complex as the vascular system begins to develop. The placenta develops from the embryo and can protect the embryo from some toxicants (e.g., ivermectin, a broad-spectrum antiparasitic drug). *Metabolism* is predominantly maternal. Many metabolizing enzymes will not be expressed until later in development. However, some metabolic enzymes have been detected as early as 8 weeks gestation (Hines 2008). *Elimination* is predominantly maternal with the exchange of wastes occurring via the placenta. The excretory system in the embryo is rudimentary. Embryos develop three sets of kidneys but only one pair turns into the adult kidneys. Urine production does not really begin until the fetal period. Production of meconium also does not occur until the fetal period. Exhalation of air will not occur until after birth although the lungs are already forming during the embryonic period.

2.2.2 *Embryos and Etretinate*

2.2.2.1 Toxicokinetics

Etretinate is a retinoid approved by the FDA in 1986 to treat severe psoriasis. Retinoids (Fig. 2.1) are chemically related to retinol (i.e., Vitamin A). Retinol is needed throughout development (Blaner 2013) but is teratogenic in high doses. To prevent birth defects retinol levels are normally tightly regulated. Protoretinoids, naturally occurring compounds such as betacarotene, are converted into retinol after being consumed. In general proretinoids are not teratogenic. Synthetic retinoids (e.g., etretinate, isotretinoin) are sometimes teratogenic.

Etretinate is taken orally. Etretinate *absorption* is improved by being taken with lipid rich foods (Wishart et al. 2006). Etretinate *distribution* includes being ~99 % bound to plasma lipoproteins. Etretinate concentrates in body fat and lipid rich organs (e.g., liver) (Judson et al. 2012). Etretinate can cross the placenta. Etretinate *metabolism* is primarily hepatic and involves the formation of active and inactive metabolites. Etretinate *elimination* occurs in two phases. With a single dose etretinate's half-life is ~ 8–13 h. With repeated dosing etretinate's half-life increases to ~120 days, one of the longest half-lives of any drug. Etretinate has been detected in human plasma over 2 years after the last dose was taken (Judson et al. 2012).

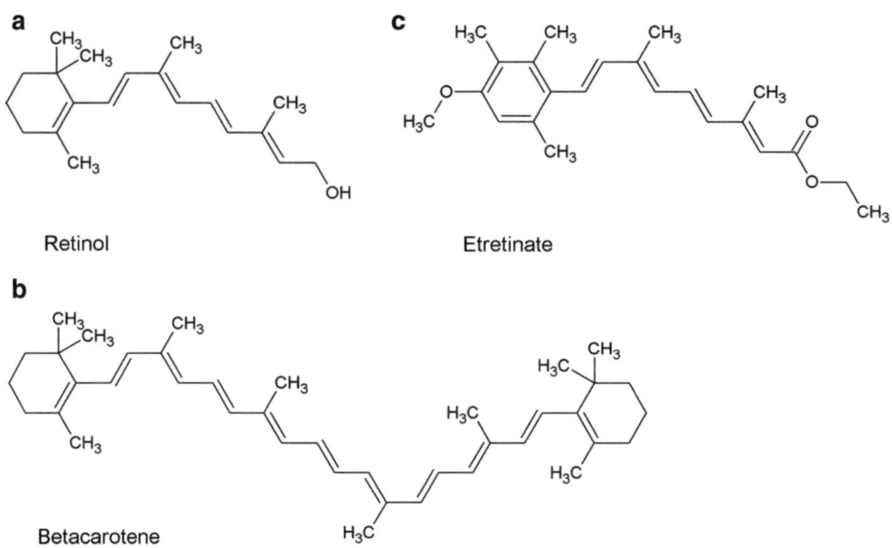

Fig. 2.1 Retinoids. Retinol (**a**) is formed in the body from protoretinoids such as Betacarotene (**b**), Etretinate (**c**), a synthetic retinoid, is a potent teratogen

2.2.2.2 Toxicodynamics

Etretinate's primary *molecular* targets are the retinoic acid receptors (RAR). Etretinate is ~100 times more potent than retinol (Teratology Society 1987). Etretinate was designed to treat skin conditions but the retinoic acid receptor is widely expressed. The major *organs* associated with retinoic acid receptor toxicity are the ears, brain, heart and mouth. In the most severe cases pregnancy loss and stillbirths occur. The retinoic acid syndrome window of susceptibility closes after the embryonic period. Etretinate has even been used successfully to treat an infant born with congenital ichthyosis (Lawlor and Peiris 1985).

2.2.2.3 Comparison with Adults

Adults are protected from the developmental toxicities associated with etretinate embryonic exposures. Reported adult toxicities tend to be reversible such as muscle damage (Judson et al. 2012).

2.2.2.4 Translational Toxicology

Etretinate is so potent and etretinate's half-life is so long (Fig. 2.2) that cases have been reported of women giving birth to infants with retinoid related birth defects even though they stopped taking etretinate months before becoming pregnant (Judson et al. 2012). Etretinate's ability to cause birth defects months after dosing ends has made etretinate unique among toxicants in being associated with a lifetime ban on blood donation (American Red Cross 2014). *Prevention* efforts include drug labeling and prescription restrictions. Etretinate is no longer approved to treat psoriasis. Its approval was lost in the United States and Canada in 1998 and 1996 respectively. Etretinate is used to treat lymphomas but is only available to women who are on effective forms of birth control (Wishart et al. 2006).

2.2.3 Embryos and Thalidomide

2.2.3.1 Toxicokinetics

Thalidomide was originally developed as anti-nausea medication. Thalidomide is currently used as treatment for leprosy and some cancers. Thalidomide is taken orally and is *absorbed* slowly through the gastrointestinal tract. Thalidomide *distribution* includes ~55 % protein binding. Thalidomide crosses the placenta. Thalidomide *metabolism* in the liver is limited and non-enzymatic hydrolysis is the major cause of thalidomide breakdown. Thalidomide is *eliminated* primarily in the urine as metabolites. The half-life of thalidomide is up to 7 h (Wishart et al. 2006).

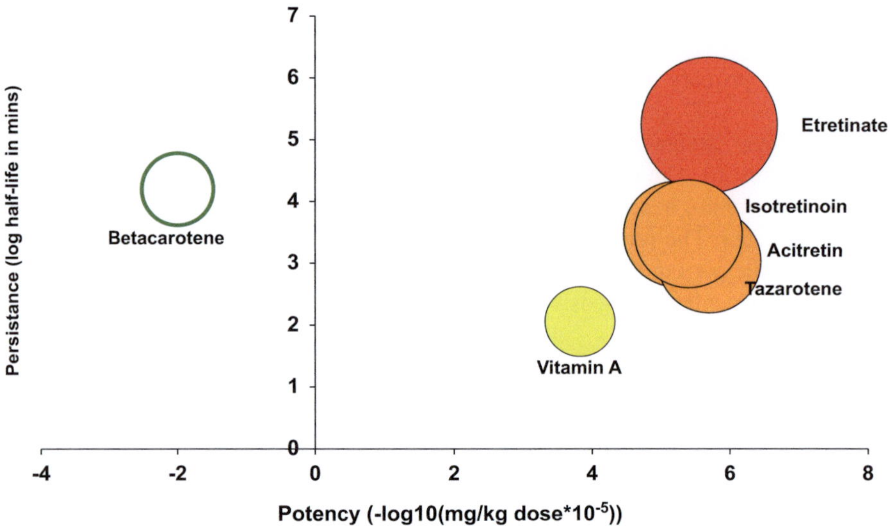

Fig. 2.2 The impact of potency and persistence on retinoid teratogens. The values on the x-axis scale represent the logarithmically transformed lowest effective doses on the *right* and the highest ineffective doses on the *left* (i.e., potency). The values on the y-axis scale represent the logarithmically transformed half-life in minutes (i.e., persistence). The *size of the bubbles* represents the product of the persistence and the potency. A *green* ring represents compounds that are not teratogenic. *Yellow* represents retinoids that are teratogenic but are approved for blood donation. *Orange* represents teratogenic retinoids that result in time-limited deferrals for blood donation. *Red* represents compounds that are teratogenic and result in lifetime bans on blood donation. This classification scheme is derived from the potency calculations developed to display genotoxicity (Waters et al. 1991)

2.2.3.2 Toxicodynamics

Thalidomide's *molecular* targets are still being identified but much of its developmental toxicity is associated with its blocking the growth of developing vascular *tissue* (Therapontos et al. 2009). Early exposures are strongly associated with limb defects. Defects in multiple *organs* including the heart, eyes, ears, genitals, urinary and digestive tracts have also been reported. Thalidomide exposures are also associated with significant mortality ~40 % in infants born with thalidomide related defects. Thalidomide *doses* range from 50 to 200 mg depending on the current usage. No doses are considered safe for pregnant women.

Thalidomide has a narrow window of susceptibility with respect to limb defects. Major limb defects are associated with exposures whose *timing* occurred during the period from gestational week 4–7 (Selevan et al. 2000). Thalidomide has been linked to limb defects such as amelia (i.e., lack of limbs) and phocomelia (i.e. flipper limbs). Major heart and brain defects are associated with exposures from gestational week 3–6 (Selevan et al. 2000).

2.2.3.3 Comparison with Adults

Adults are protected from the limb defects caused by exposure to thalidomide. Adult toxicities include peripheral neuropathy (Kim and Scialli 2011).

2.2.3.4 Translational Toxicology

Prevention is the method of translational toxicology used for thalidomide. Frances Kelsey of the Food and Drug Administration blocked its introduction to US over concerns that the toxicity seen in adults was not adequately explained. Thalidomide continues to be utilized today but with severe restrictions on who can take thalidomide and women who may become pregnant can only take thalidomide in conjunction with effective birth control.

The current program to control the provision of Thalidomide therapeutically is called Thalomid® Risk Evaluation and Mitigation Strategy (REMS) (FDA 2013). Under REMS women must first have two negative pregnancy tests. They must also agree to either abstain from heterosexual sex or use two forms of effective birth control. Males taking thalidomide must agree to wear latex or synthetic condoms even if they have had a vasectomy as thalidomide has been detected in semen. People taking thalidomide are also banned from donating blood for a month after their last dose.

Developmental toxicity screening has been changed partly because of thalidomide (Kim and Scialli 2011). The structure in thalidomide associated with its activity was identified and is used to screen chemicals *in silico*. Experimentation has allowed for the development of closely related chemicals that possess some of the pharmacologically desirable properties of thalidomide but are not teratogenic (e.g., supidimide) (Fig. 2.3).

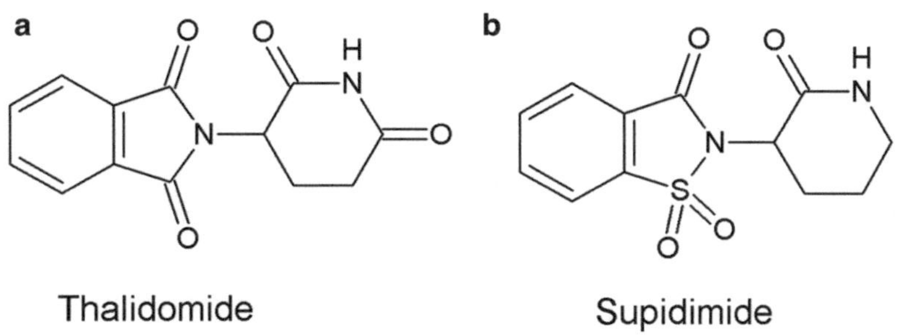

Fig. 2.3 Thalidomide SAR. Figure 2.3 shows the highly teratogenic drug Thalidomide (**a**) next to the closely related but non-teratogenic drug Supidimide (**b**)

2.2.4 *Embryos and Ethanol*

2.2.4.1 Toxicokinetics

Ethanol is the active ingredient in alcoholic drinks. Ethanol is ingested and ethanol *absorption* is rapid and complete. Ethanol *distribution* is extensive. Ethanol readily crosses the placenta and enters the embryo. Ethanol *metabolism* is primarily hepatic. Ethanol is broken down sequentially by enzymes called alcohol dehydrogenases and aldehyde dehydrogenases. The ability to metabolize alcohol varies and has a genetic component. Aldehyde dehydrogenases are polymorphic metabolizing enzymes and many individuals have reduced metabolic capacity due to their ALDH2 genotype (Judson et al. 2012).

The liver is present in the later weeks of embryonic development. However, the embryonic liver does not have many alcohol metabolic enzymes expressed in general and little is known about the expression of alcohol metabolizing enzymes early in development (Hines 2008). Most ethanol metabolism is maternal. Ethanol *elimination* is primarily maternal and metabolic. These metabolic pathways are saturable so the elimination rate of ethanol is constant above a certain dose. Roughly 95 % of ethanol is *eliminated* through metabolism while the rest is eliminated unchanged in the urine, feces, sweat and exhaled air (Judson et al. 2012).

2.2.4.2 Toxicodynamics

Ethanol has many *molecular* targets including the lipid bilayers of the cell membranes and multiple receptors. Ethanol, at high doses, damages several *organs* during development. The brain is one of the most susceptible organs to alcohol damage in embryos. Signs of fetal alcohol syndrome also include growth deficiency with weight or height under the 10th percentile. Fetal alcohol syndrome is associated with three distinct facial features; thinner upper lip, smooth space between nose and upper lip, and narrower eyes. These changes are strongly associated with brain damage and mental retardation (Lebel et al. 2011).

Fetal alcohol effects are less severe and may be due to reduced or delayed exposures. *Doses* associated with fetal alcohol syndrome are 15 or more drinks per week or 144 g of pure alcohol per day. Epidemiological evidence and animal experiments suggest that binge drinking during pregnancy is worse than regular drinking. This suggests that maximum concentration is the key factor (Maier and West 2001). Drinking ethanol during pregnancy is not advised at any *time* but is most damaging during embryonic development. Animal experiments have identified a window of toxicity that results in fetal alcohol syndrome that translates to the human embryonic timing of weeks 4 and 5 (Lipinski et al. 2012).

2.2.4.3 Comparison with Adults

Adults have a significantly greater expression of alcohol metabolizing enzymes than fetuses (Hines 2008). Adults do not suffer from the facial, intellectual or behavioral defects children develop when exposed to high levels of ethanol during the embryonic stage. Adult alcohol toxicities range from nausea to liver failure and death.

2.2.4.4 Translational Toxicology

Prevention efforts involve avoiding consumption of alcohol when attempting to get pregnant and while pregnant. Complicating these efforts are the many pregnancies that are unplanned. *Treatment* efforts are supportive in nature and are directed towards the child born with fetal alcohol syndrome.

2.3 Fetuses

2.3.1 Introduction

The fetal period occurs from 9 weeks gestation until birth which generally occurs sometime around the ninth month of gestation. During this time the fetus develops from an organism with rudimentary organs into an organism with fully functioning organs. *Absorption* can occur through the mother's lungs, skin and digestive tract. Toxicants must cross the placental blood-barrier before they enter the fetus. Maternal behavior controls exposures to toxicants. *Distribution* occurs through the circulatory system of the mother and the fetus which are linked through the placenta. *Metabolism* can be both fetal and maternal. With ingested toxicants first pass metabolism through the mother's liver can be protective. Microsomal protein levels increase as the liver develops and grows but many enzymes have not been expressed yet (Hines 2008). *Elimination* is both fetal and maternal. Developing kidneys and other excretory organs can process some wastes but ultimately fetal elimination depends upon the mother's circulatory and excretory systems.

2.3.2 Fetuses and Smoking

2.3.2.1 Toxicokinetics

Maternal smoking is associated with reduced birth weight. The evidence for the association is quite strong and between 1956 and 1987 over 100 studies were published demonstrating that women who smoke during pregnancy are more likely to give birth to low birthweight babies. Cigarette smoke contains thousands of chemicals. Talhout et al. provide a list of chemicals of concern due to both their toxic nature and their consistent and significant presence in cigarette smoke (Talhout et al. 2011). It could take an entire chapter or textbook even to cover all of these compounds adequately. Instead representative chemicals have been selected to demonstrate the wide range of toxicokinetics for cigarette smoke associated compounds.

Absorption of smoking compounds occurs through the lungs. Beryllium is poorly absorbed while nitric oxide, nicotine, benzo(a)pyrene and carbon monoxide are well absorbed. *Distribution* of smoking compounds can be primarily restricted to the surface of the lungs (e.g., beryllium) but can also be absorbed into the bloodstream and can cross the placenta and enter the fetus (e.g., benzo(a)pyrene). *Metabolism* of smoking compounds is primarily maternal but fetal metabolism can play a role. Some compounds cannot be biotransformed (e.g., beryllium) while others are detoxified (e.g., nicotine) and still others are bioactivated (e.g., benzo(a) pyrene). Most compounds in cigarette smoke are xenobiotics (i.e., foreign chemicals) however in some cases cigarettes provide exogenous sources of endogenous chemicals (e.g., ammonia, nitric oxide). In those cases they are metabolized along with the endogenous chemical (Judson et al. 2012).

Elimination of smoking compounds can occur through many different routes. Beryllium is primarily eliminated via the cilia lining the respiratory tract. The cilia bring mucus and trapped particles out of the lungs where they can be coughed up or swallowed. Cilia are damaged by cigarette smoke and this protective mechanism may become less effective the more cigarettes are smoked. Ammonia is primarily eliminated in the urine as its metabolite urea. Carbon monoxide is eliminated in the lungs. Elimination half-lives range from decades (e.g., cadmium) to seconds (e.g., nitric oxide) (Judson et al. 2012).

2.3.2.2 Toxicodynamics

Molecular targets range from selective targets (e.g., nicotine and the nicotinic receptor, carbon monoxide and hemoglobin) to non-selective targets (e.g., beryllium and any nearby proteins that can be oxidized). *Organs* affected by cigarette smoke depend on the target. Carbon monoxide can reduce the transport of oxygen to all organs. Benzo(a)pyrene causes DNA damage primarily in the organs that have the metabolic capacity to bioactivate benzo(a)pyrene (e.g., liver, lung). *Doses* vary depending on a variety factors including cigarette brand, use of filters and amount smoked. Doses also differ based on biological half-lives. Some toxicants are rapidly eliminated from the body (e.g., nitric oxide, ammonia) and have little potential for accumulation. Other toxicants remain in the body for months or even years (e.g., beryllium, cadmium).

Reduced birth weight is associated with maternal smoking but this effect is only seen if smoking occurs during the third trimester after week 30 (Wickström 2007). Smoking during the embryonic period only is not associated with growth restriction

(Wickström 2007). A possible explanation for this is that any growth restriction during the embryonic stage and early fetal stage could be overcome in the remaining gestational weeks but that after week 30 there is no longer the time to overcome any delays in growth.

2.3.2.3 Comparison with Adults

Adults are fully formed and do not have the growth restriction associated with cigarette smoking. Adults are susceptible to some of the other toxicities associated with cigarette smoking. These toxicities range from minor cosmetic changes (e.g., stained nails) to potentially lethal conditions (e.g., small-cell lung cancers).

2.3.2.4 Translational Toxicology

Prevention efforts include methods to prevent smoking in the first place. Many female smokers quit smoking once they learn they are pregnant but others struggle to quit. *Treatment* options exist to aid smoking cessation for pregnant women. There are concerns about pharmacological smoking cessation aids during pregnancy. Behavioral therapy is generally recommended while recommendations for pharmacological smoking cessation aids are often reserved for heavy (>4 cigarettes a day) smokers (Cressman et al. 2012).

2.3.3 Fetuses and Organic Mercury

2.3.3.1 Toxicokinetics

Organic mercury is created when inorganic mercury is combined with one or more carbons (Bakir et al. 1980). Unlike many pharmaceutical compounds which typically have human pharmacokinetic data generated as part of the drug approval process, environmental chemicals often lack that data. Methylmercury is one of these cases where much of the data is derived from experimental animals.

Methylmercury *absorption* is rapid and extensive. Methylmercury is readily absorbed through the skin and ~95 % of methylmercury is absorbed through the intestines (Judson et al. 2012). Methylmercury *distributes* throughout the body crossing the placenta and travels into the brain. This is accomplished by amino acid transporters. Inorganic mercury is *metabolized* by bacteria in the environment and is transformed into methyl mercury. Methylmercury bioaccumulates and becomes increasingly concentrated as it moves up the food chain and can reach high concentrations in large fish (e.g., tuna). Methylmercury conjugates with cysteine and this conjugate resembles L-lysine enough to be transported into the brain and across the placenta as if it were L-lysine (Fig. 2.4). Animal studies suggest that in humans the

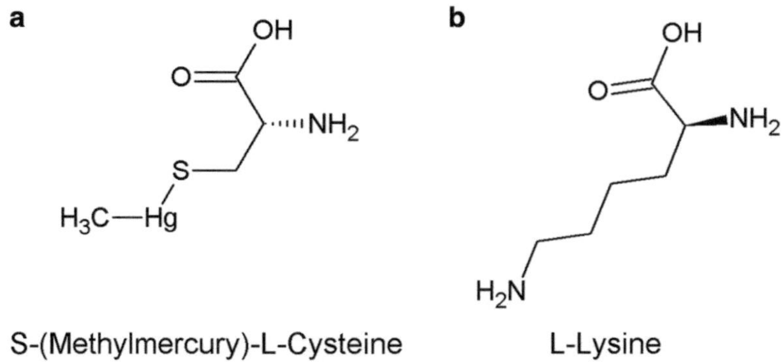

Fig. 2.4 Structural similarities between Lysine and conjugated methylmercury. S-(Methylmercury)-L-Cysteine (**a**) and L-Lysine (**b**)

elimination of methylmercury is primarily through the feces. A percentage of the mercury is also eliminated in the hair (Judson et al. 2012).

2.3.3.2 Toxicodynamics

Organic mercury targets the nervous system (Judson et al. 2012). Methyl mercury is transported into the brain using the same transport mechanisms that bring amino acids across the membrane.

2.3.3.3 Comparison with Adults

Adults are susceptible to the neurotoxicity associated with organic mercury related toxicity but not the developmental damage associated with organic mercury exposure.

2.3.3.4 Translational Toxicology

Prevention methods include standardizing hazard labels such as the recent push to implement globally harmonized Safety Data Sheets. Legislation limiting the release of inorganic mercury into the environment and guidelines limiting the consumption of contaminated fish are designed to limit the production and consumption of methymercury respectively. *Treatment* efforts range from decontamination to chelation (Judson et al. 2012).

2.4 Neonates and Infants

2.4.1 Introduction

The neonatal period is the time from birth until 30 days after birth. Infancy is from birth until 1 year of age. One of the major challenges in translation toxicology is the technical and ethical difficulties in studying very young children. Limited blood volumes in neonates and infants make pharmacokinetic studies difficult. As such data is often extrapolated from either older individuals, animals or from *ex vivo* samples (i.e., samples removed post-mortem). Compounding this relative lack of knowledge is the fact that infancy is a period of incredible growth. The average individual will have their weight double roughly four times from the time they are born until the time they reach their 20s (Fig. 2.5). These doublings correspond with infancy, toddlerhood, childhood and adolescence. Of note is that only during infancy does this doubling occur during a single year.

Birth functions as a powerful switch. Birth dramatically and rapidly alters the expression of many genes causing significant increases and decreases in many protein levels as the infant body prepares to take on functions formerly performed by the mother. *Absorption* occurs through lungs, skin and digestive tract. Infants drink more liquids and consume more food relative to adults (Selevan et al. 2000).

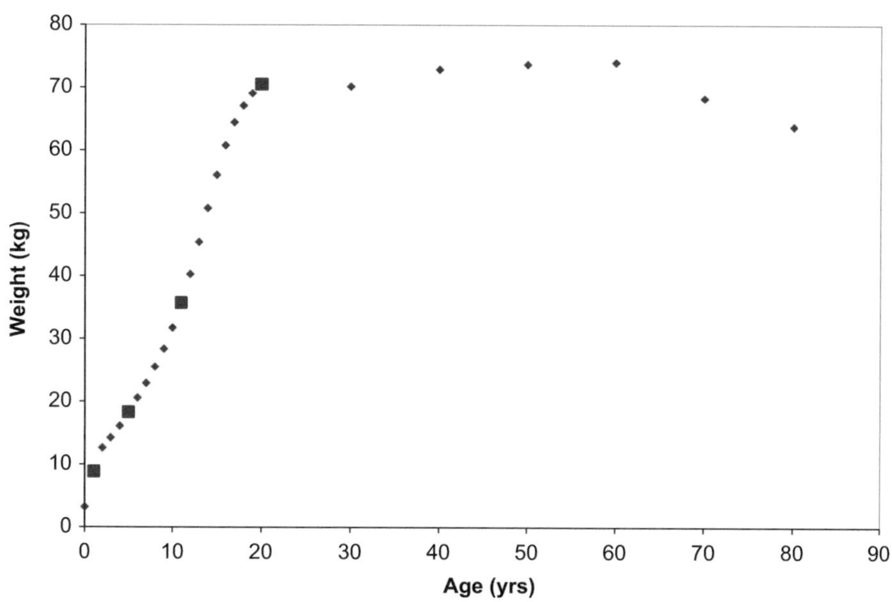

Fig. 2.5 Doubling of weight by age. The *graph* shows the weight of an average woman on the y-axis as it varies from birth until age 80. The age in years is shown on the x-axis. The *squares* mark the time when their weight doubles (Data sources are the CDC and WHO growth charts (CDC 2010))

Exposures are out of control of relatively immobile neonates. By the end of infancy the child may begin to crawl around and even "cruise" using furniture to help them stand up and take a few steps at a time. At that point the child's behavior starts to impact absorption. Crawling is the behavior that puts infants most at risk of exposures to floor dusts and contaminants (Selevan et al. 2000).

Distribution is now limited to the infant circulatory system and body. *Metabolism* is primarily hepatic. Microsomal protein is increasing but is only ~26 mg/g liver or ~65 % of adult values (Hines 2008). *Elimination* is similar to adults with the notable exception that elimination is not scheduled as infants have not learned to hold wastes in their bladders or bowels. This limits exposure time to some toxicants in those organs. After birth the kidneys become solely responsible for filtering the blood but because the renal tubules are immature the neonate kidneys function at only 20–30 % of adult values. This situation changes significantly during infancy and by 3 months of age GFR is over 60 % of adult values (Hines 2008). Also during infancy the bowels become responsible for removing solid wastes which becomes necessary as the infant begins taking in milk.

2.4.2 *Infants and Chlorpyrifos*

2.4.2.1 Toxicokinetics

Chlorpyrifos is a neurotoxic organophosphate insecticide (Fig. 2.6). Chlorpyrifos *absorption* is rapid and can occur through ingestion or dermally. Chlorpyrifos *metabolism* occurs predominantly by the cytochromes P450s (CYPs). Each chlorpyrifos metabolizing CYP forms both chlorpyrifos oxon and TCP. TCP is not neurotoxic while chlorpyrifos oxon is ~1000 times more toxic than the parent compound chlorpyrifos (Croom et al. 2010). The two major chlorpyrifos oxon producing enzymes are CYP3A4 and CYP2B6.

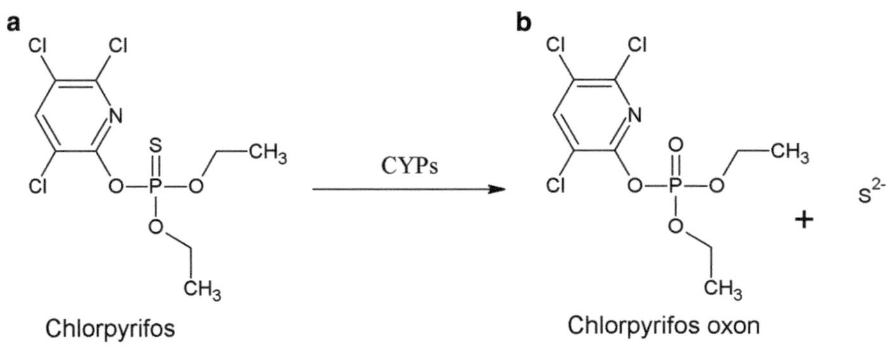

Fig. 2.6 Bioactivation of chlorpyrifos by cytochromes P450s (CYPs). Chlorpyrifos (**a**) is bioactivated into chlorpyrifos oxon (**b**) by the Cytochromes P450s (CYPs)

2.4.2.2 Toxicodynamics

The acetylcholinesterase enzymes are the molecular targets of chlorpyrifos oxon which inhibited them. Chlorpyrifos has a particular toxidrome, SLUD, which stands for Salivation, Lacrimation, Urination and Defecation. Other esterases can function to protect the nervous system by destroying chlorpyrifos oxon. The most important of these is PON1. CYP2B6, CYP3A4 and PON1 all increase after birth but appear to be regulated differently and develop at different speeds (Hines 2008; Croom et al. 2009). It is possible that many infants will have relatively high CYP levels and relatively low PON1 levels making them among the most susceptible individuals with respect to chlorpyrifos.

2.4.2.3 Comparision with Adults

Adults have fully developed PON1 and CYPs. Adults are susceptible to SLUD.

2.4.2.4 Translational Toxicology

Prevention efforts have been legislative. The use of chlorpyrifos indoors has been restricted in the United States so that only bait traps can be used in homes. These bait traps are designed so that hard plastic surrounds the bait making it difficult if not impossible for a young child to come into contact with the bait. *Treatment* of chlorpyrifos poisoning involves atropine and if necessary supportive care (Judson et al. 2012). In many parts of the world the public has greater access to insecticides. This availability results in pesticides being used more frequently in suicide attempts. When these attempts involve pregnant women there are fetal exposures. Birth can apparently be induced by the ingestion of large quantities of chlorpyrifos. When this happens a baby is born who needs to be treated immediately. At least one such newborn has been treated with atropine (Solomon and Moodley 2007).

2.4.3 Infants and Nitrates

2.4.3.1 Toxokinetics

Infants are particularly susceptible to methemoglobenemia, a condition where the nitrite formed from nitrate reacts with hemoglobin to form methemoglobin. Nitrates are ingested in the food or drinking water and *absorbed* through the digestive tract (Knobeloch et al. 2000). Infants drink relatively large quantities of water and infants who are under 3 months of age and whose formula is diluted with contaminated water may be at risk. Given the amount of water consumed by an infant and the nitrate drinking water levels associated with methemoglobenemia a dose of ~5 mg/kg/day could be dangerous. Some vegetables are naturally high in nitrates (e.g., beets) and infants ingesting large quantities of these foods can also develop methemoglobenemia (Sanchez-Echaniz et al. 2001). Nitrate *distribution* is extensive. Nitrate *metabolism* is performed by bacteria in the gastrointestinal system as well as endogenous human enzymes. The bacteria convert nitrates to nitrites. Nitrates are *eliminated* primarily unchanged in the urine (Judson et al. 2012).

2.4.3.2 Toxicodynamics

The primary *molecular* target is hemoglobin. In early infancy fetal hemoglobin predominates but by 3 months of age fetal hemoglobin is down to ~30 % of the total hemoglobin. Fetal hemoglobin is more sensitive to nitrites. Oxygen bound to hemoglobin is carried in red blood *cells* and is transported to all *organs* but the extremities are most affected by methemoglobinemia. Methemoglobinemia results in less oxygen being transported through the blood causing hypoxia often to the point of skin turning blue. Because this is primarily a problem of infants it is known as blue baby syndrome. Drinking water nitrate levels below 45 mg/L are generally regarded as safe. The percentage of hemoglobin methylated determines the outcome; 10–20 % causes limb cyanosis, 20–45 % causes central nervous system depression, 45–55 % cause coma and >60 % carries a high risk of death. Food based nitrate poisoning happens later in infancy when solid foods are added to their diet.

2.4.3.3 Comparison with Adults

Adult hemoglobin is less susceptible to nitrites than fetal hemoglobin. Adults are susceptible to methemoglobinemia but it is more often related to pharmaceutical compounds.

2.4.3.4 Translational Toxicology

Prevention efforts include testing drinking water for nitrate levels and replacing the water with a source that has lower nitrate levels. Prevention also involves limiting the amount of nitrate rich foods given to infants (e.g., sugar beets). *Treatment* efforts can include temporarily increasing the percent oxygen available.

2.4.4 Infants and Botulinum Toxin

2.4.4.1 Toxicokinetics

Absorption occurs through ingestion of *Clostridium botulinum* contaminated food. These bacteria release botulinum toxin into the food. *Metabolism* is assumed to involve proteases. Once broken down botulinum toxin is *eliminated* in the form of nitrogenous wastes although some of the amino acids will likely be reused.

2.4.4.2 Toxicodynamics

The *molecular* targets of botulinum toxin are receptors on the nerve membranes that normally bind to acetylcholine. Binding can damage the peripheral nerve *cells* causing loss of function. It takes months for new terminals to regrow. Of particular concern is loss of function of the respiratory *organs* as their loss of function poses an immediate threat to life. Botulinum toxin is extremely potent partly because it functions as an enzyme. Lethal *doses* are measured in ng/kg (Judson et al. 2012).

2.4.4.3 Comparison with Adults

In contrast to infants whose weak immune system makes them susceptible to *C.botulinum* spores adults and even toddlers can safely eat honey and be exposed to spore containing dusts and soils (Judson et al. 2012). Infant botulism represents ~65 % of the botulism cases in the United States (CDC 2014).

2.4.4.4 Translational Toxicology

Prevention includes advising parents against providing honey to infants because of their weak immune systems. The botulinum toxin is produced by bacteria that colonize food or wounds. *Treatment* always involves attempts to remove the source of the bacteria otherwise new toxin may be produced. Treatment of infants involves providing a specialized infant botulinum antitoxin (CDC 2014).

2.4.5 Infants and Heparin

2.4.5.1 Toxicokinetics

Heparin is not *absorbed* when taken orally and must be injected or infused. Heparin *distributes* within the blood and has the same volume of distribution as the blood volume (Wishart et al. 2006).

2.4.5.2 Toxicodynamics

The primary *molecular* target of Heparin is thrombin. Heparin vials have a wide range of doses and fatal mistakes have occurred where 10,000 units per ml doses have been supplied instead of 1000 units per ml to infants (Arimura et al. 2008).

2.4.5.3 Comparison with Adults

Adults are susceptible to heparin overdoses but because of their much greater size are unlikely to receive such relatively large doses.

2.4.5.4 Translational Toxicology

Prevention efforts include new labeling of heparin vials to include color differences between adult and infant doses (Monagle et al. 2012). *Treatment* efforts in overdose cases include protamine sulfate and supportive care (Judson et al. 2012).

2.5 Toddlers

2.5.1 Introduction

The toddler stage is from age 1 to 3. The toddler stage is defined by learning to walk (i.e., toddling) and the beginnings of bladder and bowel control. *Absorption* occurs through lungs, skin and digestive tract. Surface area to body mass ratio is much greater for a toddler than an adult (Fig. 2.7). Toddlers are learning to walk, fall frequently and often have abraded skin making them more susceptible to dermal

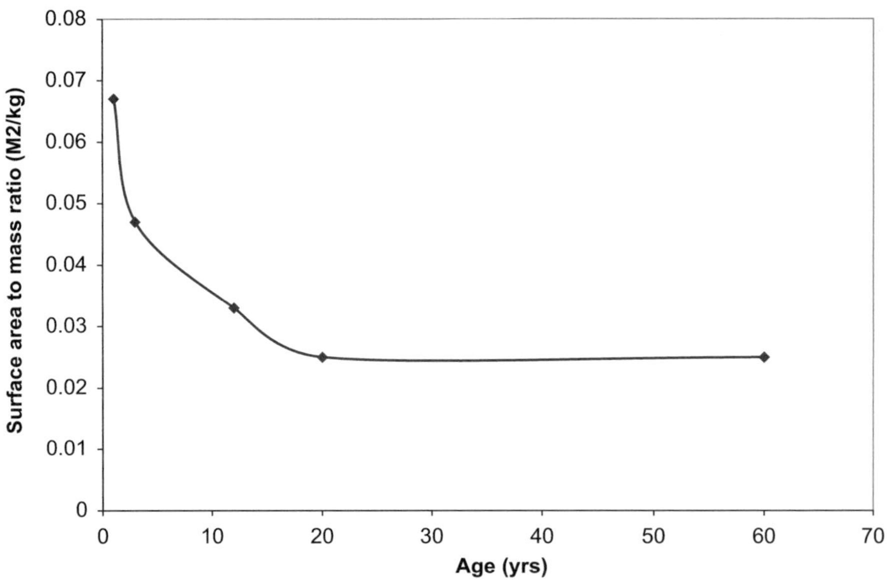

Fig. 2.7 Surface area to mass ratio by age. Age in years is shown on the x-axis. Surface area to mass ratio is shown on the y-axis. *Line* connects selected data points (Data is from (Selevan et al. 2000))

absorption. Hand to mouth behavior puts children at much greater risk of exposure (Selevan et al. 2000). Toddlers can ingest 8×as much soil as an average adult (Selevan et al. 2000). *Distribution* is through the circulatory system. Plasma protein levels are increasing during the toddler stage (Hines 2008). *Metabolism* is primarily hepatic. Microsomal protein is also increasing. *Elimination* becomes more similar to adults with age as toddlers begin to learn to hold wastes in the bladder and bowels. This increases exposure time to toxicants to those organs. Glomerular filtration rate (GFR) reaches ~80 % of adult levels (Hines 2008).

2.5.2 *Toddlers and Fluoride*

2.5.2.1 Toxicokinetics

Fluoride is a naturally occurring mineral that when consumed in drinking water can act to protect teeth from decaying. Because many sources of drinking water lack adequate levels to prevent tooth decay fluoride is added to the drinking water to prevent tooth decay. The *absorption* of the soluble fluoride in toothpaste is over 90 % (Judson et al. 2012). Normal adult behavior is to not eat toothpaste. It is normal toddler behavior to eat toothpaste. Toddlers eat sand and soil which are even grittier and less pleasant than sugary tasting toothpaste. Fluoride is rapidly *distributed* through the body by the blood but fluoride is preferentially retained in the teeth and bones. Fluoride *metabolism* is limited as it is an element. One of fluoride's major reactions in the body is its binding with calcium to form calcium fluoride. Fluoride is *eliminated* primarily through the urine. Fluoride half-life in the teeth and bones is years.

2.5.2.2 Toxicodynamics

The *molecular* targets of fluoride are generally the same as calcium and fluoride can be stored in the body the same as calcium. The *organs* targeted are the teeth and the bones (Franzman et al. 2006). *Doses* of fluoride associated with mottling of tooth enamel or fluorosis range from low but chronic exposures (e.g., daily consumption of drinking water with concentrations ~>1 ppm) (Judson et al. 2012) as well as acute exposures (e.g., ingesting toothpaste with ~1000 ppm). The *timing* of fluorosis of the teeth is restricted to children who are still developing permanent teeth.

2.5.2.3 Comparison with Adults

Adults have fully formed teeth and are not at risk of fluorosis. Adults are less likely to consume large quantities of toothpaste. With adults high exposures to fluorine is more likely occupational and due to inhalation (Judson et al. 2012). Adults are more likely to consume antacids which contain metals that can reduce the amount of fluorine absorbed.

2.5.2.4 Translational Toxicology

To get toddlers used to brushing their teeth but to *prevent* them from eating fluoride special "training" toothpaste has been developed that has the same abrasives but lacks the fluoride. Fluoride is a natural occurring mineral and when levels are high bottled water is recommended for formula and drinking water for young children.

2.5.3 Toddlers and Lindane

2.5.3.1 Toxicokinetics

Lindane is a neurotoxic insecticide used to treat lice and scabies. Toddler behavior is intensely oral and they may try to ingest anything in a tube. With toothpaste the potential toxicity is limited but some other creams can have much greater toxicity. A toddler was reported to develop seizures after ingesting the contents of a tube of lindane cream (Davies et al. 1983). The greater surface area to body mass ratio and likelihood of abrasions also made toddlers more susceptible to absorbing excessive lindane concentrations when treated with these creams.

2.5.3.2 Toxicodynamics

Lindane inhibits the GABA receptor as its *molecular* target causing nerve cells to increase their firing rates. This in turn causes muscles to spasm causing seizures if the dose is sufficient (~80 mg/kg). At higher lindane doses death can occur.

2.5.3.3 Comparison with Adults

Unlike toddlers, mentally mature adults do not ingest lindane intentionally. There have been accidental cases when lindane was mistakenly added instead of sugar (Hayes 1982). In these cases seizures also resulted. Similarly adults who have abraded skin and smaller adults may absorb more lindane through their skin.

2.5.3.4 Translational Toxicology

Prevention efforts include the banning of lindane's use on children and adults weighing less than 50 kg. Additionally because lindane is a persistent organ pollutant it has been banned and is no longer being produced. *Treatment* includes lindane removal. In the case of dermal exposures a bath or shower is recommended. If ingested then a stomach pump may be used. Supportive care and pharmaceutical treatment can also be provided (Judson et al. 2012).

2.5.4 *Toddlers and Iron*

2.5.4.1 Toxicokinetics

Iron tablets are taken in orally. Iron *absorption* is dependent partly on availability of iron in the body. Percent iron absorbed can range from a low of 10 % in individuals with adequate iron stores to a high of 95 % in individuals with iron deficiency (Wishart et al. 2006). Iron *distribution* occurs in the blood. Iron *metabolism* includes conversion into stored iron and iron can function as a catalyst but as an element metabolism is limited. Iron *elimination* occurs in the feces as well as in the blood.

2.5.4.2 Toxicodynamics

Iron has many molecular targets and plays an important role in generating free radicals which cause oxidative damage to DNA, proteins and lipids. Iron overdoses damage many *organs* including the liver. Toddlers can mistake iron tablets for candy. Toddlers can eat more iron pills in one setting than a much larger adult would normally consume in a week (Morris 2000).

2.5.4.3 Comparison with Adults

Adults are susceptible to iron poisoning but do not consume iron tablets in large quantities mistaking them for candy. Adults may however routinely ingest large quantities of iron in some foods such as instant oatmeal. Adults are often partly protected from iron overdoses by regular blood loss through blood donation and menstruation. Toddlers have neither of those mechanisms of blood loss.

2.5.4.4 Translational Toxicology

Preventative measures include using unit doses and childproof containers. Unit doses have been effective in reducing iron related overdoses and fatalities (Tenenbein 2005). *Treatment* measures for acute iron poisoning include gastric lavage and chelation (Judson et al. 2012).

2.6 Children

2.6.1 *Introduction*

Childhood occurs from age four until puberty. *Absorption* occurs through lungs, skin and digestive tract. Surface area to volume ratio is slightly larger for a small child than a normal sized adult (Fig. 2.7). Children also have a tendency to have more skin abrasions than adults. *Distribution* includes plasma protein binding and transport proteins. The brain continues to develop during childhood so distribution of neurotoxic chemicals into the brain is even more problematic with children than adults. *Metabolism* can be variable at this age. Microsomal protein content continues to increase with age and individual metabolizing enzymes may be at or above adult levels. As with toddlers and infants children have relatively larger livers than adults (Hines 2008). *Elimination* approaches adult levels and GFR reaches and sometimes exceeds adult levels during childhood (Hines 2008).

2.6.2 Children and Chemotherapy

2.6.2.1 Toxicokinetics

Absorption occurs through IV and oral routes. *Distribution* depends on the chemotherapy provided. Even chemotherapeutic medicines that do not normally cross the blood brain barrier may be injected intrathecally into the brain for children being treated for brain tumors. *Metabolism* depends on the chemotherapy provided. Sometimes metabolic activation is required for the efficacy of the compound (e.g., Ifosfamide and CYPs) but this often increases the toxicity. Other times toxicity is reduced through metabolism (e.g., epirubicin and UGTs). Metabolism is often hepatic but the metabolic capacity of the cancer cells can also be important. *Elimination* is often rapid with chemotherapeutic drugs and often involves the removal of metabolites (e.g. epirubicin).

2.6.2.2 Toxicodynamics

Doses for children can be challenging with normal medications (Hines 2008). Chemotherapeutic drugs often have narrow therapeutic indexes. Typically they are provided based on weight or body surface area but doses may have to be reduced or delayed due to toxicity. *Molecular* targets vary from selective (e.g., lapatinib and HER-2 receptors) to widespread (e.g. epirubicin and DNA). Alkylating agents (e.g., ifosfamide) target proteins and DNA and are particularly damaging to rapidly dividing *cells. Organs* damaged by alkylating agents include bone marrow, skin, epithelial linings, hair and reproductive organs. Secondary cancers are a common problem with pediatric cancer patients treated with DNA damaging agents.

2.6.2.3 Comparison to Adults

The average pediatric cancer patient is more vulnerable to fertility and DNA damaging chemotherapy than the average adult cancer patient because of their age differences. Most pediatric cancer patients are pre-pubescent and have all of their potential reproductive years ahead of them while most adult cancer patients already have most of their reproductive years behind them. Younger adults have better options of preserving their fertility than prepubescent cancer patients, particularly for males (Köhler et al. 2011). They also have less time to develop additional mutations that might drive secondary cancers.

2.6.2.4 Translational Toxicology

It is not always possible to avoid treatment of potentially lethal childhood cancers. *Prevention* efforts include efforts to preserve fertility. Effective preservation methods for of prepubertal male spermatogenesis have not been developed yet. Methods of preserving prepubescent female fertility are still investigational. A disparity has been reported with female pediatric cancer patients being referred for fertility preservation far less frequently than male patients (Köhler et al. 2011). *Treatment* can include adding protective chemicals (e.g., mesna) as well as treating symptoms (e.g., nausea).

2.6.3 *Children and Radiation*

Radiation exposures of children may be therapeutic or environmental. Therapeutic exposures are provided to treat a medical condition often as part of cancer treatments. Environmental exposures can come from natural (e.g. radon) or artificial sources (e.g., nuclear fallout).

2.6.3.1 Toxicokinetics

Absorption of radiation depends on the type of radiation. Gamma rays and proton beams can penetrate all tissues and X-rays penetrate soft tissues easily while alpha and beta particle emitters are stopped by heavy clothing and must be ingested, inhaled or inserted to cause damage to internal organs. *Distribution* depends on the radiation source. Radioactive iodine preferentially accumulates in the thyroid. Gamma rays damage tissues all along their path while proton beams mainly damage their target. Brachytherapy, inserting small pieces of radioactive materials is another technique to limit distribution. *Metabolism* of radioactive sources is mostly limited to repairing damage caused by radiation although some radioactive compounds are metabolized the same as endogenous materials (e.g., radioactive iodine, technicium (^{99m}Tc) medronic acid). *Elimination* time is intentionally limited for therapeutic radiation. Technicium 99 (^{99m}Tc) has a radiological half-life of just 6 h and a biological half-life of 1 day. Environmental sources of radiation typically take longer to be removed. Plutonium 238 has a radiological half-life of 88 years and can take 50 years to be removed from bones.

2.6.3.2 Toxicodynamics

Radiation kills rapidly dividing *cells* by damaging the *molecular* target of dividing DNA. Organs with rapidly replaced cells are particularly vulnerable (e.g., mouth, skin).

2.6.3.3 Comparison with Adults

Adults are not as susceptible to radiation as rapidly growing children. For example it is common for children being treated with radiation for brain tumors below a certain age to lose some basic mathematical abilities. Adults in general with their fully developed brains lack this risk. Adults also have significantly more options in protecting their fertility than pediatric cancer patients.

2.6.3.4 Translational Toxicology

Prevention includes reducing the exposure to the pelvic region with protective aprons. For female patients whose pelvic region must be treated with radiation in some cases the ovaries have even been moved temporarily out of the way being placed higher up into the abdomen (Köhler et al. 2011). *Treatment* efforts include treating symptoms. Human growth hormone levels are lower in children treated with radiation and may be restored artificially.

2.6.4 Children and Petroleum Distillates

2.6.4.1 Toxicokinetics

Petroleum distillates are solvents created when crude oil is heated and separated into fractions. Petroleum distillates are used in many industrial and household chemicals. Even when they are not the active ingredient, they can be the most toxic part of some products. Petroleum distillates can be inhaled and ingested and some of their components can be *absorbed* through the skin. Once inside the body petroleum distillates *distribute* based on their individual properties. *Metabolism* occurs primarily in the liver. For the more volatile compounds some *elimination* through the lungs can occur. Most elimination occurs through the urine and feces (Judson et al. 2012).

2.6.4.2 Toxicodynamics

The *molecular* targets of petroleum distillates depend on the individual compounds. The liver and lungs are two of the major *organs* that can be damaged by petroleum distillates. When ingested, vomiting and aspiration can occur. If petroleum distillates enter the lungs in sufficient quantity they can break down the surfactant in the lung causing the lungs to fail. Petroleum distillates can also displace the oxygen in the air in confined spaces causing hypoxia (Judson et al. 2012).

2.6.4.3 Comparison with Adults

Adults are also susceptible to damage due to petroleum distillates. However, they are more capable of protecting themselves than children from malicious poisonings.

2.6.4.4 Translational Toxicology

Prevention includes the increasing replacement of petroleum distillates in many household products with other solvents (e.g., water, surfactants and clays). *Treatment* efforts include removing patients to areas with better ventilation.

2.7 Adolescents

2.7.1 Introduction

Adolescence begins with the onset of puberty and continues until adulthood. Adulthood legally occurs at 18 but developmentally may not end until later. *Absorption* occurs through lungs, skin and digestive tract. Surface area to volume ratio of adolescence overlaps with adult surface area to volume ratios. However, adolescents with acne are more vulnerable to dermally absorbed toxicants. Exposures are often under the control of adolescents but adolescents often experiment and may have a sense of invulnerability. This can be seen with behaviors related to toxicants particularly illegal drugs. *Distribution* is similar to adults although the relative lipid content of adolescents can change rapidly during growth spurts and puberty. Microsomal protein content is similar to adult levels and *metabolizing* enzymes have mostly reached adult levels (Hines 2008). *Elimination* in adolescents is similar to adults and GFR is at adult levels (Hines 2008).

2.7.2 Adolescents and Inhalants

2.7.2.1 Toxicokinetics

Huffing is intentionally inhaling glue fumes, aerosol propellants or other fumes to experience euphoria and other effects. *Absorption* is through inhalation. Inhalants are small volatile molecules and they can rapidly *distribute* throughout the entire body. *Metabolism* depends on the gas inhaled. Some are rapidly metabolized by the liver (e.g., benzene) while others are barely metabolized (e.g., nitrous oxide).

Elimination can occur through exhalation. Breakdown products and parent compounds may also be eliminated through urine.

2.7.2.2 Toxicodynamics

Inhalants can displace the air in the lungs and cause hypoxia. This lack of oxygen creates a "high" but also runs the risk of causing permanent brain damage. Some inhalants are associated with damage to the myelin sheath lining the nerve *cells* (Baird and Furek 2012).

2.7.2.3 Comparison with Adults

Adults are less likely to experiment with huffing. Adult brains are fully formed while adolescent brains continue to develop putting them at greater risk of permanent loss of cognitive function (Baydala 2010).

2.7.2.4 Translational Toxicology

Prevention efforts include educational campaigns, the addition of bitterants, and the use of bag-on-valve cans that use air as a propellant. *Treatment* efforts include supportive care for acute exposures and addiction treatment.

2.7.3 Adolescents and Anabolic Steroids

2.7.3.1 Toxicokinetics

Anabolic steroids are synthetic compounds related to the male hormone testosterone. While they may be prescribed for some men and play a major role in intentional female to male sex-change they have serious risks and are often drugs of abuse. Anabolic steroids are often taken orally but can be injected intramuscularly and can also be readily *absorbed* through the skin. Once absorbed they are *distributed* throughout the body. *Metabolism* of anabolic steroids is primarily hepatic and involves the CYPs. *Elimination* of anabolic steroids and their metabolites occurs through the excretion of metabolites.

2.7.3.2 Toxicodynamics

The *molecular* targets of androgens are the androgen receptors. Anabolic steroids may also target enzymes involved in the breakdown of androgens (Wu 1997). Many *organs* are damaged by anabolic steroid use including the liver, heart, skin and reproductive organs. The brain is also altered with anabolic steroid use and rage is a common side effect. The response to anabolic steroid abuse differs by sex. Adolescent girls can have irreversible masculinization. Adolescent males can have feminization including gynecomastia (i.e., breast growth) and reduced testicular volume. There is a linear dose response for natural and synthetic androgens. Adolescence is a period of great growth and the final normal doubling of body weight occurs during this period (Fig. 2.5). When anabolic steroids are taken they can prematurely solidify the growth plates in the long bones reducing the body weight. The younger the adolescent the more likely growth restriction will occur.

2.7.3.3 Comparison with Adults

Adults suffer from most of the same side-effects but have completed growing so they avoid the growth restriction associated with anabolic steroid usage.

2.7.3.4 Translational Toxicology

Prevention efforts include laws restricting anabolic steroid usage. *Treatment* efforts include drug addiction treatment and treatment of symptoms.

2.8 Pregnant Women and Adolescents

2.8.1 Introduction

Pregnancy can occur anytime from puberty until menarche. *Absorption* occurs through lungs, skin and digestive tract. Absorption of some toxicants may be limited due to increased sensitivity to some chemicals resulting in vomiting. Absorption generally increases for oral exposures as pregnancy progresses due to slower intestinal mobility (Selevan et al. 2000). *Distribution* of toxicants may also include distribution to the embryo and fetus. Lipophilic toxicants sequestered before and during pregnancy may also be distributed into breast milk after pregnancy as fat stores are used to create milk. Blood volume increases but plasma protein content tends to decrease. *Metabolism* changes are enzyme specific. *Elimination* patterns are altered as internal organs are displaced by growing fetus but elimination volumes and demand increase as pregnancy progresses. Maternal elimination involves both maternal and fetal wastes.

2.8.2 *Pregnant Women and Adolescents and Iron*

2.8.2.1 Toxicokinetics

Many women already suffer from anemia and start pregnancy with an iron deficiency. During pregnancy the blood volume increases significantly and there is an increased need for iron to accommodate this increased blood production and to supply iron to the developing fetus and placenta.

2.8.2.2 Toxicodynamics

Iron is associated with vomiting and constipation during pregnancy even when used at the prescribed dose. Vomiting due to iron becomes less of a problem as pregnancy progresses. This is hypothesized to occur as a decline in a defense mechanism against teratogens. In contrast constipation increases as pregnancy progresses due to the increased pressure on the intestines as the fetus and uterus continue to grow.

2.8.2.3 Comparison with Adults

Pregnant adolescents are similar to pregnant adults in most respects. Adolescent behavior does not automatically change with pregnancy. Adolescents may be more inclined to hide their pregnancy than adults and may be more prone to avoiding prenatal vitamins.

2.8.2.4 Translational Toxicology

Different versions of iron exist and some may be easier to handle than traditional iron tablets. There is also the possibility of separating the delivery of iron with the delivery of other more important prenatal vitamins (e.g. folate). Additionally, research suggests that smaller prenatal vitamins may result in less stimulation of the gag reflex and that using smaller pills could decrease urge to vomit (Koren and Pairaideau 2006).

2.9 Adults

2.9.1 *Introduction*

In general adulthood is the least susceptible of the developmental stages to toxicants. The maximal average microsomal protein content of 40 mg/g liver is reached by the age of 30 and kidney function is also at its maximal (Hines 2008). The behavior of adults which makes them most susceptible to toxicants is working. In every country adults work and are potentially put at risk of occupational exposures. In contrast children and adolescents are protected from harmful occupational exposures in many countries and ideally would be protected from them worldwide.

2.9.2 Adults and Ethinyl Estradiol

2.9.2.1 Toxicokinetics

Estrogens have long been known as an occupational hazard. Gynecomastia has been associated with occupational exposures of estrogenic creams to male morticians (Bhat et al. 1989). Ethinyl estradiol *absorption* is rapid and extensive whether the exposure is oral or dermal (Wishart et al. 2006). Ethinyl estradiol *distribution* includes 97 % protein binding and 43 % bioavailability. Ethinyl estradiol is *metabolized* primarily in the liver by the CYPs including CYP3A4. Because of the role of CYP3A4 in ethinyl estradiol metabolism many compounds found to increase the expression of CYP3A4 have also been found to increase the metabolism of Ethinyl estradiol (e.g., amoxicillin) (Wishart et al. 2006). Ethinyl estradiol is *eliminated* primarily in the as hydroxylated and conjugated metabolites. Ethinyl estradiol's half-life is 36 ± 13 h.

2.9.2.2 Toxicodynamics

The main *molecular* targets for ethinyl estradiol are the estrogen receptors which are expressed in many reproductive *organs* as well as the mammary glands, hypothalamus and pituitary gland. These receptors are expressed in both women and men but have different effects based on sex (Judson et al. 2012). Contraceptive factory workers in one plant were exposed to high levels of estrogens occupationally. Gynecomastia among male workers and altered menstruation for female workers were reported (Harrington et al. 1978).

2.9.2.3 Comparison with Adults

NA.

2.9.2.4 Translational Toxicology

Prevention efforts include improved industrial hygiene (e.g., masks). *Treatment* can include surgical treatment for gynecomastia cases.

2.10 Elderly Adults

2.10.1 Introduction

The age at which adults are defined as becoming senior adults or elderly continues to change. It varies from country to country and is often tied to retirement age. According to the WHO this transition starts after 60 years of age. This fits in with the decline in average weight that occurs after age 60. *Absorption* occurs through lungs, skin and digestive tract. Skin becomes thinner. *Distribution* changes as bodies in general become more lipid rich with age as muscle mass (i.e., protein content) and water content decrease. In general *metabolism* decreases with age and microsomal protein content decreases to an average of 31 mg/g liver (Hines 2008).

Elimination becomes less effective with age as kidneys shrink and the number of functional nephrons decline. Behavior can make seniors different from other adults. A major concern is how many prescription medications are taken by older adults. This makes drug-toxicant interactions more likely. The metabolism of toxicants may be significantly reduced in seniors whose enzymes are already dealing with prescription drugs. Risk for developing adverse drug reactions requiring hospitalization is 10.7 % for elderly adults vs. 5.3 % for the general population. And 50 % of community dwelling people over 65 years of age use five or more OTC or prescription medications a week and 12 % use ≥ 10 (Wooten 2012).

2.10.2 Elderly Adults and Hydroxyzine

2.10.2.1 Toxicokinetics

Hydroxyzine is taken orally and has rapid and nearly complete *absorption.* Hydroxyzine *distribution* includes being 93 % bound to plasma proteins. Hydroxyzine, like many first generation antihistamines, readily crosses into the brain (Wishart et al. 2006). The volume of distribution of hydroxyzine has been studied in children and adults and for both populations the mean reported volume of distribution was over 15 L/kg (Judson et al. 2012). Hydroxyzine *metabolism* is extensive and primarily hepatic but poorly studied in humans. Hydroxyzine is rapidly converted into its active metabolite ceritizine. Hydroxyzine *elimination* occurs primarily through the bile and feces. Hydroxyzine has a half-life of 20–25 h.

2.10.2.2 Toxicodynamics

The major molecular targets of hydroxyzine are the histamine H1 receptors (Wishart et al. 2006). Mucus membranes are some of the *tissues* affected by hydroxyzine. The *organs* affected by hydroxyzine include the eyes, mouth, intestines. Hydroxyzine also can also affect the brain in the form of drowsiness and confusion.

2.10.2.3 Comparison with Adults

NA.

2.10.2.4 Translational Toxicology

Prevention efforts for hydroxyzine include being added to Beer's list to avoid its usage with elderly patients. *Treatment* efforts could include temporary measures to relieve symptoms of excessive dryness (e.g. eye drops, laxatives).

2.11 Warfarin During Development

2.11.1 Toxicokinetics

Warfarin is an anticoagulant. Warfarin is taken orally and its *absorption* is almost complete. Providing oral medications to young children can be challenging. Warfarin *distribution* includes being highly bound by plasma proteins and ~99 % is bound to albumin. Warfarin crosses the placenta and is *distributed* into the embryo and fetus. Warfarin *metabolism* is performed by the CYPs, primarily in the liver. The most active form of Warfarin, S-Warfarin, is primarily metabolized by CYP2C9 (Wishart et al. 2006). CYP2C9 levels remain low before birth (Fig. 2.8) and most metabolism of Warfarin during pregnancy will be maternal (Koukouritaki et al. 2004).

Warfarin *metabolism* may be reduced in elderly adults as they tend to have lower microsomal protein levels than younger adults (31 mg/g liver vs 40 mg/g liver) (Hines 2008). However, elderly adults tend to take more medications than younger adults increasing the risks of drug-drug interactions (Wooten 2012).

Warfarin *elimination* is almost entirely through the elimination of metabolites in the urine (Wishart et al. 2006). The urinary organs (i.e., kidneys, bladder), begin forming late into embryogenesis but urine production, storage and elimination mostly begin in the fetal period. Because of this Warfarin will predominately leave the embryo through umbilical cord travel to the maternal liver in the blood, be metabolized by CYPs into water soluble metabolites and be eliminated in the maternal urine. The half-life of S-Warfarin depends on genotype, diet, additional medications and health. Depending on the status of the mother Warfarin could have a half-life ranging from <20 to >50 h (Wishart et al. 2006).

2.11.2 Toxicodynamics

Warfarin has vitamin K reductase as its *molecular* target. This lowers the levels of many clotting factors and increases the risk of bleeding. Warfarin increases bleeding risks in all *organs*. The embryo is susceptible to Warfarin syndrome which is

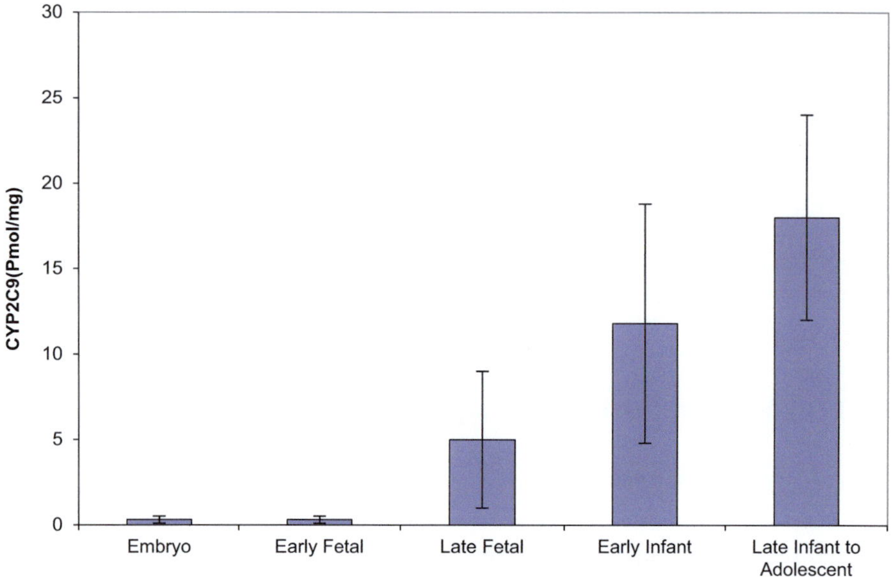

Fig. 2.8 CYP2C9 expression during childhood. CYP2C9 content is shown on the y-axis in pmol CYP2C9 per mg liver microsomal protein. *Bars* represent mean ± S.D. CYP2C9 expression increases three times during development; at 25 weeks in the late fetal stage, at birth, and at 6 months (Data from: (Koukouritaki et al. 2004))

associated with damage to the mouth, brain, skeleton, face and lungs (Abadi et al. 2002). Maternal *doses* associated with these defects range from <0.05 to >0.2 mg/kg. Warfarin exposures from 6 to 9 weeks gestational age have been associated with Warfarin syndrome (Judson et al. 2012).

Warfarin is primarily given 6 months after birth. Bleeding risks increase after coagulation system matures (Monagle 2008). Adults have fully formed organs and are not susceptible to the organ damage associated with embryonic warfarin exposures. (Judson et al. 2012). Warfarin exposures throughout pregnancy can result in increased bleeding potentially leading to loss of the pregnancy and the death of the mother. Warfarin is not part of Beer's list. At birth the coagulation system is immature. Warfarin therapy is rarely recommended for neonates. By 6 months of age the coagulation system is similar to the adult system and warfarin can be used (Monagle 2008). Typical warfarin infant, child and adolescent *doses* are 0.2 mg/kg (Monagle 2008). Warfarin doses are often based on genotype, medications and diet and response to treatment (i.e, INR) as much as they are on body mass.

2.11.3 Comparison with Adults

NA.

2.11.4 Translational Toxicology

Prevention of warfarin syndrome is encouraged through the classification of warfarin as a class X drug to be avoided while pregnant (Wishart et al. 2006). There is strong evidence that category X drugs can cause damage to the developing embryo or fetus. Pregnant women or women who are likely to become pregnant may be switched over to a safer anti-coagulant such as low molecular weight heparin which does not cross the placenta (Abadi et al. 2002). Prevention of warfarin overdoses is encouraged through the use of regular monitoring of INR levels and limiting the initial warfarin doses (Monagle 2008). *Treatment* of women who become pregnant while taking warfarin includes switching over to a safer anti-coagulant such as low molecular weight heparin which does not cross the placenta (Abadi et al. 2002). Treatment of overdose includes vitamin K treatment.

2.12 Conclusions

Development in humans takes decades and occurs through many stages. There are many toxicants that can disrupt normal development and one of the goals of translational toxicology is to either eliminate or mitigate these exposures. One of the major challenges is that human behavior plays a major role in the extent of the exposures and doses of the toxicant received (Table 2.1).

Children are not just small adults. They have many behavioral and biological differences that often make them more susceptible to toxicants. The toxicokinetics of a toxicant can be dramatically different between an adult and an infant. Small children are capable of *absorbing* more of a toxicant than an adult through most routes of exposure because they breathe more, eat and drink more, and have a greater surface area to weight ratio than the average adult.

The distribution of toxicants varies widely from chemicals that have little to no distribution (e.g., beryllium in the lungs) to toxicants that are widely distributed (e.g., hydroxyzine). Heparin fits in the middle because it is distributed within and contained by the blood vessels. Of critical importance during development is whether or not a toxicant can cross the placenta and distribute into the developing fetus or embryo.

The metabolism of toxicants ranges from rapidly and extensively metabolized toxicants (e.g., chlorpyrifos) to essentially unmetabolizeable toxicants (e.g. beryllium). Metabolism is generally hepatic and enzymatic but non-enzymatic reactions

Table 2.1 Risky behaviors of each developmental stage

Developmental stage	Characteristic behaviors and risk factors
Embryos	Maternal diet and drug use
Fetuses	Maternal diet and drug use
Infants	Crawling
Toddlers	Learning by mouthing and eating
Children	Developing cancers
Adolescents	Illegal drug use
Pregnancy	Pregnancy "morning" sickness
Adults	Occupation
Elderly	Prescription drug use

can predominate (e.g., thalidomide). At its most complicated metabolism can be fetal, placental and maternal.

Elimination of toxicants varies greatly in route and rate. In the beginning maternal elimination predominates. In all developmental stages the doses and the chemical properties of the toxicant largely determine how long it might take to eliminate the toxicant. Toxicant half-lives range enormously (Table 2.2) from mere seconds (e.g., nitric oxide) to decades (e.g., cadmium). The shorter the half-life of a teratogen the sooner it will be removed from body and the sooner it will be safe for a pregnancy to occur. The example of etretinate is a cautionary tale about the need to understand persistence of chemicals particularly for teratogens. The possibility exists that some new pharmaceutical or even environmental chemical will be both a potent teratogen and a persistent chemical.

Potency varies enormously between toxicants. There is over a billion-fold difference in the potency of botulinum toxin where ng/kg concentrations can cause infantile botulism and ethanol where pregnant women must drink g/kg quantities to cause fetal alcohol syndrome. Potency also depends on the presence of molecular targets. Embryos are particularly vulnerable to damaging toxicants. There are many examples of defects that only occur when exposures occur during the embryonic stage (Table 2.3). A notable exception is with smoking and reduced birth weight where the fetal period is the critical point.

Warfarin is an example of how developmental changes can dramatically alter the efficacy and toxicity of a compound. Warfarin is dangerous to the embryo and fetus, ineffective toward the neonate, but can beneficial towards older infants, children, adolescents, adults (unless they are pregnant). The ability to metabolize warfarin generally increases with age but also depends on genotype, diet and drug interactions. Warfarin highlights the need with some compounds to account for both the life stage and the lifestyle of the individual patient.

One of the greatest challenges in developmental toxicology is predicting the toxicokinetics and toxicodynamics of toxicants. This is due to our relatively limited knowledge on the developmental expression of many molecular targets and metabolizing enzymes. *In silico*, *in vitro* and *in vivo* developmental models are often only as good as the data used in their development. In many ways the growth

Table 2.2 Half-lives of representative toxicants

Toxicant	Exposure route	Half-life
Ammonia	Inhalation	Minutes
Beryllium	Inhalation	Months
Cadmium	Inhalation	Decades
Carbon monoxide	Inhalation	Hours
Ethinyl estradiol	Oral	Hours
Etretinate	Oral	Months
Hydroxyzine	Oral	Hours
Nitric oxide	Inhalation	Seconds
Nitrous oxide	Inhalation	Minutes
Thalidomide	Oral	Hours
Warfarin	Oral	Days

Table 2.3 Role of timing on birth defects

Toxicant	Effect	Embryo	Fetus
Etretinate	Brain, heart, ear and jaw defects	Y	N
Ethanol	Brain and face defects[a]	Y	N
Ethanol	Learning and behavioral changes[b]	Y	Y
Thalidomide	Limb defects	Y	N
Smoking	Reduced birth weight	N	Y

[a]These "fetal" alcohol syndrome related defects are due to embryonic exposures
[b]These fetal alcohol effects are due to embryonic and fetal exposures

of knowledge derived from *ex vivo* samples remains a rate limiting step in developmental toxicology.

These representative toxicants reveal how knowledge of toxicokinetics and toxicodynamics can be used by translational toxicology to prevent and treat many potentially devastating exposures. They also demonstrate how the toxicokinetics and toxicodynamics of chemicals towards the different developmental stages should be considered when new chemicals and products are being developed. The "dose makes the poison" but the timing makes the teratogen.

References

Abadi S, Einarson A, Koren G (2002) Use of Warfarin during pregnancy. Can Fam Physician 48(4):695–697

American Red Cross (2014) Blood donor eligibility: blood pressure, pregnancy, disease & more|American Red Cross. [online] Available at: http://www.redcrossblood.org/donating-blood/eligibility-requirements/eligibility-criteria-alphabetical-listing. Accessed 10 July 2014

Arimura J, Poole R, Jeng M, Rhine W, Sharek P (2008) Neonatal heparin overdose-a multidisciplinary team approach to medication error prevention. J Pediatr Pharmacol Ther 13(2):96–98

Baird C, Furek M (2012) Adolescents and inhalant abuse: how huffing affects the myelin sheath. J Addict Nurs 23(2):129–131

Bakir F et al (1980) Clinical and epidemiological aspects of methylmercury poisoning. Postgrad Med J 56(651):1–10
Baydala L (2010) Inhalant abuse. Paediatr Child Health 15(7):443–454
Bhat N, Rosato E, Gupta P (1989) Gynecomastia in a mortician. A case report. Acta Cytol 34(1):31–34
Blaner W (2013) The fat-soluble vitamins 100 years later: where are we now? J Lipid Res 54(7):1716–1718
Brust J (2010) Ethanol and cognition: indirect effects, neurotoxicity and neuroprotection: a review. Int J Environ Res Public Health 7(4):1540–1557
CDC (2010) Growth charts 2010. http://www.cdc.gov/growthcharts/cdc_charts.htm. Accessed 3 July 2014
CDC (2014) Botulism. http://www.cdc.gov/nczved/divisions/dfbmd/diseases/botulism/. Accessed 4 July 2014
Cressman A, Pupco A, Kim E, Koren G, Bozzo P (2012) Smoking cessation therapy during pregnancy. Can Fam Physician 58(5):525–527
Croom E, Stevens J, Hines R, Wallace A, Hodgson E (2009) Human hepatic CYP2B6 developmental expression: the impact of age and genotype. Biochem Pharmacol 78(2):184–190
Croom E, Wallace A, Hodgson E (2010) Human variation in CYP-specific chlorpyrifos metabolism. Toxicology 276(3):184–191
Davies J, Dedhia H, Morgade C, Barquet A, Maibach H (1983) Lindane poisonings. Arch Dermatol 119(2):142–144
FDA (2013) THALOMID® Risk evaluation and mitigation strategy (REMS)
Franzman M, Levy S, Warren J, Broffitt B (2006) Fluoride dentifrice ingestion and fluorosis of the permanent incisors. J Am Dent Assoc 137(5):645–652
Harrington J, Stein G, Rivera R, de Morales A (1978) The occupational hazards of formulating oral contraceptives – a survey of plant employees. Arch Environ Health 33(1):12–15
Hayes W (1982) Pesticides studied in man, 1st edn. Williams & Wilkins, Baltimore, pp 211–228
Hines R (2008) The ontogeny of drug metabolism enzymes and implications for adverse drug events. Pharmacol Ther 118(2):250–267
Judson R, Martin M, Egeghy P, Gangwal S, Reif D, Kothiya P, Wolf M, Cathey T, Transue T, Smith D, and others (2012) Aggregating data for computational toxicology applications: the US environmental protection agency (ePA) Aggregated Computational toxicology Resource (ACtoR) system. Int J Mol Sci 13(2):1805–1831
Kim J, Scialli A (2011) Thalidomide: the tragedy of birth defects and the effective treatment of disease. Toxicol Sci 122(1):1–6
Knobeloch L, Salna B, Hogan A, Postle J, Anderson H (2000) Blue babies and nitrate-contaminated well water. Environ Health Perspect 108(7):675
Köhler T, Kondapalli L, Shah A, Chan S, Woodruff T, Brannigan R (2011) Results from the survey for preservation of adolescent reproduction (SPARE) study: gender disparity in delivery of fertility preservation message to adolescents with cancer. J Assist Reprod Genet 28(3):269–277
Koren G, Pairaideau N (2006) Compliance with prenatal vitamins. Patients with morning sickness sometimes find it difficult. Can Fam Physician 52(11):1392–1393
Koukouritaki S, Manro J, Marsh S, Stevens J, Rettie A, McCarver D, Hines R (2004) Developmental expression of human hepatic CYP2C9 and CYP2C19. J Pharmacol Exp Ther 308(3):965–974
Lawlor F, Peiris S (1985) Progress of a harlequin fetus treated with etretinate. J R Soc Med 78(Suppl 11):19
Lebel C, Roussotte F, Sowell E (2011) Imaging the impact of prenatal alcohol exposure on the structure of the developing human brain. Neuropsychol Rev 21(2):102–118
Lipinski RJ et al (2012) Ethanol-induced face-brain dysmorphology patterns are correlative and exposure-stage dependent. PLoS One 7(8):e43067
Maier S, West J (2001) Drinking patterns and alcohol-related birth defects. Alcohol Res Health 25(3):168

Monagle (2008) Antithrombotic therapy in neonates and children. Chest 133.6 suppl (2008):887S
Morris C (2000) Pediatric iron poisonings in the United States. South Med J 93(4):352–358
Monagle P, Studdert D, Newall F (2012) Infant deaths due to heparin overdose: time for a concerted action on prevention. J Paediatr Child Health 48(5):380–381
Sanchez-Echaniz J, Benito-Fern'andez J, Mintegui-Raso S (2001) Methemoglobinemia and consumption of vegetables in infants. Pediatrics 107(5):1024–1028
Selevan S, Kimmel C, Mendola P (2000) Identifying critical windows of exposure for children's health. Environ Health Perspect 108(Suppl 3):451
Solomon G, Moodley J (2007) Acute chlorpyrifos poisoning in pregnancy: a case report. Clin Toxicol 45(4):416–419
Stevens J, Hines R, Gu C, Koukouritaki S, Manro J, Tandler P, Zaya M (2003) Developmental expression of the major human hepatic CYP3A enzymes. J Pharmacol Exp Ther 307(2):573–582
Talhout R, Schulz T, Florek E, Van Benthem J, Wester P, Opperhuizen A (2011) Hazardous compounds in tobacco smoke. Int J Environ Res Public Health 8(2):613–628
Tenenbein M (2005) Unit-dose packaging of iron supplements and reduction of iron poisoning in young children. Arch Pediatr Adolesc Med 159(6):557–560
Teratology Society (1987) Teratology society position paper: recommendations for vitamin A use during pregnancy. Teratology 35:269–275
Therapontos C, Erskine L, Gardner E, Figg W, Vargesson N (2009) Thalidomide induces limb defects by preventing angiogenic outgrowth during early limb formation. Proc Natl Acad Sci 106(21):8573–8578
Waters M, Stack H, Garrett N, Jackson M (1991) The genetic activity profile database. Environ Health Perspect 96:41
Wickström R (2007) Effects of nicotine during pregnancy: human and experimental evidence. Curr Neuropharmacol 5(3):213
Wishart D, Knox C, Guo A, Shrivastava S, Hassanali M, Stothard P, Chang Z, Woolsey J (2006) DrugBank: a comprehensive resource for in silico drug discovery and exploration. Nucleic Acids Res 34(suppl 1):668–672
Wooten J (2012) Pharmacotherapy considerations in elderly adults. South Med J 105(8):437–445
Wu FC (1997) Endocrine aspects of anabolic steroids. Clin Chem 43(7):1289–1292

Part II
Toxicant Modes of Action and Biomarkers

Chapter 3
Mutational Effects

Edward L. Croom

Abstract Mutagens are agents that cause permanent genetic changes (i.e., mutations). These mutations can be limited to a single cell or may become part of an entire population of organisms. Chemicals as well radiation can cause mutations. Mutagens can be divided into direct and indirect acting mutagens. Direct acting chemical mutagens are often electrophilic and chemically reactive. Indirect acting chemical mutagens often require metabolic activation. Some agents (e.g., ionizing radiation) can act as both. Radiation induced mutational damage is related to the size and speed of the particles with larger particles causing more damage but having less penetrating power. Mutations can be limited to a single base or can involve multiple chromosomes. Mutations can occur in coding and non-coding regions of the genome. Mutations in the coding regions of the genome range in severity from generally harmless silent mutations to lethal vital protein destroying mutations. Mutations in the non-coding regions can sometimes cause functional changes (e.g., splice enhancer mutations). The earlier during development a mutation occurs the greater the potential for it to spread and impact the entire organism. Mutations that occur in the germ cells can be passed on to future generations and are of the greatest concern. Mutation frequencies differ depending on the bases being substituted and the mutagens involved. Translational toxicology includes efforts to limit exposures to potent chemical mutagens (e.g., aflatoxin) and mutagenic radiation (e.g., radon gas, UV light).

Keywords Mutagens • Mode of action • Radiation • DNA repair

3.1 Introduction

Deoxyribonucleic acid (DNA) is the molecule that allows biological information to be passed down from generation to generation. DNA provides the instructions on how many and what types of proteins and ribonucleic acids (RNAs) are made in

E.L. Croom (✉)
VP Chemistry, BOV Solutions, INC, Statesville, NC 28677, USA

230 Union St N, Concord, NC 28025, USA
e-mail: elcroom@gmail.com

C.L. Hughes, M.D. Waters (eds.), *Translational Toxicology*, Molecular and Integrative Toxicology, DOI 10.1007/978-3-319-27449-2_3

each cell. It is these proteins and RNAs that perform most of the functions of the cell and by extension the functions of the entire organism.

RNAs are short-lived molecules as are most proteins. These molecules are designed to be replaced over time and generally will last in the range of seconds to months. In contrast the DNA that makes up chromosomes is designed to last the lifetime of the cell. In the case of nerve cells those chromosomes have to last the lifetime of the organism which for humans is decades. Although many cells are short lived and are replaced after only a couple of days (e.g., mouth epithelial cells) the underlying cells that form a constant supply of replacement cells are longer lived and are designed to be able to divide regularly as needed.

It is the overwhelmingly successful ability to faithfully copy DNA from one cell to another that allows cells and organisms to function successfully over time. In general mutations are likely to be destructive at worst or silent at best. A small percentage of mutations will convey some advantage. These advantages may even be passed down through generations and maintained if they are sufficiently beneficial to convey some survival advantage. Even mutations that convey some survival advantage tend to have harmful effects. Examples of this are show with mutations in the G6PDH and hemoglobin genes that confer increased resistance to malaria but which are also associated with favism and sickle cell disease respectively.

This chapter is organized into four parts. The first part describes the chemistry of DNA and basic mechanisms of how it is first transcribed into RNA and then next translated into protein. The second part covers the types of mutagens that exist. The third part describes the types of mutations that can occur. The forth part describes repair mechanisms. Throughout the chapter case studies will be provided that highlight clinically relevant mutagens, mutations, and mechanisms.

Because of ethical and practical restrictions on the studies of how mutagens cause human mutations much of the available information is derived from studies involving *in vitro* model systems (e.g., bacteria, blood cells) and *in vivo* models (e.g., repair deficient mice). For this reason case studies may have mechanistic data based on models even when they are based on known clinically relevant human mutations. When possible, examples of translational toxicology efforts to limit exposures to known or anticipated human mutagens will be provided.

3.2 Normal Replication

DNA is the molecule that allows biological information to be passed down from generation to generation. This occurs through a process called DNA replication. DNA is normally a doubled stranded molecule that is found in the form of a double helix. For DNA to be copied this helix must first be unwound. This unwinding action is performed by enzymes called helicases (Fig. 3.1).

Although both DNA strands can be copied at the same time they are copied at different rates. This is because DNA is only formed in one direction from 5′ to 3′ because only the 3′ end has an open hydroxyl group open to link to another phosphate and base. Because DNA can only be formed from 5′ to 3′, DNA is only read

Fig. 3.1 DNA unwound by a helicase. Image modified from RCSB PDB file 4CGZ deposited by Newman et al. (2014)

from the 3′ to 5′ end (Fig. 3.2. The two strands are identified as the lagging strand (5′ to 3′) and the leading strand (3′ to 5′). Although both strands can be read once that section of DNA is unwound the leading strand can be read continuously while the lagging strand is read in sections.

In humans DNA is organized in 23 pairs of chromosomes. DNA is read and copied by enzymes called DNA polymerases. These DNA polymerases are extremely efficient and have an error rate of less than 1 in 100,000 bases. Proofreading and repair mechanisms further reduce that rate so the overall error rate is only around 1 in a billion. This is a very high degree of fidelity and to continue the text analogy would be the equivalent of having to type roughly 1000 books worth of text but only be allowed a single typo (Wang 2008). These repair mechanisms will be examined later in Sect. 3.1.5.

DNA is made up of four bases; cytosine, guanine, thymine and adenine (Fig. 3.3). These bases differ in terms of their chemical reactivity and likelihood to cause mutations when modified. DNA is normally double stranded with cytosines pairing with guanines and adenines pairing with thymines. These DNA bases are transcribed (i.e., copied) into an RNA format. Human chromosomal DNAs are extremely long molecules, too long and too large to ever fit through the nuclear membrane. Even if the DNA could fit through nuclear membrane even the smallest chromosomes contain many genes and most of the information would need to be removed before a single gene could be transcribed.

To efficiently transfer a gene's worth of information from the nucleus to the cytoplasm where proteins are made a special form of RNA called precursor messenger

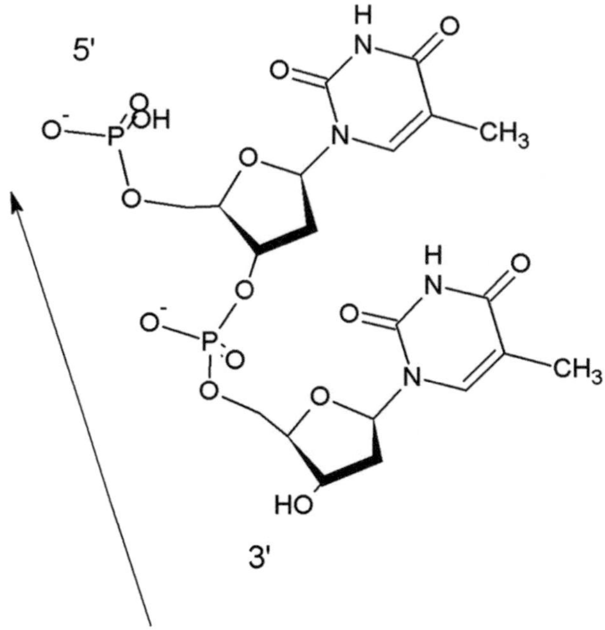

Fig. 3.2 DNA. The *arrow* next to the DNA molecule shows the direction of the order of which DNA is read

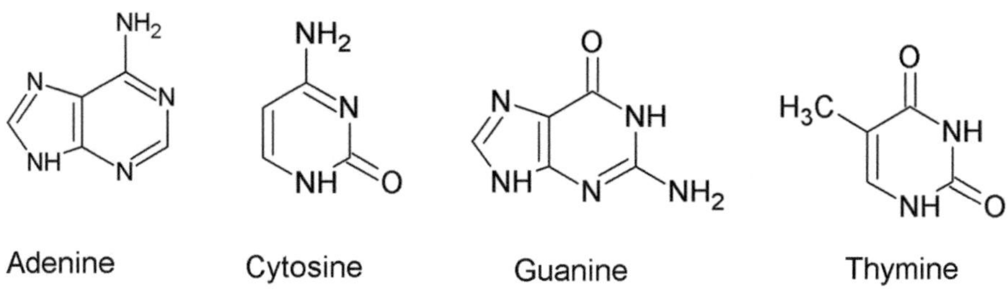

Fig. 3.3 DNA bases

RNA (pre-mRNA) is created. This pre-mRNA contains codons identical to the opposite DNA sequence except that instead of thymine, RNA has uracil. Both DNA and RNA are read in units of three base pairs called codons. These DNA codons are transcribed into mRNA and then translated into amino acids. DNA can be thought of as an instruction manual where the instructions are always provided using three letter words. For example "USE THE BAR AND PRY THE RED TOP OFF."

There is redundancy in the codon language and there are more codons than amino acids. Typically the first two bases are the same for a given amino acid and only the third base in the codon differs. This is exemplified by proline which is coded for by the four mRNA codons CCU, CCC, CCA, and CCG. An exception to

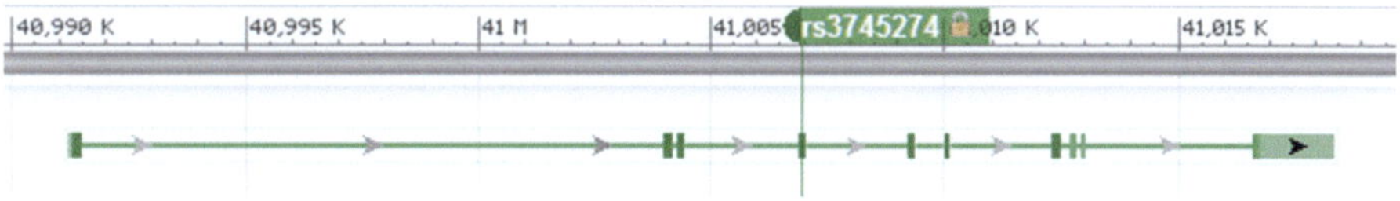

Fig. 3.4 CYP2B6 exons and introns. CYP2B6 exons are shown as the *vertical green bars*. The introns are shown as the *light green lines* between the exons. The *scale above* shows the position of the gene on the chromosome which ranges roughly 25,000 bp (Source rs3745274, http://www.ncbi.nlm.nih.gov/SNP/snp_ref.cgi?rs=3745274)

redundancy in coding is methionine. Methionine, the start codon, is coded for by just one codon, AUG. Leucine, Arginine and Serine also differ from this rule by having more than four codons each.

DNA is organized into genes that form proteins and surrounding portions of DNA that do not form proteins although they may play a role in regulating gene expression. Genes also contain coding and non-coding portions. Exons are the coding parts and introns are the non-coding parts (Fig. 3.4. Both introns and exons are transcribed into pre-RNA but only exons are taken out of the nucleus as mRNA and translated into protein. Introns generally take up most of the gene sequence while the exons only take up a small amount.

DNA is located inside the nucleus while proteins are created in the cytoplasm. In order to get the instructions on how to make those proteins from the nucleus to the cytoplasm an intermediary, messenger RNA (mRNA), is used. RNA polymerases use a DNA strand as a template the bases forming mRNA. This mRNA contains all of the codons but loses the introns and is further modified with the addition of a cap on the 5′ end and poly A (i.e., adenosine) tail on the 3′ end. This effectively shortens the pre-mRNA from thousands of bases down to just hundreds of bases for mRNA. A special collection of proteins combine to form the spliceosome. The spliceosome is responsible for taking the original transcript, called pre-mRNA, and modifying it by removing the introns and attaching the remaining exons into a much shorter messenger RNA.

Translation occurs in the ribosomes. The ribosomes are where the transfer RNA (tRNA) anticodons are paired with the mRNA codons. This positions the amino acids in the proper order and proximity to allow them to be joined together by the ribosome forming a polypeptide chain. The first amino acid in this polypeptide chain is always methionine however there can be many modifications before the final protein is formed and methionine may not be the first amino acid in the finished protein.

Human DNA is organized into 23 pairs of chromosomes. During most of the cell-cycle these chromosomes are generally in an open state and available for transcription. In this state open state the chromosome is referred to as euchromatin. However the DNA has to be condensed and packed tightly in order for mitosis and meiosis to occur. Proteins called histones function as tiny spools wrapping up the DNA tightly around them. There are many histones per chromosome and in this tightly wrapped form the chromosome is referred to as heterochromatin. There are always small portions of the chromosome that are heterochromatin even during interphase. These parts are known as constitutive heterochromatin. Even when the X-chromosome is selectively deactivated forming Barr bodies some euchromatin is still accessible and some transcription still occurs. There is very strong evidence of this in the different phenotypes of XO females who exhibit the symptoms of Turner syndrome (e.g., short stature, webbed neck).

Because the enzymes involved in normal replication are not perfect with enough replication events some mistakes will be made even without external exposures. Even without the action of external forces DNA has a slight tendency to change spontaneously. DNA is chemically reactive and can be damaged by ionizing radiation, free radicals and electrophilic chemicals. The next section focuses on the types of chemicals and radiation capable of damaging DNA.

3.3 Types of Mutagens

Mutagens can be chemicals or radiation. Chemicals mutagens can be organic (e.g., benzo(a)pyrene) or inorganic (e.g., arsenic, chromium, nickel). Not all inorganic chemical mutagens are metalloids or metals. Nitric oxide is a mutagenic inorganic compound. Mutagenic radiation can be non-ionizing (e.g., UV) or ionizing (e.g., X-rays). There are many different types of mutagens but one of the most important ways of classifying them is as direct and indirect mutagens.

3.3.1 Direct Acting Mutagens

Direct acting mutagens interact directly with DNA. Direct acting mutagens are either electrophilic chemically reactive molecules or are a form of radiation containing enough energy to change DNA directly.

3.3.1.1 Direct Acting Chemical Mutagens

The electrophilic theory of chemical carcinogenesis was developed decades ago to explain the observation that human carcinogens tended to be mutagens that either started out as electrophiles or were activated to become electrophiles (Miller and Miller 1981).

Nitrogen mustards are an example of direct acting chemical mutagens. Nitrogen mustards were originally developed as chemical warfare agents. Their ability to alkylate DNA made them deadly chemicals on the battlefield. Their ability to damage DNA and kill rapidly dividing cells also led to them being used as an early form of chemotherapy.

Alkylating agents remain an important part of chemotherapeutic agents. Melphalan (Fig. 3.5 is an example of a direct acting chemotherapeutic mutagen.

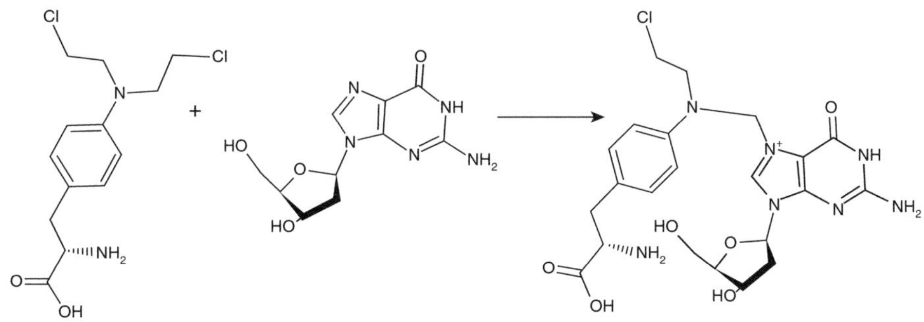

Fig. 3.5 Melphalan adduct. Melphalan is shown reacting with Deoxyadenosine to form a melphalan DNA adduct

The N7 nitrogens in DNA purine bases are not involved in hydrogen bonding making them good targets for direct acting mutagens.

Platinum based chemotherapeutic drugs (e.g., cisplatin, carboplatin) are another group of direct acting mutagens that interact with N7 nitrogen atoms of purines (Burnouf et al. 1990). Platinum based chemotherapeutic drugs form intrastrand adducts by linking two adjacent purines (Fig. 3.6).

Another way mutagens can directly interact with DNA is through intercalation. Intercalating chemicals stick inside the double helix disrupting the structure of DNA. DNA contains a minor groove and major groove. Intercalating agents interact with either the major or minor grooves (Fig. 3.7). Intercalating mutagens generally have a ring structure that creates a planar part of the molecule that can slide between the bases.

Case Study 1: Chemotherapeutic Mutagens Mutagens can damage rapidly dividing cells which can make them useful as chemotherapeutic agents. However, mutagenic chemotherapeutic drugs are also often quite toxic to normal healthy cells causing significant damage to organs and limiting their use. Translational toxicology includes efforts to either develop chemoprotective drugs or more selective chemotherapeutic mutagens. In the case of doxorubicin, an intercalating agent, the major organ affected is the heart. In the case of cyclophosphamide, an alkylating agent, a major organ affected is the bladder. Melphalan, an alkylating agent, often reduces blood cell counts.

To help overcome the cardiotoxicity of doxorubicin a similar intercalating drug, epirubicin, was created. Epirubicin, can be as damaging to cancer cells as doxorubicin but is more readily metabolized and removed by normal cells. To help overcome the bladder toxicity associated with cyclophosphamide mesna, a chemoprotective drug, is added to scavenge the urotoxic cyclophosphamide metabolite acrolein. There is some evidence that in addition to protecting the bladder, mesna can reduce the mutagenicity of cyclophosphamide. While mesna has been reported to not reduce the mutagenicity of cyclophosphamide *in vitro*, there is

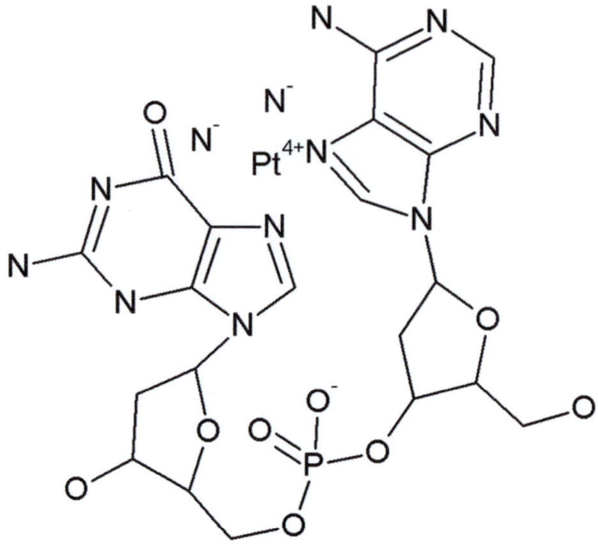

d(ApG)/cis-diamminedichloroplatinum(II)

Fig. 3.6 Cisplatin based DNA adduct (Structure source Burnouf et al. 1990)

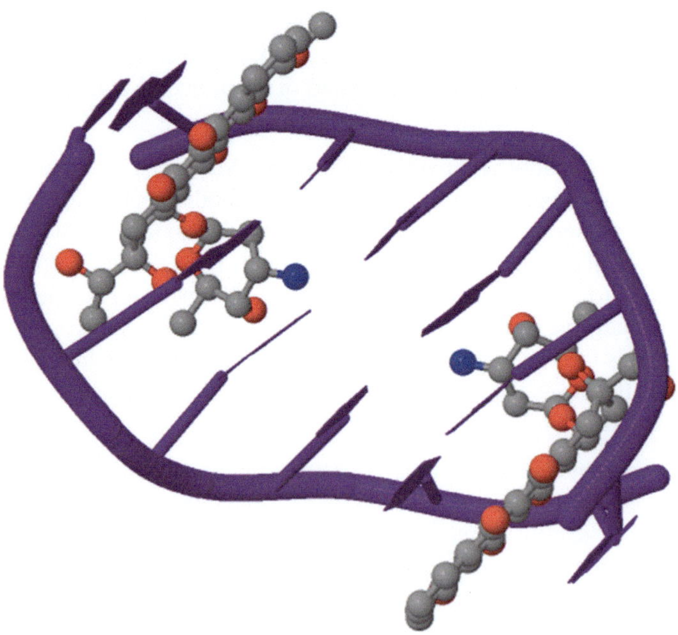

Fig. 3.7 DNA and intercalating chemotherapeutic agents. Daunorubicin is shown in red, gray and blue. The DNA is shown in purple (Source RCSB PDB file 1DA0 deposited by Wang et al. 1987)

evidence that it can reduce the mutagenicity of cyclophosphamide *in vivo* as it was shown to reduce the amount of urinary mutagens in rodents treated with cyclophosphamide (Pool et al. 1988).

A third method of reducing the mutagenicity of chemotherapeutic drugs has been to develop prodrugs that are selectively activated by cancer cells. In effect this turns direct acting mutagens into targeted indirect acting mutagens. A recent example of this is the investigational drug melphalan-flufenamide, a pro-drug version of the alkylating chemotherapeutic agent melphalan. Melphalan-flufenamide has a peptide bond connecting melphalan to an amino acid which cleaved by a peptidase that is commonly overexpressed in cancer cells (Chauhan et al. 2013).

3.3.1.2 Direct Acting Radiological Mutagens

Radiation can be directly mutagenic. Most electromagnetic radiation is not mutagenic. Visible light lacks sufficient energy to mutate cells as do the types of electromagnetic radiation with even longer wavelengths (e.g., infrared, microwaves). As wavelengths shorten the available energy increases. Only the shortest waves on the end of the electromagnetic spectrum found in ultraviolet (UV) radiation, x-rays, gamma rays and cosmic rays contain enough energy to directly mutate DNA.

UV radiation exposure primarily comes from sunlight. UV radiation is an important source of vitamin D in small exposures for many people but UV radiation causes DNA damage that can result in mutations. UV radiation is associated with cyclobutane lesions. The most common of these is a thymine dimer (Fig. 3.8).

Thymine dimers are one form of cyclobutane lesions. Cyclobutane lesions are more stable when the bonds holding the thymines together are shorter (Fig. 3.9). While the double helix structure of DNA is normally very symmetrical, cyclobutane lesions interferes with structure and reshape the molecule causing it to bend around them (Fig. 3.10).

Case Study 2: UV Radiation Exposure to UV radiation can be limited through the use of protective equipment, clothing and sunscreens. UV radiation is divided into UVA and UVB. Sunscreens contain chemicals designed to absorb most of the energy in UV radiation. Early sunscreens were designed primarily to absorb UVB radiation but increasingly they contain multiple active ingredients that when combined are designed to absorb both UVA and UVB radiation. The presence of at least one benzene ring is a common characteristic of many UV absorbing chemicals (Fig. 3.11).

3.3.2 Indirect Acting Mutagens

Indirect acting mutagens interact first with another molecule and then a product of that interaction interacts with DNA. Indirect acting mutations can be chemical or radiological.

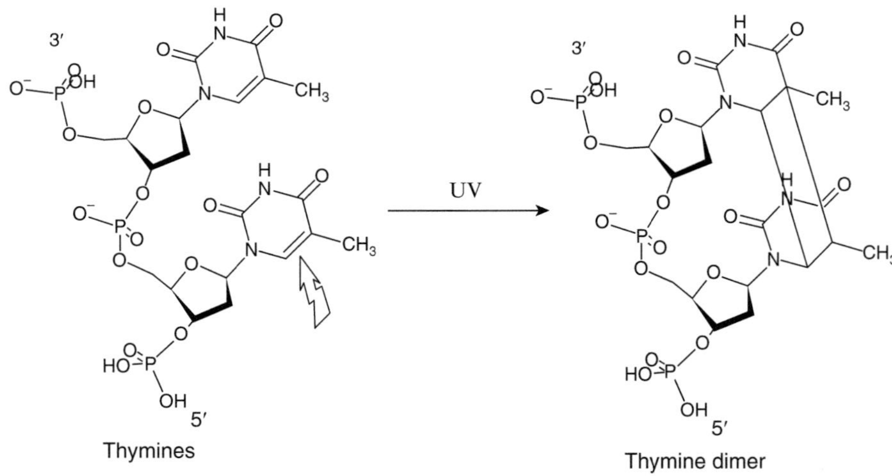

Fig. 3.8 Thymine dimers. Thymine dimers form when UV light excites the double bonds in a thymine. When there is an adjacent thymine this extra energy makes it possible for the two double bonds to open up forming instead two new covalent bonds linking the thymines together

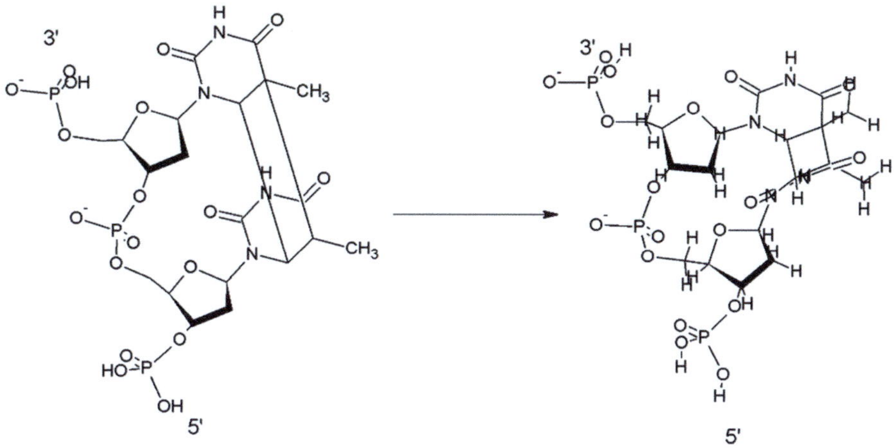

Fig. 3.9 Cyclobutane lesions. When cyclobutane rings are created as in the formation of the thymine dimer shown tension is created. It takes a great deal of energy to have a cyclobutane ring stretched out and the ring will contract pulling the bases together

Fig. 3.10 DNA with a thymine dimer is shown above. Note in the ends of the figure how the bases are neatly stacked upon one another and are directly across from their partner on the opposing DNA strand. In the middle of the strand is a thymine dimer where the two thymines have been pulled together after a cyclobutane ring forms. On the opposite strand one of the adenines is shown having been released. Figure created using a structure published on RCSB PDB by Vassylyev et al. (1995)

Fig. 3.11 UV absorbing chemicals

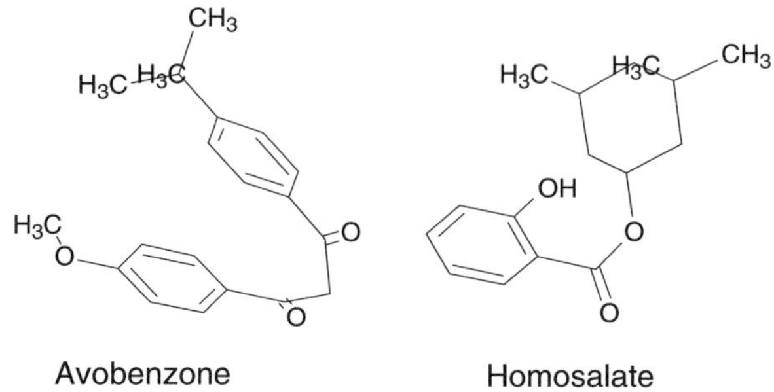

3.3.2.1 Indirect Acting Chemical Mutagens

An example of an indirect acting mutagen is the promutagen benzo(a)pyrene. Benzo(a)pyrene is an aromatic hydrocarbon commonly found in combustion byproducts. In the liver it is bioactivated by the cytochromes P450 (CYPs) forming the mutagenic metabolite Benzo(a)pyrene-trans-7,8-dihydrodiol-10,11-epoxide (BPDE). BPDE forms adducts with DNA.

CYPs are commonly involved in bioactivation but other phase I and even phase II enzymes can play a role. In some cases phase I and phase II xenobiotic metabolizing enzymes can work together in sequence to make the ultimate mutagen. For example, sulfotransferases have been implicated in the further bioactivation of hydroxylated polycyclic aromatic hydrocarbons and some amines. The classic carcinogenic activity 2-acetylaminofluorine (2-AAF) has been shown to be the result of first hydroxylation by CYPs and then conjugation by sulfotransferases to form a reactive, electrophilic sulfuric acyl ester metabolite.

Case Study 3: Aflatoxin Aflatoxin is a naturally occurring indirect acting chemical mutagen. Aflatoxin is commonly produced by the fungi *Aspergillus flavus*. Aflatoxin $B_1$is metabolized in the liver by CYPs to form the reactive compound Aflatoxin $B_1$8,9-epoxide. High doses of aflatoxin B_1 are associated with DNA adduct formation and liver cancer development. To limit exposures to aflatoxin B_1 two main strategies are deployed. The first is to alter the storage and handling of the grains and peanuts that can become contaminated with *Aspergillus flavus* under favorable growth conditions (e.g., drought then damp). The second strategy is to screen food products for aflatoxin B_1 levels. The availability of relatively inexpensive and rapid testing methods (e.g., ELISAs) has made testing easier and in some cases field ready.

3.3.2.2 Indirect Acting Radiological Mutagens

Ionizing radiation is also an indirect acting mutation. Ionizing radiation can create free radicals in molecules near the DNA and those free radicals can then bind to the DNA. This can happen when the energy of the particles that make up ionizing radiation is enough to remove an electron on the outside of an atom. Water is the most

1. Initial ionization $H_2O \longrightarrow H_2O^+ + e^-$

2. Thermalization $e^- + H_2O \longrightarrow e^-_{(aq)}$

3. Second ionization $e^-_{(aq)} + H_2O \longrightarrow H_2O^-$

4. Radical formation $H_2O^+ + H_2O \longrightarrow H_3O + {}^*OH$

$$H_2O^- \longrightarrow OH^- + H^*$$

Fig. 3.12 The first steps involved in the formation of free radicals in water by ionizing radiation

commonly molecule in the human body and not surprisingly is the most frequently ionized molecule as a result of exposure to ionizing radiation (Fig. 3.12).

The amount of energy contained in each particle depends on both the size and the speed of the particle. The larger the particle the less penetrating ability it possesses. Linear energy transfer is the covalent energy force related to the amount of energy transferred by the ionizing particle as it travels through a material by unit of length. Below a certain linear energy transfer level the body will not be damaged and mutations will not form. Linear energy transfer is closely related to stopping power.

Alpha particles are the largest particle emitted. They each contain two protons and two neutrons. This size significantly limits their ability to penetrate the body. However, when ingested or inhaled they can cause significant damage. Alpha emitters may be used medicinally but exposures may also be environmental. The high energy and low penetrance of alpha emitters make them useful for targeting tumors. The high energy and size of alpha-particles means that only a few alpha particles crossing the nucleus of a cell are enough to cause cell death. Alpha particles cause a high degree of double strand breaks.

Beta-emitters release beta particles which are electrons. Beta particles have limited linear electron transfer. It can take thousands of beta particle hits to kill a cell. The path length of a beta particle can range from ~50 to 1000 cell diameters. There a naturally occurring beta emitters (e.g., Carbon-14). Strontium-90 is a beta emitter produced when nuclear bombs are detonated. Levels of strontium-90 have declined significantly with implementation of the nuclear test ban treaty.

Gamma emitters lack the mass to directly ionize. Gamma rays are made up of photons and have significant penetrance but limited linear electron transfer. Instead they indirectly ionize by exciting nearby electrons. A single excited electron can transfer its kinetic energy in several ways. It can heat the surrounding tissue, excite nearby electrons or ionize. The ionization is what leads to mutation. A single high energy electron that is the result of absorbing a gamma or x-ray photon can produce over 1,000 low energy secondary electrons (i.e., delta rays). Eventually the excitation energy becomes less than excitation level of water (7.4 eV), and the final extra energy is transferred through vibrational, rotational and collisional energy exchanges. This can lead to chromosomal breakage and other forms of DNA damage.

X-rays have even lower energy than gamma rays and are the last form of electromagnetic radiation that can ionize. Ultraviolet radiation, the next lowest down has ~100 fold less energy than X-rays and although UV radiation is mutagenic it lacks the ability to ionize water and other molecules. In contrast X-rays can penetrate soft tissues although the lack the ability to penetrate bone.

Case Study 4: Alpha Particles The size of alpha particles limits their pathway to only ~2–10 cell diameters. This makes them ideal candidates for being conjugated to tumor selective antibodies (e.g., Bi-213). When alpha particles conjugated with tumor selective antibodies are used against small tumors they have the potential to cause significant damage to the tumors but limited damage to the surrounding tissues.

Radon is an important alpha emitter that forms during the natural radioactive decay of uranium and thorium in the earth. Because radon is a gas it can concentrate inside homes build above uranium containing rock. Radon exposure is associated with increased lung cancer rates. While alpha particles have limited penetrance the thin lung alveoli walls make them easily damaged.

Case Study 5: Beta Particles One of the properties of radioactive chemicals is that they have a tendency to accumulate in bones and teeth. Strontium-90 levels have been studied non-invasively for years by collecting baby teeth. One of the more widely used beta emitters is Iodine-131. Iodine collects in the thyroid. To prevent the accumulation of radioactive iodine in the thyroid people at risk of being exposed (e.g., living near nuclear power plant) are often supplied with potassium iodide tables). These tablets can be taken preemptively to saturate the thyroid limiting the ability of radioactive iodide to accumulate. This tendency to accumulate in the thyroid has also made Iodine-131 useful to treat thyroid diseases (e.g., Graves' disease). In order to protect others from their temporarily radioactive bodies, patients receiving Iodine-131 have to follow strict procedures to protect their family members. They particularly have to stay away from very young children.

Case Study 6: Gamma Knife Gamma radiation has significant penetrating power. Gamma rays generated from gamma emitters (e.g., Cobalt 60) have become an important therapy for tumors that would otherwise be inoperable. Stereotactic radiosurgery uses externally generated radiation that is aimed at a target to destroy diseased tissue while preserving surrounding healthy tissue. With this technique tumor tissues inside the skull can be destroyed without having to create an incision.

3.3.3 Types of Human Mutagens

Global Harmonization System of Classification and Labeling of Chemicals (GHS) is the international standard used to classify chemicals with respect to toxicity. Chemicals are classified into categories based on the weight of evidence. Mutagens are classified as Category 1, 2 or not classified based on whether or not they are regarded as human germ cell mutagens.

3.3.3.1 Category 1 Mutagens

Category 1 mutagens are human germ mutagens. They are further divided into Category 1A and Category1B. To date no chemicals have been classified as Category 1A as none have had adequate human epidemiological evidence. Category 1B human germ cell mutagens have some positive evidence towards mammalian germ cells. This evidence may be supported by evidence of mutagenicity towards somatic cells. With mammalian germ cells there may be evidence of generational inheritance. With human germ cells the evidence is limited to one generation. For example aneuploidy of the sperm cells is not going pass down to the next generation as changing the number of chromosomes will typically be lethal.

3.3.3.2 Category 2 Mutagens

Category 2 mutagens are chemicals of concern as they are considered potential human germ cell mutagens. Category 2 chemicals are positive in animals and/or in vitro assays. The in vivo assays include sister chromatid exchange assays and assays of DNA damage (e.g., comet assay). The in vitro assays include chromosomal aberration tests and salmonella reverse mutagenicity tests (Hansel et al. 1997).

3.4 Types of Mutations

3.4.1 Base Substitutions

3.4.1.1 Introduction

Base substitutions are a point mutation where a single DNA base is mutated and changed from one base into another. Transition mutations occur when a purine is mutated into another purine (e.g., A→G). Transversion mutations occur when a pyrimidine is converted into a purine or a purine is converted into a pyrimidine (e.g., A→T). Mutation frequencies differ depending on the genes, tissues and the mutagens involved. Human lung tumors have been examined for mutation frequencies. With human tumors there is no controlling the exposures unlike with animal models. However, when tumor tissues collected from the same organs from hundreds of human donors were examined trends were seen. Roughly 40 % of the tumors had TP53 mutations while <20 % of the tumors had Kras2 mutations and <2 % of the tumors had Braf mutations (Jackson et al. 2006).

For example with the Kras2 gene exposures to 2-amino-3-methylimidazo[4,5-f] quinolone and 7H-dibenzo[c,g]carbazole were associated primarily with AT>TA transversions (80 % and 92 % respectively) (Jackson et al. 2006). In contrast exposures to benzidine resulted primarily in GC>TA transversions and NNK caused Kras2 mutations in mouse lungs that were predominantly (96 %) GC>AT (Jackson et al. 2006).

3.4.1.2 Silent Mutations

Silent mutations are mutations that change the base pair but do not change the amino acid (Fig. 3.13). These mutations most often occur in the third base of a codon. For example CGA codes for Arginine but so does AGA. With a silent mutation the original message "USE BAR AND PRY THE TOP OFF END" would stay the same "USE BAR AND PRY THE TOP OFF END."

Silent mutations in coding regions can still have an effect of protein formation. This is because coding regions also can contain regulatory elements such as splice enhancers. An example of this is the silent mutation (c.321C>T, p.D107D) in the rennin receptor gene (*ATP6AP2*) that nonetheless results in functional changes due to its function as a splice enhancer mutation. Ramser et al. reported a loss of exon 4 in 50 % of the mRNA and confirmed that truncated proteins were produced that had reduced function towards ERK1/2 activation (Ramser et al. 2005).

3.4.1.3 Missense Mutations

Missense mutations are mutations that change the base pair and change the amino acid (Fig. 3.14). Clinically important missense mutations tend to change the amino acid property. Changing a large amino acid into a small amino acid or a polar amino acid into a non-polar amino acid are major changes that can disrupt the structure of the protein. Missense mutations can also result in a loss in protein expression if they also disrupt regulatory elements. An example of this is the CYP2B6*6 allele which

a

DNA:	5' -	ATG	ACT	CAC	CG**A**	GCG	CGA	AGC	TGA	- 3'
	3' -	TAC	TGA	GTG	GCT	CGC	GCT	TCG	ACT	- 5'
mRNA:	5' -	AUG	ACU	CAC	CGA	GCG	CGA	AGC	UGA	- 3'
Protein:		Met	Thr	His	**Arg**	Ala	Arg	Ser	Stop	

b

DNA:	5' -	ATG	ACT	CAC	CG**G**	GCG	CGA	AGC	TGA	- 3'
	3' -	TAC	TGA	GTG	GCC	CGC	GCT	TCG	ACT	- 5'
mRNA:	5' -	AUG	ACU	CAC	CGG	GCG	CGA	AGC	UGA	- 3'
Protein:		Met	Thr	His	**Arg**	Ala	Arg	Ser	Stop	

Fig. 3.13 Silent mutations. Both the normal strand (**a**) and the mutant strand (**b**) form Arginine in the fourth position shown in *bold*

a

```
DNA:      5' - ATG ACT CAC CGA GCG CGA AGC TGA - 3'
          3' - TAC TGA GTG GCT CGC GCT TCG ACT - 5'
mRNA:     5' - AUG ACU CAC CGA GCG CGA AGC UGA - 3'
Protein:       Met Thr His Arg Ala Arg Ser Stop
```

b

```
DNA:      5' - ATG ACT CAC GGA GCG CGA AGC TGA - 3'
          3' - TAC TGA GTG CCT CGC GCT TCG ACT - 5'
mRNA:     5' - AUG ACU CAC GGA GCG CGA AGC UGA - 3'
Protein:       Met Thr His Gly Ala Arg Ser Stop
```

Fig. 3.14 Missence mutations. Normal strand (**a**) forms Arginine while mutant strand (**b**) forms Glycine instead due to a missence mutation changing Cytosine to Guanine. Affected amino acids and nucleotides are *bolded*

involves the 516G>T, Q172H missense mutation removes a splice enhancer (Hoffman 2008). These types of missense mutations can be common and may play a major role in human variablitiy such as the common CYP2B6 polymorphisms and their impact on drug metabolism (Lang et al. 2001).

3.4.1.4 Nonsense Mutations

Nonsense mutations are mutations that create a premature stop codon (Fig. 3.15). There are three stop codons and they have been named for colors UAA (Ochre), UAG (Amber), UGA (Opal). With a nonsense mutation the instructions to make a functioning protein are stopped prematurely. In text it would be as if instead of the sentence "USE THE BAR AND PRY THE RED TOP OFF END" you were only given "USE THE BAR AND END." It is not surprising that premature stop codons often result in non-functioning proteins.

3.4.2 Small Insertions and Deletions (INDELS)

Small insertions (Fig. 3.16) and deletions (Fig. 3.17) (i.e., INDELS) are defined as insertions or deletions of a size between 1 and 10,000 bases. While INDELS are less common than base pair changes and SNPs they are still widespread. Several million INDELS have been identified in the human genome (Mulaney et al. 2010). Continuing the text example "USE THE BAR AND PRY THE RED TOP OFF END" could change to "USE THE BAR PRY THE RED TOP OFF END" with a small deletion or "USE THE BAR TOP OFF END" with a larger deletion. Similarly an insertion could change "USE THE BAR AND PRY THE RED TOP OFF END" to "USE THE BAR AND CAR FOR GET TEN PRY THE RED TOP OFF END."

a

```
DNA:     5' - ATG ACT CAC CGA GCG CGA AGC TGA - 3'
         3' - TAC TGA GTG GCT CGC GCT TCG ACT - 5'
mRNA:    5' - AUG ACU CAC CGA GCG CGA AGC UGA - 3'
Protein:      Met Thr His Arg Ala Arg Ser Stop
```

b

```
DNA:     5' - ATG ACT CAC TGA GCG CGA AGC TGA - 3'
         3' - TAC TGA GTG ACT CGC GCT TCG ACT - 5'
mRNA:    5' - AUG ACU CAC UGA - 3'
Protein:      Met Thr His Stop
```

Fig. 3.15 Nonsense mutations. Normal strand (**a**) is four amino acids longer than the shorter mutant strand (**b**) that forms as a result of a nonsense mutation when a cytosine is replaced with a thymine resulting a premature stop codon

a

```
DNA:     5' - ATG ACT CAC CGA GCG CGA AGC TGA - 3'
         3' - TAC TGA GTG GCT CGC GCT TCG ACT - 5'
mRNA:    5' - AUG ACU CAC CGA GCG CGA AGC UGA - 3'
Protein:      Met Thr His Arg Ala Arg Ser Stop
```

b

```
DNA:     5' - ATG ACT CAC CGA GGA GCG CGA AGC TGA - 3'
         3' - TAC TGA GTG GCT CCT CGC GCT TCG ACT - 5'
mRNA:    5' - AUG ACU CAC CGA GGA GCG CGA AGC UGA - 3'
Protein:      Met Thr His Arg Gly Ala Arg Ser Stop
```

Fig. 3.16 Insertion mutations. Normal strand (**a**) forms Arginine while mutant strand (**b**) forms Arginine and inserted Glycine. Affected amino acids and nucleotides are *bolded*

With insertions or deletions that occur within a gene the amount of damage caused depends on the number of bases inserted or deleted and whether or not the reading frame changes. Often insertions and deletions destroy the function of a gene but they can at times create different functions.

Frameshifts occur as a result of deletions or insertions. DNA is read three bases at a time. If a single base is deleted or inserted it shifts the entire reading frame (Fig. 3.18). This is a major mutation that generally destroys the gene. Insertions and deletions are equally deleterious. In a text example a deletion of the letter e in the first word would change the reading from frame shifting the instructions from “USE THE BAR AND PRY OFF THE RED TOP OFF END” to “UST HEB ARA NDP RYT HER EDT OPO FFE NDN OWT…”

a

DNA: 5' - ATG ACT CAC **CGA** GCG CGA AGC TGA - 3'
3' - TAC TGA GTG GCT CGC GCT TCG ACT - 5'
mRNA: 5' - AUG ACU CAC CGA GCG CGA AGC UGA - 3'
Protein: Met Thr His **Arg** Ala Arg Ser Stop

b

DNA: 5' - ATG ACT CAC GCG CGA AGC TGA - 3'
3' - TAC TGA GTG CGC GCT TCG ACT - 5'
mRNA: 5' - AUG ACU CAC GCG CGA AGC UGA - 3'
Protein: Met Thr His Ala Arg Ser Stop

Fig. 3.17 Deletion mutations. Normal DNA strand (**a**) has its last four amino acids as Arginine, Alanine, Arginine, and Serine while the mutated DNA strand (**b**) has lost the first Arginine so its last four amino acids are Histidine, Alanine, Arginine and Serine

a.

DNA: 5' - ATG ACT CAC CG**A** GCG CGA AGC TGA - 3'
3' - TAC TGA GTG GCT CGC GCT TCG ACT - 5'
mRNA: 5' - AUG ACU CAC CGA GCG CGA AGC UGA - 3'
Protein: Met Thr His **Arg Ala Arg Ser Stop**

b.

DNA: 5' - ATG ACT CAC CG**G CGC GAA GCT GAT** - 3'
3' - TAC TGA GTG GCT GCG CTT CGA CTA - 5'
mRNA: 5' - AUG ACU CAC CGG CGC GAA GCU GAU - 3'
Protein: Met Thr His Arg **Glu Gly Asp**

Fig. 3.18 Frameshift mutations. Normal DNA strand (**a**) has its last three amino acids as Alanine, Arginine, and Serine while the mutated DNA strand (**b**) has had its reading frame shifted due to the deletion of an adenine. This replaces the stop codon with Aspartic Acid and the other three previous amino acids with Arginine, Glutamic acid and Glycine

3.4.3 Chromosomal Alterations

Chromosomal changes are another type of mutation. Chromosomes alterations come in two major forms; numerical and structural. Numerical changes involve the loss or gain of chromosomes. Polyploidy is when there are more than two copies of each chromosome. In humans polyploidy is not survivable at the organism level. Roughly 1–3 % of pregnancies are triploid but these pregnancies rarely result in term. Some infants are born triploid with a very limited survival time (Boghossian et al. 2012).

Aneuploidy occurs when there is a loss or gain in one or more chromosomes. Monosomy occurs when there is only one chromosome instead of two. Monosomy of the X-chromosome results in Turner's syndrome. Monosomies of the other chromosomes are lethal at the organismal level with the exception of the y-chromosome that is normally paired with the much larger x-chromosome. Trisomy occurs when there are three chromosomes instead of two. Trisomy 21 results in Down Syndrome. Trisomy in humans at the organism level is rarely survivable and survivability is inversely proportional to the size of the chromosomes. The most common trisomies that make it to term are trisomy 13, 18 and 21 and their hospital mortality rates have been reported as 92 %, 89 % and 33 % respectively (Boghossian et al. 2012).

Aneuploidy is different for the sex chromosomes. Males develop normally with only one X-chromosomes and females develop normally without at y-chromosome. This is because the y-chromosome has few genes, none that are needed by females, while the x-chromosome is selectively deactivated so that only one X-chromosome per cell is open during most of the cell-cycle while the other X-chromosome is largely deactivated through the process of lionization. These chromosomes are also smaller than most allowing them to be added or deleted with far less clinical impact than larger chromosomes. This is also true with XXY males who have Klinefelter syndrome (Strachan and Read 1999).

Chemicals and energy particles that break chromosomes are called clastogens. Benzene and arsenic represent organic and inorganic chemical clastogens respectively. Non homologous end joining (NHEJ) is a repair mechanism for chromosomal damage. When multiple chromosomes are broken it is possible to have different parts of different chromosomes joined together. This can result in the creation of acentric (i.e., without centromeres) chromosomes and dicentric (i.e., with two centromeres) chromosomes. Small acentric chromosomes are also known as micronuclei and are tested for as a sign of chromosome damage.

Chromosome breaks can alter the expression of a gene even if they occur outside of the gene. This is because a chromosome break can move a gene away from a regulatory element, toward a regulatory element or can move the DNA from normally active euchromatin that is available for transcription into condensed heterochromatin that is protein bound and inaccessible for transcription.

Chromosomal abnormalities are considered unbalanced if there is a net loss of chromosomal material and balanced if there is no net loss of chromosomal material. Inversions occur when a piece of a chromosome breaks off and is reattached in the opposite direction. Abnormal chromosomes can be passed down through cell division if they have the proper number of centromeres. Chromosomes without a centromere or with two centromeres cannot be properly moved to one of the new daughter cells during division and will be lost. The exception to this is with Robertsonian translocations where only a small portion of DNA is lost and the two centromeres are so close together that they can function as one centromere.

Chromosome specific dyes can be used to identify which parts of which chromosomes have been joined together. These breakages and joinings can result in pathological changes. A type of trisomy 21 is the result of the large arm of chromosome 21 being added to the large arm of chromosome 14. The Philadelphia chromosome is an example of what happens when a gene that normally has low expression is combined with a promoter that is normally associated with a gene that has high expression.

Mutations that occur in somatic cells are limited to that individual. Mutations may be limited to a single cell or they may be spread throughout the body. Some common mutations occur throughout the body because of extensive damage. Thymine dimers caused by UV exposure are examples of this type of widespread mutation. Other mutations may spread throughout the body because they occur early in development.

This is most significant during the embryonic stage when a mutation can be passed on to many different cells. Mosaics are examples of this where they have been seen. This is most apparent with mutations in skin color. A striking example of this involves a reported case of identical twins who shared mostly identical genetics but differed in terms of the neurofibromatosis-1 (NF-1) gene (Kaplan et al. 2010). These twins had their monozygosity confirmed. The mutation status of the NF-1 gene was tested the fibroblasts, lymphocytes and buccal cells of the twins and their relatives. The twins shared a non-sense mutation in the NF-1 gene but only one twin had the phenotype of neurofibromatosis.

Neurofibromatosis type 1 is a progressive condition that affects the skin and nerves. Only one twin had the café-au-lait macules. Every cell tested in the affected twin possessed the NF-1 nonsense mutation. In contrast many of the cells tested in the unaffected twin lacked the NF-1 nonsense mutation. The unaffected twin was identified as a mosaic. It is hypothesized that the NF1 mutation occurred after fertilization but before the zygote split into two separate zygotes. The affected twin developed from part of the zygote that only possessed mutated cells while the unaffected twin developed from a part of the zygote that possessed both mutated cells and cells lacking the NF-1 mutation.

Constitutional mutations are present in all of the cells of the body. Constitutional mutations must occur in the sperm, egg or fertilized egg to be spread throughout the entire organism as it develops. However, constitutional mutations that are only present in one copy may not be passed on to the next generation as during meiosis the diploid cells divide to form haploid gametes that pass down only one copy of each chromosome. For example an X-linked constitutional mutation present in the father would not be passed down to a son but would be passed down to a daughter.

Another way mutations can spread is through the growth and metastasis of cancer cells. The clonal theory of expansion for cancers describes how a single cell is mutated and that cell grows forming clones. Genomic instability is a common characteristic of cancer cells and cancer cells generally lack adequate repair mechanisms. Over time these cells may be exposed to other mutagens and will be involved in new cell divisions. Genomic analysis of late stage cancers reveals that it can take decades for mutations to accumulate and a single patient can develop over a dozen different clones. Analysis of pancreatic cancer metastatic tumors demonstrated that different clones from a primary tumor can form different metastases. Those metastatic cells can continue to evolve developing new clones and those clones can even form their own metastatic cells and produce new metastatic tumors (Campbell et al. 2010).

Comparisons between the germline DNA of the patient can be compared to the DNA collected from the primary tumor and metastastic tumors to determine when and where a mutation develops. These analysis reveal that many of the mutations that appear to provide a selective advantage to clones include mutations that damage the cellular repair mechanisms. One of the most consistently mutated genes in tumor samples is P53 (Table 3.1).

Table 3.1 TP53 mutations by cancer cell type

Type	Mutation (%)
Renal cell carcinoma	2
AML	9
Breast adenocarcinoma luminal	24
Endometrial carcinoma	28
Glioblastoma multiforme	30
ALL	41
Bladder adenocarcinoma	51
Lung adenocarcinoma	52
Colorectal adenocarcinoma	58
Lung squamous-cell	72
Breast adenocarcinoma basal	80
Ovarian carcinoma	94

Source. Hoadley et al. (2014)

Mutations that occur in germ cells may be passed on to the next generation. An example of this is with retinoblastoma related rb mutations where roughly half of the mutations are inherited.

Most mutagens are associated with environmental exposures. When these exposures are limited or eliminated it may be possible to eliminate or limit the mutations associated with these exposures. There are some mutagens that are unavoidable. For example, iron and oxygen are two vital elements that are also involved in the creation of reactive oxygen species which can damage DNA causing mutations.

In humans trisomy is only survivable in some cases. When survivable it is limited to the smaller chromosomes (e.g., 21). Tetraploidy is even more limited in humans and is predominantly limited to the X-chromosome. In individual cells the number of chromosomes can increase dramatically. Similarly loss of chromosomes on an organism wide scale is almost always fatal with the notable exception of the sex chromosomes. This is a condition known as aneuploidy. Cancer cells can have anywhere from over 100 chromosomes to under 38 chromosomes. Monosomy is limited at the organism level to the X-chromosome. This is because in humans the X-chromosome is selectively inactivated so that only one copy is active in both women and men. All of the other chromosomes have evolved to produce the gene products from both parents.

There are many instances of haplo-insuficiency where both copies of the gene are needed to produce an adequate amount of the gene product. Examples of haplo-insufficiency are seen with Williams and Marfan syndromes. With Marfan syndrome a single mutated copy of fibrillin-1 is enough to cause widespread tissue weakening (e.g., aortic dilation). With Williams syndrome a deletion of ~26 genes from chromosome 7 results in cognitive, physical and metabolic defects.

3.4.4 Mutation Penetrance

Mutations can be highly penetrant or of limited penetrance. For many syndromes with known genetic causes it is common to for individuals to have wildly different phenotypes even when they share the same mutation. These cases highlight the complexity of the human body and environment and how genes interact with other genes and the environment. An example of this can be seen with syndromes known to reduce height. Among the affected individuals there is often still a wide range of heights as many other factors impact body height including health, diet, sex, and parental height.

What area of the genome is mutated is critical. Only some areas of the genome are expressed. Some areas of the genome are hypervariable and mutations in these regions may have limited impact. At the other end of the spectrum some areas are highly conserved. Mutations in some regions are more likely to result in cancers. Tumor suppressor genes (e.g., TP53) and oncogenes (e.g., RAS) are two major examples of genes whose mutation is often involved in carcinogenesis. Mutations involving large chromosomal changes alter many genes at once. When expressed over the entire organism they are often lethal.

What part of the body is mutated is critical. Cells in the lining of the digestive tract and the outer layers of the skin are routinely sloughed off. Genetic damage to these cells may be of limited impact if they are removed. In contrast the cells that are below these layers and are frequently dividing to form new replacement cells are potentially dangerous cells to mutate. Many of the most common primary cancer sites involve these kinds of linings (e.g., glandular epithelial tissue).

The period of development is also important. Mutations are more problematic earlier in development. At the earliest stage they are more likely to be carried throughout the body during cell divisions. However even with younger adults mutations are likely to be more problematic than a similar mutation occurring with an elderly adult. This is because it often takes decades for an initiating mutation to lead to tumor formation. Growth during development also lends itself towards susceptibility towards mutagens because of the widespread cell divisions. Regrowth is also an issue as the same genes that are turned on for cellular repair can become involved in tumorogenesis.

3.5 Repair Mechanisms

3.5.1 Overview of Repair

The fidelity in DNA replication is significantly improved by the presence of multiple DNA repair mechanisms. In general the smaller the lesion the easier and more likely is the repair. Repair can restore the fidelity of the DNA or can result in a mutation but still allow replication to occur. Repair enzymes may be at low levels until significant DNA damage is detected. If DNA damage is detected then a global response known as the SOS response can be initiated. During the SOS response cells block replication until repair is completed. Lower levels of damage can result in repair genes increases in response to the damage.

Similar to the immune system which has multiple ways of responding to different types of infections and injuries, humans have multiple ways of responding to DNA damage. Some pathways (e.g., homologous repair) require replication to be occurring. Other pathways (e.g., global genomic nucleotide excision repair) are not dependent on replication. Humans have a range of enzymes and other proteins involved in DNA repair and their importance in health can be seen with their association with mutation related diseases (e.g., XPA and xeroderma pigmentosa).

3.5.2 Base Repair

Damaged DNA bases can be excised and replaced through the use of DNA polymerases through a process known as nucleotide excision repair. Because DNA is double stranded in the case of an abasic site the opposing strand can be used as the template (Fig. 3.19). This type of repair restores the original DNA copy. Humans have multiple DNA polymerase types only some of which have significant DNA repair ability.

Cyclobutane lesions (Fig. 3.11) which are caused by UV radiation are repaired by photolyase enzymes. Photolyases use photon energy to drive photochemical reactions. This process is known as photoreactivation (Cooper 2000).

O^6-methylguanine methyltransferase is another DNA repair enzyme that specializes in the repair of alkylated DNA that has been modified at the O^6 residue of guanine.

3.5.3 Chromosomal Repair

Ligases can reassemble broken chromosomes. Repair can involve either joining the ends together or by capping the end of the chromosome with a telomere. With homologous recombination ends are joined together that have been matched up with an extensive homologous sequence to guide the repair. With non-homologous

Fig. 3.19 Abasic site. DNA is shown missing a base. This figure was created using the structure published by Serre et al. (2002) on RCSB PDB

recombination the repair is not guided by an extensive homologous sequence. Small sequences (i.e., microhomologies) that are only a few bases long can guide non-homologous end joining. This can lead to accurate repair but often results in the loss of a few bases as the pieces are not perfectly lined up.

3.5.4 Cell Death

The last resort is cell death which can be viewed on an organismal scale as a form of repair as it can allow for the replacement of damaged cells with new healthy ones. This cell death can be uncontrolled death where the cell contents spill out and can damage surrounding cells and tissues or it can be programmed and controlled through the process of apoptosis. Apoptosis limits the damage to surrounding cells and tissues. Apoptosis is the primary mechanism used to remove normal but unneeded tissue during development such as the webbing between fingers in the developing embryo. Apoptosis is largely directed by checkpoint regulator p53 which has been nicknamed "guardian of the genome." When extensive DNA damage is detected p53 can interact with other proteins to stop cell division and initiate repair. If the genetic damage is too severe to allow for repair then apoptosis may be initiated. A common cause of this is double stand breaks.

Case Study 6: Xeroderma Pigmentosa Xeroderma Pigmentosa is a condition where the normal repair mechanism for UV radiation caused pyrimidine dimers is broken. As a consequence the patients are unable to repair the damage to their skins cells and can develop skin lesions and skin cancers at an early age. Some patients, if diagnosed early and extensively shielded from UV radiation avoid this profound skin damage. These efforts include use of protective clothing and sunscreen as well as a nocturnal lifestyle. Efforts continue to develop effective therapies to limit the development of skin cancers which remain a leading cause of death for patients with Xeroderma Pigmentosa. Recently there have been preclinical studies to develop methods to provide gene therapy. The goal of these gene therapy studies is to insert repair genes into skin cells so they can begin expressing repair enzymes (e.g., photolyases).

3.6 Conclusion

In order to alter DNA mutagens must have the ability to interact with electrons on the DNA either directly or through an intermediary. Mutagens can be chemicals or radiation. Chemicals are generally either direct or indirect mutagens. Indirect mutagens need to be activated before they can react with DNA. Radiation can be both indirect and direct mutagens. When acting as direct mutagens the radiation particles ionize the DNA. When acting as indirect mutagens the radiation ionizes a neighboring molecule that then reacts with the DNA.

Mutagens can damage a small part of a single base or can break apart an entire chromosome. The larger the radiation particle the more damage it can cause. However, larger particles have limited penetrating ability. This makes alpha emitters less able to cause damage with dermal exposures but they can still be extremely dangerous if inhaled or ingested. Mutagens are most damaging at the earliest stages of development. At the organism level minor mutations are more survivable. Mutagens that activate proto-oncogenes, inactivate tumor suppressor genes or damage DNA repair genes can lead to the development of uncontrollable cell-growth, new mutations and cancers.

Mutagens are further classified based on their likelihood of causing mutations towards germ cells. Mutagens that affect the germ cells are the most troubling as they can be passed on to future generations. For this reason the Global Harmonizing System classifies mutagens based on their ability or expected ability to cause mutations in human germ cells.

References

Boghossian N, Horbar J, Carpenter J, Murray J, Bell E (2012) Major chromosomal anomalies among very low birth weight infants in the Vermont Oxford Network. J Pediatr 160(5):774–780

Burnouf D, Gauthier C, Chottard J, Fuchs R (1990) Single d (ApG)/cis-diamminedichloroplatinum (II) adduct-induced mutagenesis in Escherichia coli. Proc Natl Acad Sci 87(16):6087–6091

Campbell P, Yachida S, Mudie L, Stephens P, Pleasance E, Stebbings L, Morsberger L, Latimer C, McLaren S, Lin M et al (2010) The patterns and dynamics of genomic instability in metastatic pancreatic cancer. Nature 467(7319):1109–1113

Chauhan D, Ray A, Viktorsson K, Spira J, Paba-Prada C, Munshi N, Richardson P, Lewensohn R, Anderson K (2013) In vitro and in vivo antitumor activity of a novel alkylating agent, melphalan-flufenamide, against multiple myeloma cells. Clin Cancer Res 19(11):3019–3031

Cooper G (2000) DNA repair in the cell: a molecular approach, 2nd edn. Sinauer Associates, Sunderland. Available from http://www.ncbi.nlm.nih.gov/books/NBK9900/

Hansel S, Castegnaro M, Sportouch M, De Méo M, Milhavet J, Laget M, Duménil G (1997) Chemical degradation of wastes of antineoplastic agents: cyclophosphamide, ifosfamide and melphalan. Int Arch Occup Environ Health 69(2):109–114

Hoadley K, Yau C, Wolf D, Cherniack A, Tamborero D, Ng S, Leiserson M, Niu B, McLellan M, Uzunangelov V, Zhang J, Kandoth C, Akbani R, Shen H, Omberg L, Chu A, Margolin A, van't Veer L, Lopez-Bigas N, Laird P, Raphael B, Ding L, Robertson A, Byers L, Mills G, Weinstein J, Van Waes C, Chen Z, Collisson E, Benz C, Perou C, Stuart J (2014) Multiplatform analysis of 12 cancer types reveals molecular classification within and across tissues of origin. Cell 158:929–944

Hofmann M, Blievernicht J, Klein K, Saussele T, Schaeffeler E, Schwab M, Zanger, U (2008) Aberrant splicing caused by single nucleotide polymorphism c.516G>T [Q172H], a marker of CYP2B6*6, is responsible for decreased expression and activity of CYP2B6 in liver. J Pharmacol Exp Ther [online] 325(1):284–292. Available at: http://dx.doi.org/10.1124/jpet.107.133306. Accessed 16 Aug 2014

Jackson MA (2005) Genetic alterations in cancer knowledge system: analysis of gene mutations in mouse and human liver and lung tumors. Toxicol Sci 90(2):400–418

Kaplan L, Foster R, Shen Y, Parry D, McMaster M, O'Leary M, Gusella J (2010) Monozygotic twins discordant for neurofibromatosis 1. Am J Med Genet A 152(3):601–606

Lang T, Klein K, Fischer J, Nussler A, Neuhaus P, Hofmann U, Eichelbaum M, Schwab M, Zanger U (2001) Extensive genetic polymorphism in the human CYP2B6 gene with impact on expression and function in human liver. Pharmacogenet Genomics 11(5):399–415

Miller E, Miller J (1981) Searches for ultimate chemical carcinogens and their reactions with cellular macromolecules. Cancer 47(10):2327–2345

Mullaney J, Mills R, Pittard W, Devine S (2010) Small insertions and deletions (INDELs) in human genomes. Hum Mol Genet 19(R2):R131–R136

Newman JA, Savitsky P, Krojer T, Von Delft F, Arrowsmith CH, Edwards A, Bountra C, Gileadi O. Crystal structure of the Bloom'S syndrome helicase Blm in complex with DNA. Pdb file 4CGZ. http://www.rcsb.org/pdb

Pool B, Bos R, Niemeyer U, Theuws J, Schmähl D (1988) In vitro/in vivo effects of Mesna on the genotoxicity and toxicity of cyclophosphamide a study aimed at clarifying the mechanism of Mesna's anticarcinogenic activity. Toxicol Lett 41(1):49–56

Ramser J, Abidi F, Burckle C, Lenski C, Toriello H, Wen G, Lubs H, Engert S, Stevenson R, Meindl A et al (2005) A unique exonic splice enhancer mutation in a family with X-linked mental retardation and epilepsy points to a novel role of the renin receptor. Hum Mol Genet 14(8):1019–1027

Serre L, Pereira de J'esus K, Boiteux S, Zelwer C, Castaing B (2002) Crystal structure of the Lactococcus lactis formamidopyrimidine-DNA glycosylase bound to an abasic site analogue-containing DNA. EMBO J 21(12):2854–2865

Strachan T, Read AP (1999) Chapter 2: Human molecular genetics, 2nd edn, Chapter 2. Chromosomes in cells. Wiley-Liss, New York

Vassylyev D, Kashiwagi T, Mikami Y, Ariyoshi M, Iwai S, Ohtsuka E, Morikawa K (1995) Atomic model of a pyrimidine dimer excision repair enzyme complexed with a DNA substrate: structural basis for damaged DNA recognition. Cell 83(5):773–782

Wang AHJ et al (1987) Interactions between an anthracycline antibiotic and DNA: molecular structure of Daunomycin complexed to D(Cpgptpapcpg) At 1.2-.ANG. resolution. Biochemistry 26(4):1152–1163

Wang Z (2008) DNA damage and mutagenesis. In: Molecular and biochemical toxicology, 4th edn. Edited by Robert C Smart, Ernest Hodgson. Wiley, Hoboken, pp 441–491

Chapter 4
Ligand-Mediated Toxicology: Characterization and Translational Prospects

Rais Ansari, Claude L. Hughes, and Kazim Husain

Abstract In biochemistry a ligand is a molecule that binds to a receptor or other biomolecule that forms a complex and produces a biological effect. We live in a world of exposure to exogenous ligands and there are three long-established strands of biological research that have investigated the actions of biologically active dietary or environmental compounds on animals, including humans.

First, in agricultural and animal husbandry research, there is extensive documentation of the many thousands of phytochemicals that mediate plant-animal interactions many of which produce toxic effects at certain levels of exposure in terms of dose, duration and window of sensitivity. In a broader biological context, many of these relationships have been formalized by concepts of co-evolution of plants and animals.

Second, there is a long history of toxicology research showing a wide range of adverse effects that often result from exposures to hormones and other modulators of endocrine function, especially during developmentally sensitive intervals. In recent decades much of this research has been encapsulated by a single term of "endocrine disruptors" even though many ligand-mediated actions are not endocrine *per se* or definitively adverse.

R. Ansari (✉)
Department of Pharmaceutical Sciences, College of Pharmacy, Nova Southeastern University, 3200 S University Drive, Fort Lauderdale, FL 33328, USA
e-mail: ra557@nova.edu

C.L. Hughes
Therapeutic Science & Strategy Unit, Quintiles, Inc., 5927 South Miami Blvd., Morrisville, NC 27560, USA

Department of Obstetrics and Gynecology, Duke University Medical Center, Durham, NC, USA

Department of Mathematics, North Carolina State University, Raleigh, NC, USA
e-mail: claude.hughes@quintiles.com

K. Husain
Department of Physiology, Pharmacology and Toxicology, Ponce School of Medicine and Health Sciences, 7004, Ponce, PR 00732, USA

C.L. Hughes, M.D. Waters (eds.), *Translational Toxicology*, Molecular and Integrative Toxicology, DOI 10.1007/978-3-319-27449-2_4

Third, nutritional and functional foods research have demonstrated a wide range of health benefits produced by consumption of many phytochemical ligands through mechanisms that are not truly distinct from those that seem to mediate the adverse effects noted in either the animal husbandry or developmental toxicology research arenas.

We are confident that environmental and public health can be enhanced by defining how such exposures to exogenous signaling molecules (ligands) pose both risks and benefits. Therefore, it is time to modify the scientific perspective and nomenclature to encompass these several streams of investigation such that both risks and benefits of exposure to dietary and environmental ligands may be characterized and translated into sound strategies to address environmental and public health concerns. Fetal alcohol spectrum disorder (FASD) is an illustrative case for translational toxicology-informed interventions where there is a viable prospect for a multi-component nutritional health benefit to reduce occurrence or severity of FASD. Health promoting interventional options must include reduction(s) in exposures to ligands that incur health risks but when exposures cannot be adequately reduced or preempted, we should also define and employ active mechanistically-based mitigative interventions.

Keywords Translational toxicology • Endocrine disruption • Epigenetics • Ligand • Receptor mediated toxicity • Xenosensors

4.1 Living in a World of Ligands: Translating Exposures to Exogenous Signaling Molecules into Human Health Benefits and Risks

In biochemistry a ligand is generally defined as a molecule (usually small), that binds to a receptor or other biomolecule that forms a complex and produces a biological effect. Our complex environment is laced with ligands to which animals, including humans, are continuously and episodically exposed. Endogenous ligands are "part and parcel" of our *milieu interieur* while naturally-occurring and man-made exogenous ligands are major elements of our *milieu exterieur*. At all times across the lifespan, growth impacts and functional/homeostatic effects of exposure to such exogenous ligands may occur. Effects of such exposures on differentiation events might also occur at any time in an animal's life, but such organizational impacts are more likely to be robust during developmentally-sensitive intervals ("developmental windows"). Our challenge for human health is not merely to document the occurrence of such effects but to also to characterize the risk – non-effect – benefit profile of such exposures and further to ascertain how prevention or intervention may be used to enhance environmental and public health while reducing exposures to ligands that incur health risks. We will consider how current toxicologic approaches might be reoriented into a more translational scientific endeavor by reflecting on the broad range of exposures from a perspective that we will call ligand-mediated toxicology.

4.2 Traditional Toxicology

Toxicology studies the potential adverse effects of exposures, especially xenobiotics, on various forms of life including human beings. Modern toxicology goes beyond documenting adverse effects to the study of molecular toxicology, performance of safety evaluations and risk assessments as ways to translate basic toxicology into public health or public policy interventions. Two main principles underlie all descriptive toxicity testing: (1) the adverse effects produced by an agent in laboratory animals are applicable to humans and (2) high dose exposures of a toxicant to animals may be required in order to discover possible hazards in humans. The first toxicity test performed on a new chemical is acute toxicity which gives a quantitative estimate of lethality (lethal dose or lethal concentration that kills 50 % of exposed animals, LD_{50} or LC_{50}). Acute studies are used to compare the toxicity of agents, to identify target organs and other clinical manifestations of acute toxicity, to establish the reversibility of the toxic response and to provide dose-ranging guidance for other studies. Subacute toxicity testing is performed to obtain toxicity data for a chemical after repeated exposure as well as to establish doses for subchronic toxicity. Subchronic exposure can be of different durations but 90 days is the most common duration. Long term or chronic toxicity studies are performed similarly to subchronic toxicity studies except that the duration of exposure is greater than 90 days and up to the lifetime of the animal. Chronic toxicity studies are performed to assess the cumulative toxicity of chemicals (Eaton and Gilbert 2010).

4.3 Mechanisms of Toxicity

The quantitative characterization of how toxic effects occur is important for an evaluation of the potential hazard posed by a chemical. It is valuable to understand the mechanisms responsible for the manifestation of toxicity. The understanding of mechanism of toxicity is of both practical and theoretical importance. Selective or altered toxicity may be due to different or altered exposure and delivery, target molecules, biochemical alterations, cellular and molecular repair and the mechanism of circulatory and thermoregulatory reflex in the adaptation of the toxic events. The significance of the chemistry of a toxicant for its delivery and reaction with the target molecule as well as the biochemistry, cell and molecular biology, immunology and physiology of the exposed organism in its response to the action of the toxicant are equally important. An organism has specific mechanisms to counter the delivery of the toxicant, to detoxify the toxicant, to reverse the toxic injury by repair mechanisms and to develop an adaptive response. There are several fundamental steps in any mechanism of toxicity such as delivery from the site of exposure to the target involving absorption (versus presystemic elimination), distribution to and away from the target, excretion versus reabsorption and activation as a part of toxication versus detoxication and, ultimately, reaction of the active toxicant(s) with the target molecule(s). Various types of reactions may occur including effects of toxicants on target molecules and toxicity not initiated by reaction with target molecules; cellular dysfunction and resultant toxicities involving toxicant-induced cellular dysregulation and toxic alteration of cellular maintenance including repair or dysrepair involving molecular repair, cellular repair, tissue repair, repair failure and toxicity resulting from dysrepair (Gregus 2010).

4.4 Translational Toxicology

The nature of toxicology has always been driven by a translational sense of purpose with respect to human health. The earliest toxicological studies carried out were to assess the effects of plant and animal toxins on human subjects (Gallo 2008). However there are still gaps in the transformation of human toxicology into a translational discipline in the contemporary biomedical sense of the term. There is a gap between laboratory toxicology research and testing for cellular, molecular and pathological outcomes in cell systems and animal models and epidemiological and ecosystem studies, which examine exposed populations for disease. The current approach may or may not correlate the exposure to a toxicant with a clinical outcome, but this almost never leads to identification of a specific protective, corrective or compensatory therapy that may be used in currently intoxicated individuals or those who may be exposed in the future. The claims of association of cause-and-effect are often no better than a tentative association of exposure and effect. If there is inadequate evidence to show either adverse effects due to exposure to a potential toxicant or that there is no plausible risk due to exposure, then there is little or no prospect for intervention of any sort beyond application of the "precautionary principle." Therefore in strategic terms, the main objectives of translational toxicology are to advance environmental health sciences in order to: (1) transform research and development programs to better link exposures to disease occurrence or outcomes; (2) link target tissue/organ/system outcomes in animal models to animal biomarkers to human biomarkers to human target organ outcomes; and (3) enhance predictive use of animal and human biomarker data by development of computational models (Hale et al. 2009). Therefore we will reconsider how current knowledge from toxicological research and epidemiological studies can be seen as a contemporary translational biomedical sub-discipline of ligand-mediated toxicology in which developmental exposures may impact health unfavorably or favorably across the human lifespan.

4.5 Translating Biomarkers

In clinical toxicology potential biomarkers that may have utility in translational toxicology research include chemical and molecular markers such as clinical chemistry and hematology profiles, and hormone levels (Hale et al. 2009) as well as imaging techniques such as radiology, ultrasound, computerized tomography (CT), magnetic resonance imaging (MRI), and positron emission tomography (PET). In preclinical toxicology, histopathology remains a key component in assessing target tissue responses to exposures while new genomic, transcriptomic, proteomic and metabolomics technologies as well as next generation sequencing methods generate data sets that can serve as biomarkers in toxicological investigations across the lifespan (West et al. 2010). Global metabolite profiling allows the measurement of numerous chemicals in minimal amounts of biological material (Patti et al. 2012) and is applicable to translational toxicity research. Within genomics and RNA signaling, micro RNAs that are known to be key regulators of metabolism are available in accessible human blood to quantify as either signaling molecules or disease biomarkers (Rottiers and Naar 2012). Additionally, other molecular and cellular biomarkers such as the levels of leukocyte mitochondrial DNA copy number can also be used in translational studies (Kim et al. 2012).

For biomarkers to be useful in predicting and monitoring drug safety in a clinical population, temporal response must be compared clinically and preclinically. Thorough preclinical qualification ideally includes studies of acute and chronic dosing, a range of toxicants with varying mechanisms of injury in the target organ, negative control compounds, and several toxicants known to affect other organs to assess the specificity of the biomarker. Limitations exist to how we can study a biomarker's behavior in a clinical population. Histological examination of tissue is not possible in most circumstances; scheduled sampling is more complex in an outpatient population; and ethical issues prevent the deliberate administration of toxicants to humans. However, strategies for qualifying biomarkers in humans can take advantage of marketed therapeutics with known toxicities. For example, a clinical study for qualifying nephrotoxicity biomarkers considered several clinical populations with known kidney injury before deciding to monitor cardiac catheterized patients with an unfortunately high event rate of contrast dye–induced nephropathy (Fisch 2012). Considered were patients with shock trauma, aminoglycoside-treated cystic fibrosis, and cisplatin-chemotherapy. The goal of toxicology testing in pharmaceutical development is to identify and predict safety issues to inform the risk–benefit decision about using a drug for a particular indication. The safety assessment is not always fully predictive, and limitations exist in extrapolating findings from animal studies to clinical studies. Therapeutic dosing prior to clinical trials is the accepted paradigm of safety testing. This depends on the ability to extrapolate to humans from the drug's activity and dose-response in the animal model.

Although, specific pharmacokinetic and pharmacodynamic parameters and a standard battery of toxicological end points are routinely measured, many factors that affect the comparative disposition of the drug in preclinical model versus human are rarely fully elucidated. Post-sacrifice microscopic histopathology plays a central role in identifying drug-induced lesions in diverse tissues in animal studies; extrapolating these findings to clinical trials requires a noninvasive, monitorable end point (Euling et al. 2008). However, studies in healthy, genetically homogeneous animals can hardly be expected to predict low-incidence toxicities that arise clinically as a result of individual human differences in response due to genetic variations in metabolism or drug–target pathways or to coexisting conditions and subsets of disease phenotypes. Extrapolation from preclinical to clinical following the working principles of translational toxicology research is required.

It has also been proposed that peroxisome proliferation could be used as a biomarker of exposure to a variety of pollutants in environmental pollution assessment such as polycyclic aromatic hydrocarbons (PAHPAHs), polychlorinated biphenyls (PCBs), chlorinated herbicides, chlorinated solvents, perfluorinated acids, phthalate ester plasticizers, phenoxyacetic acids, alkyl carboxyl acids and tetrazoles (Woods et al. 2007). The increase in the volume and number of peroxisomes is usually accompanied by induction of specific peroxisomal enzymes including all three enzymes of the peroxisomal β-oxidation pathway, namely, acyl-CoA oxidase (ACO), peroxisomal hydratase-dehydrogenase-isomerase or multifunctional enzyme, and thiolase. Enzyme induction occurs through transcriptional activation of the corresponding genes by peroxisomal proliferator activator receptors (PPARs). The mechanism by which humans and nonhuman primates are resistant to peroxisome proliferation is not known. Humans possess a functional PPARα, indicating that the lack of receptor is not responsible for the lack of peroxisome proliferation upon treatment with fibrate drugs.

4.6 Receptor Mediated Toxicity: Xenosensors, Nuclear Hormone Receptor Activation, Actions of Hormonal Ligands

Living organisms including humans are exposed to thousands of exogenous chemicals, both man-made (synthetic, xenobiotic) and of natural origin (prominently phytochemicals). For survival, organisms have developed systems for handling these potentially harmful exogenous substances. When a subject (human, wildlife, farm animal, etc.) is exposed to a chemical, these processes typically either eliminate and/or transform the chemical within the exposed subject. If a chemical is modified by the exposed subject, it interacts with sensors called xenosensors which prepares the subject for the elimination of the chemical from the system. The exposure to chemicals leads to activation of enzymes referred to collectively as xenobiotic metabolism (monoxygenases, conjugation enzyme and transporters) which creates increased centers of hydrophilicity, followed by subsequent conjugation with either glucuronic acid or sulfate to prepare the drug or chemical for its excretion from cell

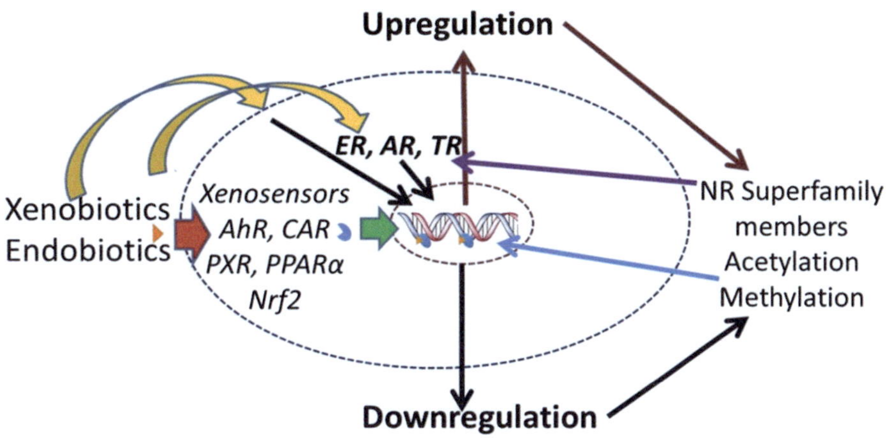

Fig. 4.1 Xenosensing and xenobiotics mediated possible effects via gene regulation. *AhR* Aryl hardrocarbon receptor, *CAR* Constitutive androstane receptor, *PXR* Pregnane X receptor, *PPARα* Peroxisomal proliferator activated receptor-alpha, *Nrf2* Nuclear factor (erythroid-derived-2) like 2 (NFE2L2)

using multi-drug resistant transport processes. During the process of metabolism, the chemicals either become less toxic or may acquire characterstics which allow further interaction with cellular constituents. Compounds like PAH and halogenated hydrocarbons (HHC) bind to aryl hydrocarbon receptor (AhR) which not only serves as a xenosensor but also acts as receptor and activates the monoxygenase system for creating hydrophilic centers and conjugation with hydrophlic compounds for elimination. One mode of action of endocrine disruption is the ability of some chemicals to either directly or indirectly disrupt such nuclear hormone receptor signaling. In addition to man-made chemicals, nature also produces a number of compounds which possess the ability to bind and either disrupt or activate nuclear receptor superfamily members. Figure 4.1 depicts this general type of mechanism of receptor mediated toxicity.

A number of receptors sense the presence of xenobiotics in the cell and induce a cascade of events leading to neutralization and excretion. However, in many cases the metabolism of xenobiotics can give rise to toxic metabolites or free radicals/ reactive oxygen species (ROS) that can cause further oxidative injury to the cells. The metabolism of xenobiotics can also perturb production and metabolism of certain hormones in the body. Therefore in some instances receptors that induce the elimination of toxicants may themselves mediate some of their toxic effects.

Certain environmental toxicants can also bind to receptors other than the xenosensors (Ruegg et al. 2009). In particular the receptors for the female sex steroids (estrogen receptors, ERs) have been identified as targets for many synthetic and natural compounds. Interactions of exogenous substances with hormone receptors influence a number of other steps in physiological signaling which have been termed "endocrine disruption." One common and widely-used definition was formulated as follows:

> An environmental endocrine disruptor was broadly defined as an exogenous agent that interferes with the production, release, transport, metabolism, binding, action or elimination of natural hormones in the body responsible for the maintenance of homeostasis and the regulation of developmental processes. (Kavlock et al. 1996)

Note that this definition did not require consideration of whether any effects that may result from a compound capable of this mode of action be assessed as adverse or even potentially beneficial. Additionally over time and with ongoing usage, the designation of "endocrine disruptor" has broadened to such an extent as to often not relate to endocrine (hormonal) actions at all, and we accept that this is the established usage of the term. However, given the abiding need to consider both risks and/or benefits of exposures to dietary and environmental chemicals and the utilitarian value of including a broad spectrum of highly specific but non-hormonal modes of action, we will use "ligand-mediation" as the broader core concept.

Endocrine disruption has been associated with major health threats in the western world, such as development of diabetes and obesity, decreased fertility, and with the occurrence of hormone-associated cancers. Several of these chemicals of concern are persistent organic pollutants (POPs) because they are resistant to environmental degradation. They are thus persistent over a long time in the environment, can contaminate drinking water and food, and can be biomagnified via the food chain. Organochlorine pesticides, like DDT (dichloro diphenyl trichloroethane), and gamma hexane are quite persistent. A number of the man-made environmental chemicals identified as ER disruptors were pesticides. Symptoms in humans working with the manufacture of DDT led to the identification of DDT as an ER agonist. Other pesticides that activate ER are the DDT metabolites DDD (dichlorodiphenyldichloroethane) and DDE (dichlorodiphenyldichloroethylene), methoxychlor, and dieldrin (Lemarie et al. 2006). In addition, DDE and the fungicide vinclozolin can also affect the function of other hormone receptors by acting as androgen receptor (AR) antagonists (Kavlock and Cummings 2005), while vinclozolin can further antagonize the activity of progesterone receptor (PR), glucocorticoid receptor (GR), and mineralocorticoid receptor (MR) in vitro (Molina-Molina et al. 2006). Chemicals intentionally produced besides pesticides include biocides, plasticizers, and food additives and cosmetics. Many of these substances are structurally related to steroid hormones and may thus act on the respective hormone receptor(s). Non-persistent compounds such as phthalates and bisphenol A are additives used in the plastics industry. These compounds leach from plastic containers to water and liquids. These chemicals may also interact with nuclear receptors ER and AR while bisphenol A can also bind to thyroxine receptor (TR).

Endocrine-disrupting chemicals (EDCs) can have effects at low doses that are not predicted by effects at higher doses (Vandenberg et al. 2012). Low-dose effects are those that occur within the range of human exposures or those observed at doses below the range used in traditional toxicological studies. These effects may depend upon non-monotonic dose–response relationships. A nonlinear relationship may be seen between dose and effect where the slope of the curve changes sign somewhere within the range of doses examined. It is noteworthy that non-monotonic responses and low-dose effects are remarkably common in studies of natural hormones and EDCs. In most instances where there is a scientifically valid health concern, the developmental toxicology concept of endocrine disruption must still be "put to the test" and advanced to an active translational (protective, mitigative or compensatory) step in which: (a) human exposure(s) occur, (b) there is a plausible mode of action such that the exposure might elicit developmental effects, (c) human health outcomes are assessed and (d) both passive [reduce exposures] and active [nutritional, sociobehavioral or pharmaceutical interventions] health risk intervention strategies can be developed for exposed individuals.

4.7 Endocrine Disruption and Epigenetic Changes After Exposure to Chemicals

The range of mechanisms of endocrine disruption is not highly constrained. Studies with EDCs like vinclozolin, bisphenol A and polychlorinated biphenyls demonstrate disruption of the hypothalamic-pituitary axis (Vandenberg et al. 2012; Bergman et al. 2012; Zoeller et al. 2012), to disruption of nuclear hormone receptor signaling (Bergman et al. 2012), to epigenetic changes transmitted from exposed individuals to their fetuses and even to subsequent generations. This latter mechanism of endocrine disruption, epigenetic changes, (Bollati and Baccarelli 2010) is gaining prominence as an important mode of action in developmental toxicology. Epigenetics is the study of heritable changes in gene expression without changes in the DNA sequence of the genome. Modification of DNA by methyltransferases leads to decreased charge on DNA causing compaction of chromatin which is not suitable for active transcription. It is known that methylation of CpG islands which are present in promoters of genes results in gene silencing. In addition chromatin associated proteins, i.e., histones, can be modified by acetylation, methylation, phosphorylation, sumoylation etc. Acetylation of histones is known to relax the chromatin and activate transcription. *In utero* exposures to bisphenol A, a contituent of polycarbonate plastics, has been implicated in breast and prostate cancer (Bollati and Baccarelli 2010). Experimentally, it has been demonstrated that bisphenol A exposure causes hypomethylation. Exposure of bisphenol A to pregnant mice results into altered methylation in CpG islands of the forebrain. In addition, it leads to shifting of agouti coat color as a result of decreased DNA methylation (Walker and Gore 2011). In rats, bisphenol A exposure increases methylation of several prostate specific genes while decreasing the methylation of phosphodiesterase 4 variant 3 (PDE4D3). PCBs have also been demonstrated to decrease DNA methylation enzyme activities (Dnmt1 in hypothalamus and liver and Dnmt1 and methylation of 16 genes in liver) in postnatal day 21 in offspring (Walker and Gore 2011).

Transfer of epigenetic changes has been demonstrated by animal experimentation. It has been demonstrated that exposure to EDCs *in utero* leads to epigenetic changes in fetuses which leads to altered gene expression and these epigentic changes may be passed on to the next generation (Walker and Gore 2011). Thus exposure of fetuses in pregnant female rats to vinclozin during the sex determination phase results in transgenerational spermatogenic defects and prostate enlargement (Anway et al. 2005, 2006a, b; Anway and Skinner 2006; Chang et al. 2006). Other environmental exposures may activate endogenous toxicologic mechanisms that could lead to epigenetic changes. Studies have demonstrated that workers who are exposed to particulate matter exhibit reduced CpG methylation of inducible nitric oxide synthase. Many chemicals when detoxified produce oxidative stress which in turn may cause DNA damage and this can in turn hinder methyltransferase and binding of methyl groups to DNA. Under mild oxidative stress DNA methylation machinery malfunctions, leading to hypomethylation of genes (Collotta et al. 2013) and this can lead to altered methylation of cytosine residues in DNA. Hypothetically, DNA methylation and chromatin modification can change the activity of vital genes. Correcting methylation via chemicals may prove a good prospect to counteract environmental chemical-induced epigenetic effects. In the case of hypomethylation, methyl donors like folic acid can be used to block epigenetic changes. Genistein, a phytohormone which blocks BPA-induced hypomethylation might also be employed. These are only preliminary findings but preclinical and clinical trials can be envisoned. We need to determine if some nutritional or other generally-regarded-as-safe (GRAS) agents could potentially limit environmental chemical-induced epigenetic changes and whether such effects could have a salutory impact on human health.

Changes in acetylation of histones which can open chromatin and increase the transcriptional activity of genes have also been observed with EDCs including persistent organic pollutants, arsenic metal and pesticides (herbicides, insecticides and fungicides). Increased acetylation of histone and neurodegeration was observed with dieldrin in N27 cells (Song et al. 2010). Another herbicide, paraquat was also observed to increase acetylation of histone *in vitro* using N27 cells (Song et al. 2011). Histone modification in brain is linked to sexual differentiation of brain. In males, the bed nucleus of the stria terminalis in the hypothalamus is larger compared to females. Increasing acetylation by inhibiting the deacetylase activity with valproic acid in androgenized female results in femanization of the bed nucleus of stria terminalis (Walker and Gore 2011). A number of agents which alter the activity of bromodomain proteins (which sense epigentic changes) are being developed and may be able to reverse and rectify epigenetic changes due to various chemical exposures including EDCs.

In addition to methylation and acetylation as epigenetic changes, gene regulation at the transcriptional level is also affected by micro-RNAs (miRNA) (Yang and Zhou 2013). miRNAs are short 22–24 nucleotide long having sequence complementarity to the 3′- untranlasted sequences of mRNA. miRNAS binds to 3′-untranslated sequences as well as other regions of mRNA to regulate mRNA level and thus controlling the gene expression. miRNA is an evolving field and it is believed that one miRNA can regulate several mRNAs and several miRNA can regulate one gene. A good amount of data documents changes in miRNAs expression patterns in diseases and after exposure to chemicals (Berezikov 2011; Wang et al. 2009). It is proposed that miRNAs can serve as signalling molecules besides their role as transcriptional regulators (Russo et al. 2014). If changes in the expression pattern of miRNAs are transferred from mother to fetus due to exposure to environmental chemicals, the pattern of key gene expression may affect the development of the offspring. By performing miRNA arrays after exposure of porcine kidney epithelial cells with the organophosphate pesticide, dichlorvos, alterations in miRNA expression were observed (Li et al. 2011). Changes in miRNA expression profiles were observed in zebrafish after exposures to several pesticides (Wang et al. 2010). Changes in miRNA profiles have been linked to development of cancer. Cancer causing chemicals including the fungicides triadimefon and propiconazole, and arsenic have been found to alter miRNA expression profiles in mouse liver and in human lymphoblastoid cells. To our knowledge, studies on transgenerational changes in miRNA due to exposure with chemicals including ECDs have not been preformed. Thus it is imperative to document the miRNA expression patterns (over 1000 miRNAs in humans are known) in unexposed subjects for comparative purposes. Then the possibility of effects of altered miRNA mediated toxicity can be visualized and a translational approach can be imagined.

4.8 Broadening the Translational Perspective About Exogenous Ligands (Exogenous Signaling Molecules): Risks, Benefits, Neither or Both?

While there are numerous well-documented hormone receptors, the range of potential biomolecules with which ligands may interact in highly specific ways is far greater, exceeding two thousand different intracellular molecular species. The full range of potential modes of action is captured in the "Guide to Receptors and Channels (GRAC)" (Alexander et al. 2011; Pawson et al. 2014). This compilation of the major pharmacological targets is divided into seven sections: G protein-coupled receptors, ligand-gated ion channels, ion channels, catalytic receptors, nuclear receptors, transporters and enzymes. The most straightforward access to this resource is the website, http://www.guidetopharmacology.org/ which was originally created as a collaboration between The British Pharmacological Society (BPS) and the International Union of Basic and Clinical Pharmacology (IUPHAR), and is now supported with funding from the Wellcome Trust. Within this searchable database, in addition to listing of thousands of known ligands, thousands of biomolecules that are known to interact with small molecular ligands are listed. The categories of biomolecules include the following:

- G protein-coupled receptors
- Ion channels
- Nuclear hormone receptors

- Kinases
- Catalytic receptors
- Transporters
- Enzymes
- Other protein targets

All of these biomolecular targets present potential mechanisms by which ligand-mediated toxicity could occur. While some targets such as enzymes might seem less likely to robustly affect developmental events, numerous other targets in the several categories could plausibly impact development and functional status over the lifespan in either adverse or beneficial ways.

The current list of ligands (Pawson et al. 2014) within the website includes multiple categories of chemicals as follows:

- Approved [drugs] (current or previously approved for clinical use in humans)
- Synthetic organic chemicals
- Metabolites
- Natural products
- Endogenous peptides
- Other peptides
- Antibodies
- Labeled (for research applications)

Especially when we add that many compounds are known to have pleiotropic actions, the potential number of ligand-biomolecule signaling effects are astronomical, and we must accept that we must link exposures of interest to attendant modes of action and with resultant "phenotypes" that are a definable functional or structural outcome plausibly implying adverse or beneficial health consequences.

4.9 Dietary and Environmental Ligands: Simple Insights from Animal Husbandry and Co-evolution of Plants and Animals

When we consider potential effects of exogenous molecules that act as ligands on human development, it makes biological and pharmacological sense to consider that some such effects may be either adverse (Hughes and Tansey 1998) or beneficial (Adlercreutz 1998) as is well-documented for livestock (Cheeke 1989) and wild animals (Belovsky and Schmitz 1991). A paradigm for considering the human situation regarding such a Janus-like consequence of developmental exposures to exogenous ligands may derive from consideration of co-evolution of plants and animals (Harborne 1982; Thompson 1994). The chemical warfare contest between plants and their animal predators captures the two-faced nature of animal exposures to plant derived ligands. On one hand, massive amounts of scientific data demonstrate that thousands of phytochemicals produced by plant secondary metabolism limit or deter herbivores via specific and general mechanisms of toxicity (Cheeke 1989; Belovsky and Schmitz 1991). On the other hand, many of the same classes of phytochemicals have been shown to provide health promoting benefits when present at modest levels in the diet of animals and humans (Adlercreutz 1998; Hughes and Tansey 1998; Polya 2003). Particularly for omnivores like humans (Hughes and Dhiman 2002), dietary diversity has been convincingly shown to be health-promoting especially regarding diverse intake of colorful and flavorful plant-based foods which are rich in biologically active phytochemicals. Therefore, when we consider the actions of xenobiotic exogenous ligands and their potential effects on developmental events in humans, we must assess their modes of action as well as manifestations of persistent changes in biological function that can be judged as adverse, beneficial or nil (inconsequential). In this context we include as xenobiotic exogenous ligands all man-made compounds with specificity of actions on mammalian signaling pathways. Thus it is essential that all exposures including environmental contaminants, pharmaceuticals, over-the-counter medications, herbal and supplement products, lifestyle drugs of use or abuse, household products, etc. merit thorough assessment. Furthermore, mode of action alone cannot be sufficient to determine presence or absence or degree of risk to human development. Understanding what exposures (in terms of dose, duration, window of exposure etc.) incur human developmental risk is the critical basis upon which alternative protective strategies must be based. For adverse exposures that can be obviated, exposure should be reduced or eliminated. For the many exposures which are unforeseen or unavoidable, thorough understanding of the effects of the antecedent ligand exposure and the induced persistent functional change(s) will be essential for the translational steps of proposing, assessing and implementing mitigative or restorative therapeutic treatments.

Early in life humans are exposed to numerous environmental factors that may influence developmental processes and subsequently health across the entire ensuing lifespan. During developmentally sensitive windows such as *in utero*, neonatal and peripubertal periods, the young human may be and sometimes is inevitably exposed to complex exogenous factors such as psychosocial stressors, maternal metabolic *in utero* environment, hypertensive disorders of pregnancy which may subject the fetus to placental insufficiency, inadequate or inappropriate nutrition, and other mixtures for which the toxicopathologic pathways are poorly understood. The translational prospects for developing mitigative or restorative interventions for adverse effects that may have been induced by those categories of exposures is certainly daunting because the specific molecular, cellular and organizational perturbations are not well characterized. In contrast to the staggering complexities posed by several of these developmental exposures, other exposures to compounds that act as ligands in interactions with biomolecules offer some prospects for more specific countervailing interventions. In brief, if exposure to an exogenous ligand acts through a particular molecular signaling pathway then the prospect for identifying a therapy that may antagonize or circumvent the induced signal should be more straightforward. If the signaling perturbation during a developmentally sensitive window is known with a high degree of precision and certainty, then therapeutic mitigation of that signal during that developmental window can be specifically assessed and possibly implemented. Additionally if ligand exposure during a sensitive developmental window is shown to have occurred, understanding of the affected signaling pathway may also allow more precise compensatory therapy at some later stage in life.

As mentioned earlier, one area of developmental toxicology that has been extraordinarily controversial over the last two decades is the field defined as "endocrine disruptors" (Kavlock et al. 1996; Olsson et al. 1998). There are core elements to the controversy that relate to the prospects of adverse developmental effects of variable levels of exposure to hormone like compounds via interactions with highly specific hormone receptors and non-monotonic dose response curves.

On one hand, there seems to often be limited recall regarding decades of historical studies that investigated potential toxicities attributed to hormonal exposures which built the foundation for our more contemporary research regarding hormone mediated developmental effects (McLachlan 1980, 1985; Kincl 1990) and show the extent to which adverse effects can be manifested. On the other hand, the designation of "endocrine" has often been generalized to such an extent that any apparent mode of action, even those that may be totally unrelated to the long-established "action at a distance" notion that underpins endocrinology as a field of research and medicine has been included as endocrine disruption rather than simply and more generally, developmental toxicology. Additionally, some use of nomenclature in this field suggests that any compound that affects any signaling biomolecule is "endocrine" and *de facto* is hazardous. It seems likely that some transient perturbations in signaling pathways may be inconsequential depending upon whether they occur and persist in a critical developmental window and whether or not there is an adverse impact on functional status especially later in life. It is our viewpoint that the field is now confounded by the way in which current nomenclature is used by both those who appear to hold entrenched positions in this controversy as well as the many objective open-minded scientists and physicians who continue basic, epidemiologic and clinical research. Accordingly, we propose that the study of developmental effects of hormone and hormone-like compounds should be seen as a subtopic encompassed within the broader field of ligand mediated pharmacology which we call "ligand-mediated developmental toxicology."

4.10 Nutrition and Functional Foods: Intentional Exposures to Dietary Ligands

In our present effort to consider both the health risks and benefits of exposures to exogenous ligands, the dynamic field of functional foods research appears to be carrying nutritional research into the realm of therapeutics by assessing intentional exposures to enhanced levels of dietary ligands with the anticipation of health benefits (Martirosyan and Liu 2014; Martirosyan 2014). A number of definitions of "functional food" have been advanced and debated by investigators, institutions and regulatory bodies in several countries. Functional food development is driven by efforts to prevent, mitigate or treat various diseases and is based on anticipated actions of bioactive compounds (predominantly ligands). Widely-recognized examples of health benefits of bioactive compounds include the following:

- Lutein and zeaxanthin for cataract prevention and for macular degeneration
- Beta-carotene and lycopene for skin protection from ultraviolet radiation injury
- Lutein and lycopene for cardiovascular risk reduction
- Lycopene for prostate cancer risk reduction

Many more therapeutic uses are currently under active research and development (Martirosyan and Liu 2014).

Some modes of action of bioactive compounds in functional foods include activation of antioxidant defenses, modulation of signal transduction pathways, changes in cell survival-associated gene expression, changes in cellular proliferation and differentiation and enhancement of mitochondrial integrity and function (Martirosyan 2014).

Given the breadth and depth of knowledge that is being developed in the field of functional foods, we think that there is great potential for defining generally-regarded-as-safe (GRAS) interventions for mitigation of adverse developmental effects that would be attributable to ligand-mediated toxicant actions.

4.11 Illustrative Therapeutic Opportunities in Ligand-Mediated Toxicology

"Fetal therapy" is a treatment administered to the fetus or via the mother with a primary indication to improve perinatal or long-term outcomes for the fetus or newborn (Paulis et al. 2013). Due to concern about unknown consequences of fetal exposures, it is often presumed that no pharmaceutical interventions should be given to pregnant women for the purpose of enhancing lifelong well-being of the mother's offspring. Besides immunizations, a few obstetrical therapies are clearly "fetal therapy" including care of maternal acute and chronic diseases as well as specific obstetrical conditions such as pre-eclampsia and gestational diabetes, and this includes administration of antenatal corticosteroids to enhance fetal lung maturity when premature delivery is anticipated. There are a few additional treatments that are often provided to pregnant women where a primary aim is to effect fetal benefit that would obviously or likely manifest later in the lifespan (Hughes et al. 2013a, b). What are the prospects for developing others to address other clinically important disorders that result from adverse effects of developmental exposures? We will consider two general cases where ligand mediated signaling mechanisms may present opportunities to impact developmentally-driven metabolic disorders in offspring, and one specific exposure as a "strawman" entity that may illustrate the prospect for pairing of exposure reduction with protective/mitigative intervention.

4.12 Childhood Effects of Maternal Metabolic Milieu on the Fetus

It is now well demonstrated that the maternal metabolic milieu in which the fetus develops has a profound effect on the subsequent risks of several childhood diseases (Hughes et al. 2013a) including obesity, hypertension, early onset of type II diabetes mellitus, cardiovascular disease and immune-mediated disorders such as asthma. A common theme and potential organizing principle is that activated inflammatory pathways during fetal life seem central in mediating many of these adverse effects. Accordingly, signaling pathways that may modify inflammation are worthy candidates for investigation as potentially protective interventions. One such prospect is liver X Receptor α, a nuclear transcription factor that regulates lipid metabolism. Recently, it has been shown that activation of LXRα with synthetic ligands has anti-inflammatory effects in atherosclerosis and chemical-induced dermatitis.

Obesity is a worldwide problem and its prevalence is increasing rapidly (Danaei et al. 2011). It is caused by the storage of excess calories as triglycerides in adipose tissue (Popovich et al. 1997), which is associated with insulin resistance; type 2 diabetes, hypertension, hyperlipidemia, cardiovascular disease, stroke and non-alcoholic steatohepatitis (Saltiel et al. 2001; Visscher and Seidell 2001). LXRs are potential drug targets for obesity, dyslipidemia and atherosclerosis. Earlier studies have shown that the synthetic LXR agonist GW3965 lowers cholesterol levels in both serum and the liver, inhibits the development of atherosclerosis in mouse models (Joseph et al. 2002; Kratzer et al. 2009), and improves glucose tolerance in diet-induced obesity and insulin resistant mice by regulating genes involved in glucose metabolism in the liver and the adipose tissue (Lafitte et al. 2003). On the other hand, LXR antagonists, such as 5a, 6a-epoxycholesterol-3-sulfate, block the formation of plaques of atherosclerosis by inhibiting LXR function (Song et al. 2001).

Remarkably the liver-X-receptors have shown anti-inflammatory ability in several animal models of respiratory disease. The effect of LXR agonist in allergen-induced airway remodeling in mice was studied by Shi et al. 2014. Ovalbumin-sensitized mice were chronically challenged with aerosolized ovalbumin for 8 weeks. The LXR agonist failed to attenuate the inflammatory cells and Th2 cytokines in bronchoalveolar lavage fluid. But the application of LXR agonist reduced the thickness of airway smooth muscle and the collagen deposition. It is concluded that LXRs may attenuate the progressing of airway remodeling, providing a potential treatment of asthma (Shi et al. 2014). The effect of the LXRα agonist T0901317 on lung inflammation in a rodent model of hemorrhagic shock shows that it increased nuclear LXRα expression and DNA binding while also inhibiting activation of NF-κB, a pro-inflammatory transcription factor, in the lung. A study suggests that LXRα is an important modulator of the inflammatory response and lung injury after severe hemorrhagic shock, likely through the inhibition of the NF-κB pathway (Solan et al. 2011).

In summary, LXR-mediated signaling pathways illustrate the potential for identifying and developing therapies that could mitigate both metabolic and immunologic diseases of childhood.

4.13 Prospective Interventions for Free Radial-Mediated Pathways

Various antioxidants have been shown or presumed to have a wide range of health benefits but broad public health recommendations have often been confounded by contradictory results or unexpected outcomes in studies. Nonetheless we think that the biology of free radicals as mediators of injury remains critically important to understand and for development of protective interventions. We wish to illustrate such prospects by reference to current research with one natural product (kuding tea made from the leaf of Broadleaf Holly which belongs to the *Folium Llicis Latifoliae* family) and an established pharmaceutical drug (N-acetylcysteine).

Recently, several clinical studies have focused on kuding tea effects on lipid lowering, body weight reduction and blood glucose lowering in patients with metabolic syndrome. Animal studies have shown that the phenolic constituents and phenyl ethanoid glycosides of kuding tea exhibit significant antioxidant activities *in vitro* (Zhu et al. 2009). The extracts of kuding tea prevent the development of obesity, hyperlipidemia and glucose tolerance in high-fat diet-fed C57BL/6 mice, and inhibit the transactivities of LXRβ (Fan et al. 2012). Kuding tea has both preventive and therapeutic roles in metabolic disorders in mice induced with high-fat diets (Fan et al. 2012). The effects appear to be mediated through the antagonism of LXRβ transactivity. Kuding tea may be a useful dietary therapy and a potential source for the development of novel anti-obesity and lipid lowering drugs in humans.

N-acetylcysteine (NAC) is a low molecular weight compound administered to neutralize the deleterious effects of free radicals. NAC raises the intracellular concentration of cysteine and hence of reduced glutathione (GSH), which acts as an important endogenous antioxidant. Moreover, NAC has direct scavenging properties *in vitro* against hydroxyl radicals and hypochlorous acid (Brunet et al. 1995). NAC is rapidly absorbed following oral dosing and reaches its peak concentrations in the blood within 1 h. After passing through the intestine and liver, NAC is transformed into a variety of metabolites. Consequently, NAC has been shown to have multiple therapeutic benefits (Marthler and Keresztes 2004). NAC has been used in clinical practice since the 1960s when it was introduced as a mucolytic agent for the treatment of respiratory diseases, such as chronic bronchitis and cystic fibrosis. Since the late 1970s, NAC has been administered as an antidote for the therapy of acute acetaminophen intoxication (Van Shooten et al. 2002). During the last decade, numerous *in vitro* and *in vivo* studies have suggested that NAC has beneficial medicinal properties including inhibition of carcinogenesis, tumorigenesis, and mutagenesis, as well as the inhibition of tumor growth and metastasis (Kumamoto et al. 2001). Currently, NAC is being studied for use in many disorders, such as chronic obstructive pulmonary disease (COPD), contrast-induced nephropathy, influenza, idiopathic pulmonary fibrosis and polycystic ovary syndrome (Millea 2009). Recent study showed that treatment with NAC may contribute to the restoration of non-enzymatic antioxidant reserves when administered to lead-exposed workers (Kasperczyk et al. 2014). When administered to workers chronically exposed to lead, NAC reduced lipid peroxidation in a dose-dependent manner. The simultaneously elevated concentrations of alpha-tocopherol may enhance the beneficial effects of NAC. However, the influence of NAC on the levels of uric acid, bilirubin, albumin and ferric reducing ability of plasma (FRAP) seems to be limited.

4.14 Fetal Alcohol Syndrome (FAS) and Fetal Alcohol Spectrum Disorder (FASD) as a Prototype "Endocrine Disruptor" Syndrome: Reduce Exposure and Mitigative Intervention

Alcohol (ethanol) is a well-known developmental toxicant that exerts its effects via mechanisms that fall within the definition of "endocrine disruptor." The primary public health intervention to prevent FAS/FASD is and must be to avoid fetal exposure. Nevertheless, presently in the industrialized world, even in the socioeconomic "middle class," 2.4–4.8 % of children are victims of FAS/FASD (May et al. 2014). This reality compels us to ask if there are mitigative therapies that could be given in pregnancy or after birth to minimize the adverse developmental effects or that could enhance or accelerate post-natal compensatory processes.

4.15 Antioxidant Therapy for Reduction of Oxidative Stress and Application of Modulators of Bromodomain Proteins to Alter Transcription Pattern for Reversing FAS/FASD

It has long been known that children exposed to alcohol *in utero* exhibit a range of abnormalities. The abnormalities include mental, cranofacial and immunological disorders. The severity of the symptoms relates to the amount, duration and timing (in fetal life) of alcohol consumption by the pregnant mother. A milder and lower level of exposure can lead to hyperactivity, anxiety and depression which will affect overall quality of future life. With severe exposures, deformities, mental retardation and impaired metabolic and immune function may be obvious (Momino et al. 2012).

Studies in animals also have established that fetal exposure with ethanol leads to fetal alcohol spectrum disorder (FASD). A number of mechanisms have been proposed to pinpoint how FASD occurs. Two proposed mechanisms are dysregulation of hypothalamic-pituitary-adrenal (HPA) axis function (Mead and Sarkar 2014) and epigenetic changes (Kleiber et al. 2014). Endocrine dysfunction relates to how HPA regulation is altered by alcohol by a loop mechanism (Mead and Sarkar 2014). Experimental studies with rats demonstrate that proopiomelanocortin (POMC) producing neurons which serves as precursor of a number of biologically active peptide were found to be affected (Boyadjieva et al. 2009; Hellemans et al. 2008; Sarkar et al. 2007). In rats, replacing β-endorphin-POMC producing cells has been shown to ameliorate alcohol-mediated effects. This observation suggests the need to measure the circulating levels of POMC in unexposed mothers who bear normal children. Consideration could then be given to whether rectification of possible FASD could be attained by use of recombinant POMC in exposed pregnant mothers to protect the fetus. In the present era of biotechnology, production of POMC for partial treatment and amelioration of FAS/FASD should face no major hurdle at a time when an ever expanding spectrum of biomolecules are being produced as therapeutic drugs (Smith et al. 2009). Finally, in any program to develop a fetal protective therapy, maternal genetic heterogenity will likely be an important consideration since the number of pregnant mothers drinking alcohol is quite high compared to number of children who are diagnosed with FASD. Alcohol metabolism is linked to single nucleotide polymorphism of alcohol dehydrogenase (ADH) and that impacts alcohol metabolism.

Another mechanism for induction of FASD which is gaining support is the inheritable epigenetic changes to fetuses due to alcohol exposure through mothers. A number of processes involving DNA modification including methylation and acetylation as well as synthesis and secretion of micro RNAs (miRNA) have been shown to be affected by ethanol exposure. The genome is protected from changes in nucleotide sequence (mutations) by DNA repair enzymes to maintain integrity of the genome. Similarly, the epigenome is protected by bromodomain proteins (as mentioned earlier) which sense the changes in methylation and acetylation pattern (Sanchez et al. 2014). Methylation of CpG and CpG islands in promoters and histone acetylation regulate gene expression. Increased methylation of POMC promoter in fetal pups in rats and decreased acetylation have been observed with the promoter of POMC leading to decreased circulating levels (Govorko et al. 2012). This leads to a feedback effect on the hypothalamus-pituitary axis. An increase in POMC by feedback increases the POMC derived peptides, like endorphins and adrenocortical releasing hormone (ACTH) leading to increased cortisol, an indicator of stress. Experimentally, it has been demonstrated that effects of alcohol exposure to pregnant mothers is transmitted to the third generation. Modifying the activity of bromodomain proteins may be able to reverse alcohol-mediated epigenetic changes. In order to reverse alcohol-induced FASD, either the physiological level of POMC could be maintained by adminstration *per se* or the activity of bromodomain proteins could be modified to upregulate the synthesis and secretion of PMOC or a combination of both could be attempted. Currently, a number of prospective agents which can modify the activity of bromodomain proteins are being developed and could be considered for use in clinical trials for mitigating occurrence of FASD (Sanchez et al. 2014).

Alcohol metabolism by Cytochrome P450 2E1 is induced in chronic alcoholics and produces superoxides and increases oxidative stress. In addition, alcohol metabolism by alcohol dehydrogenase (ADH) is decreased by fasting and malnutrition due to degration of ADH and metabolism by CYP2E1, thereby increasing the oxidative stress in alcoholics and possibly among pregnant mothers who consume alcohol. As regards fetal effects, antioxidants are a good prospect for inhibiting some alcohol toxicity but it is unsettled as to which antioxidants may be most effective for this therapeutic application. Given the plausiblity of benefit with likely safety, further research into use of antioxidants for FAS/FASD should be pursued in animal models and hopefully could advance into clinical trials. This general point-of-view has recently been robustly encouraged by a recent comprehensive and insightful review (Young et al. 2014). Young et al. detailed the range of metabolic and nutritional perturbations that appear to be involved in the occurrence of FASD. These investigators considered studies in animal models as well as human FASD-related research and propose that a modest set of nutritional supplementations including vitamin A, docosahexaenoic acid, folic acid, zinc, choline, vitamin E, and selenium could plausibly prevent or alleviate the development of FASD. These investigators (Young et al. 2014) emphasized that avoidance of exposure is the primary protective intervention; however, they also argued that further research should be conducted to assess the potential for reducing the occurrence or severity of FASD by determining optimized amounts of these nutrients to reduce specific detrimental outcomes of FASD and to define the most effective timing of such supplementation, e.g., first, second, third or all trimesters and/or into the early postnatal and childhood intervals.

Another class of biomolecules receiving great attention relative to FASD is micro RNAs (miRNA) as mentioned earlier. Changes in miRNA with FASD as related to neuronal development have been observed in rodents (Wang et al. 2009; Miranda 2012). It is possible that expression of miRNA could be altered via epigenetic genetic changes due to exposure to ethanol (Miranda 2012). Alteration in miRNA expression could significantly contribute towards FASD among children of women who abuse alcohol during pregnancy. Experimentally, alteration in miRNAs of miR-29c and miR-204 has been demonstrated in adult rat frontal cortex due to ethanol while changes in miR-29c and miR-30c were observed in fetal brains exposed through maternal ethanol. It is anticipated that these miRNAs can affect the HPA axis, thereby producing FASD. As aforementioned, miRNA based gene regulation is a recently discovered area, only a few agents have been developed for targeted therapy for controlling the activity of miRNA. Antagomirs are the chemicals which target miRNA and modify its activity. Only a few antagomirs have been developed and their safety for use in humans remains to be determined. There is a timely need for the development of agents which can modulate miRNAs involved in FAS/FASD.

4.16 Summary

If we set aside the design and testing of pharmaceutical compounds as exogenous chemical agents to intentionally influence mammalian physiology, there are three other long-established strands of biological research that have investigated the actions of biologically active dietary or environmental compounds on animals including humans.

First, in agricultural and animal husbandry research there is extensive documentation of the many thousands of phytochemicals that mediate plant animal interactions, many of which produce toxic effects at certain levels of exposure in terms of dose, duration and window of sensitivity. In a broader biological context, many of these relationships have been formalized by concepts of co-evolution of plants and animals.

Second, there is a long history of developmental toxicology research showing a wide range of adverse effects that often result from exposures to hormones and other modulators of endocrine function during developmentally sensitive intervals. In recent decades much of this research has been encapsulated by the single term of endocrine disruptors.

Third, nutritional and functional foods research have demonstrated a wide range of health benefits produced by consumption of many phytochemical ligands through mechanisms that are truly not distinct from those that seem to mediate the adverse effects noted in the animal husbandry and developmental toxicology research arenas.

Humans (and animals) live in a world of exposure to ligands. We are confident that environmental and public health can be enhanced by defining how such exposures to exogenous signaling molecules (ligands) pose both risks and benefits. It is timely to modify scientific perspectives and nomenclature to encompass these several streams of investigation such that risks and benefits of exposure to dietary and environmental ligands may be translated into validated strategies to improve environmental and public health. Health promoting interventional options must include exposure reduction(s) to ligands that incur health risks but we should also define and employ active mechanistically-based mitigative interventions when exposures cannot be adequately reduced or preempted.

References

Adlercreutz H (1998) Chapter 14: Human health and phytoestrogens. In: Korach KK (ed) Reproductive and developmental toxicology. Marcel Dekker, Inc, New York, pp 299–371

Alexander S, Mathie A, Peters J (2011) Guide to receptors and channels (GRAC), 5th edn. Br J Pharmacol 164(Suppl 1):S1-324

Anway M, Skinner M (2006) Epigenetic transgenerational actions of endocrine disruptors. Endocrinology 147(6 Suppl):S43–S49

Anway M et al (2005) Epigenetic transgenerational actions of endocrine disruptors and male fertility. Science 308(5727):1466–1469

Anway M, Leathers C, Skinner M (2006a) Endocrine disruptor vinclozolin induced epigenetic transgenerational adult-onset disease. Endocrinology 147(12):5515–5523

Anway M et al (2006b) Transgenerational effect of the endocrine disruptor vinclozolin on male spermatogenesis. J Androl 27(6):868–879

Belovsky G, Schmitz O (1991) Chapter 1: Mammalian herbivore optimal foraging and the role of plant defenses. In: Palo R, Robbins C (eds) Plant defenses against mammalian herbivory. CRC Press, Inc, Boca Raton, pp 1–28

Berezikov E (2011) Evolution of microRNA diversity and regulation in animals. Nat Rev Genet 12(12):846–860

Bergman A, Jobling S, Kidd K, Zoeller R (2012) State of science of endocrine disrupting chemicals. WHO Press, World Health Organization, United Nations Environmental Programme, Geneva

Bollati V, Baccarelli A (2010) Environmental epigenetics. Heredity (Edinb) 105(1):105–112

Boyadjieva N et al (2009) Beta-endorphin neuronal cell transplant reduces corticotropin releasing hormone hyperresponse to lipopolysaccharide and eliminates natural killer cell functional deficiencies in fetal alcohol exposed rats. Alcohol Clin Exp Res 33(5):931–937

Brunet J et al (1995) Effects of N-acetylcysteine in the rat heart reperfused after low-flow ischemia: evidence for a direct scavenging of hydroxyl radicals and a nitric oxide-dependent increase in coronary flow. Free Radic Biol Med 19(5):627–638

Chang H et al (2006) Transgenerational epigenetic imprinting of the male germline by endocrine disruptor exposure during gonadal sex determination. Endocrinology 147(12):5524–5541

Cheeke P (1989) Toxicants of plant origin, vol I–IV. CRC Press, Inc, Boca Raton

Collotta M, Bertazzi P, Bollati V (2013) Epigenetics and pesticides. Toxicology 307:35–41

Danaei G et al (2011) National, regional, and global trends in fasting plasma glucose and diabetes prevalence since 1980: systematic analysis of health examination surveys and epidemiological studies with 370 country-years and 2.7 million participants. Lancet 378(9785):31–40

Eaton D, Gilbert S (2010) Chapter 2: Principles of toxicology. In: Klaassen CD, Watkins JB III (eds) Casarett & Doull's essentials of toxicology, 2nd edn. McGraw-Hill, New York

Euling S et al (2008) Role of environmental factors in the timing of puberty. Pediatrics 121(Suppl 3):S167–S171

Fan S et al (2012) Extract of Kuding tea prevents high-fat diet-induced metabolic disorders in C57BL/6 mice via liver X receptor (LXR) beta antagonism. PLoS One 7(12):e51007

Fisch G (2012) Nosology and epidemiology in autism: classification counts. Am J Med Genet C: Semin Med Genet 160C(2):91–103

Gallo M (2008) Chapter 1: History and scope of toxicology. In: Klaassen CD (ed) Casarett and Doull's toxicology. McGraw-Hill, New York

Govorko D et al (2012) Male germline transmits fetal alcohol adverse effect on hypothalamic proopiomelanocortin gene across generations. Biol Psychiatry 72(5):378–388

Gregus Z (2010) Chapter 3: Mechanism of toxicity. In: Klaassen CD, Watkins JB III (eds) Casarret & Doull's essentials of toxicology, 2nd edn. McGraw-Hill, New York

Hale G et al (2009) Atypical estradiol secretion and ovulation patterns caused by luteal out-of-phase (LOOP) events underlying irregular ovulatory menstrual cycles in the menopausal transition. Menopause 16(1):50–59

Harborne J (1982) Introduction to ecological biochemistry. Academic, London

Hellemans K et al (2008) Prenatal alcohol exposure increases vulnerability to stress and anxiety-like disorders in adulthood. Ann N Y Acad Sci 1144:154–175

Hughes C, Dhiman T (2002) Dietary compounds in relation to dietary diversity and human health. J Med Food 5(2):51–68

Hughes C, Tansey G (1998) Chapter 13: Phytoestrogens and reproductive medicine. In: Korach KK (ed) Reproductive and developmental toxicology. Marcel Dekker, Inc, New York, pp 277–298

Hughes C, Waters M, Allen D, Obasanjo I (2013a) Translational toxicology: a developmental focus for integrated research strategies. BMC Pharmacol Toxicol 14(1):51. doi:10.1186/2050-6511-14-51
Hughes C, Waters M, Obasanjo I, Allen D (2013b) Translational developmental toxicology: prospects for protective therapeutic obstetrical and neonatal interventions. J Neonatal Biol 2:122. doi:10.4172/2167-0897.1000122
Joseph S et al (2002) Synthetic LXR ligand inhibits the development of atherosclerosis in mice. Proc Natl Acad Sci U S A 99(11):7604–7609
Kasperczyk S, Dobrakowski M, Kasperczyk A, Zalejska-Fiolka J, Pawlas N, Kapka-Skrzypczak L, Birkner E (2014) Effect of treatment with N-acetylcysteine on non-enzymatic antioxidant reserves and lipid peroxidation in workers exposed to lead. Ann Agric Environ Med 21(2):272–277. doi:10.5604/1232-1966.1108590
Kavlock R, Cummings A (2005) Mode of action: inhibition of androgen receptor function – vinclozolin-induced malformations in reproductive development. Crit Rev Toxicol 35(8–9):721–726
Kavlock R, Daston G, DeRosa C et al (1996) Research needs for the risk assessment of health and environmental effects of endocrine disruptors: a report of the US EPA sponsored workshop. Environ Health Perspect 104(suppl 4):715–740
Kim J, Im J, Lee D (2012) The relationship between leukocyte mitochondrial DNA contents and metabolic syndrome in postmenopausal women. Menopause 19(5):582–587
Kincl F (1990) Hormone toxicity in the newborn. Springer, Berlin
Kleiber M et al (2014) Long-term genomic and epigenomic dysregulation as a consequence of prenatal alcohol exposure: a model for fetal alcohol spectrum disorders. Front Genet 5:161
Kratzer A et al (2009) Synthetic LXR agonist attenuates plaque formation in apoE–/– mice without inducing liver steatosis and hypertriglyceridemia. J Lipid Res 50(2):312–326
Kumamoto M et al (2001) Effects of pH and metal ions on antioxidative activities of catechins. Biosci Biotechnol Biochem 65(1):126–132
Laffitte B et al (2003) Activation of liver X receptor improves glucose tolerance through coordinate regulation of glucose metabolism in liver and adipose tissue. Proc Natl Acad Sci U S A 100(9):5419–5424
Lemaire G et al (2006) Activation of alpha- and beta-estrogen receptors by persistent pesticides in reporter cell lines. Life Sci 79(12):1160–1169
Li S et al (2011) microRNA and mRNA expression profiling analysis of dichlorvos cytotoxicity in porcine kidney epithelial PK15 cells. DNA Cell Biol 30(12):1073–1083
Marthaler M, Keresztes P (2004) Evidence-based practice for the use of N-acetylcysteine. Dimens Crit Care Nurs 23(6):270–273
Martirosyan D (2014) Introduction to functional food science, 2nd edn. Food Science Publisher, Dallas
Martirosyan D, Liu S (2014) Discovery, utilization, and control of bioactive components and functional foods. Food Science Publisher, Dallas
May P et al (2014) Prevalence and characteristics of fetal alcohol spectrum disorders. Pediatrics 134:855–866
McLachlan J (1980) Estrogens in the environment. Elsevier, New York
McLachlan J (1985) Estrogens in the environment II: influences on development. Elsevier, New York
Mead E, Sarkar D (2014) Fetal alcohol spectrum disorders and their transmission through genetic and epigenetic mechanisms. Front Genet 5:154
Millea P (2009) N-acetylcysteine: multiple clinical applications. Am Fam Physician 80(3):265–269
Miranda R (2012) MicroRNAs and fetal brain development: implications for ethanol teratology during the second trimester period of neurogenesis. Front Genet 3:77
Molina-Molina J et al (2006) Steroid receptor profiling of vinclozolin and its primary metabolites. Toxicol Appl Pharmacol 216(1):44–54

Momino W et al (2012) Maternal drinking behavior and Fetal Alcohol Spectrum Disorders in adolescents with criminal behavior in southern Brazil. Genet Mol Biol 35(4 suppl):960–965
Olsson P-E, Borg B, Brunstrom B, Hakansson H, Klasson-Wehler E (1998) Endocrine disrupting substances, Report 4859. Swedish Environmental Protection Agency, Stockholm
Patti G, Yanes O, Siuzdak G (2012) Innovation: metabolomics: the apogee of the omics trilogy. Nat Rev Mol Cell Biol 13(4):263–269
Paulis G et al (2013) Long-term multimodal therapy (verapamil associated with propolis, blueberry, vitamin E and local diclofenac) on patients with Peyronie's disease (chronic inflammation of the tunica albuginea). Results of a controlled study. Inflamm Allergy Drug Targets 12(6):403–409
Pawson A, Sharman J, Benson H, Faccenda E, Alexander S, Buneman O, Davenport A, McGrath J, Peters J, Southan C, Spedding M, Yu W, Harmar A, NC-IUPHAR (2014) The IUPHAR/BPS Guide to PHARMACOLOGY: an expert-driven knowledgebase of drug targets and their ligands. Nucl Acids Res 42(Database Issue):D1098–D1106
Polya G (2003) Biochemical targets of plant bioactive compounds. Taylor & Francis, New York
Popovich N, Wood O (1997) Drug therapy for obesity: an update. J Am Pharmcol Assoc (Wash), NS37(1):31–39, 56
Rottiers V, Naar A (2012) MicroRNAs in metabolism and metabolic disorders. Nat Rev Mol Cell Biol 13(4):239–250
Ruegg J et al (2009) Receptors mediating toxicity and their involvement in endocrine disruption. EXS, 99:289–323
Russo F et al (2014) A knowledge base for the discovery of function, diagnostic potential and drug effects on cellular and extracellular miRNAs. BMC Genomics 15(Suppl 3):S4
Saltiel A, Kahn C (2001) Insulin signalling and the regulation of glucose and lipid metabolism. Nature 414(6865):799–806
Sanchez R, Meslamani J, Zhou M (2014) The bromodomain: from epigenome reader to druggable target. Biochim Biophys Acta 1839(8):676–685
Sarkar D et al (2007) Alcohol exposure during the developmental period induces beta-endorphin neuronal death and causes alteration in the opioid control of stress axis function. Endocrinology 148(6):2828–2834
Shi Y et al (2014) A liver-X-receptor ligand, T0901317, attenuates IgE production and airway remodeling in chronic asthma model of mice. PLoS One 9(3):e92668
Smith K et al (2009) Rapid generation of fully human monoclonal antibodies specific to a vaccinating antigen. Nat Protoc 4(3):372–384
Solan P et al (2011) Liver X receptor alpha activation with the synthetic ligand T0901317 reduces lung injury and inflammation after hemorrhage and resuscitation via inhibition of the nuclear factor kappaB pathway. Shock 35(4):367–374
Song C, Hiipakka R, Liao S (2001) Auto-oxidized cholesterol sulfates are antagonistic ligands of liver X receptors: implications for the development and treatment of atherosclerosis. Steroids 66(6):473–479
Song C et al (2010) Environmental neurotoxic pesticide increases histone acetylation to promote apoptosis in dopaminergic neuronal cells: relevance to epigenetic mechanisms of neurodegeneration. Mol Pharmacol 77(4):621–632
Song C et al (2011) Paraquat induces epigenetic changes by promoting histone acetylation in cell culture models of dopaminergic degeneration. Neurotoxicology 32(5):586–595
Thompson J (1994) The coevolutionary process. The University of Chicago Press, Chicago
Van Schooten F et al (2002) Effects of oral administration of N-acetyl-L-cysteine: a multi-biomarker study in smokers. Cancer Epidemiol Biomarkers Prev 11(2):167–175
Vandenberg L et al (2012) Hormones and endocrine-disrupting chemicals: low-dose effects and nonmonotonic dose responses. Endocr Rev 33(3):378–455
Visscher T, Seidell J (2001) The public health impact of obesity. Annu Rev Public Health 22:355–375

Walker D, Gore A (2011) Transgenerational neuroendocrine disruption of reproduction. Nat Rev Endocrinol 7(4):197–207
Wang L et al (2009) Ethanol exposure induces differential microRNA and target gene expression and teratogenic effects which can be suppressed by folic acid supplementation. Hum Reprod 24(3):562–579
Wang X et al (2010) Effect of triazophos, fipronil and their mixture on miRNA expression in adult zebrafish. J Environ Sci Health B 45(7):648–657
West P et al (2010) Predicting human developmental toxicity of pharmaceuticals using human embryonic stem cells and metabolomics. Toxicol Appl Pharmacol 247(1):18–27
Woods C et al (2007) Sustained formation of alpha-(4-pyridyl-1-oxide)-N-tert-butylnitrone radical adducts in mouse liver by peroxisome proliferators is dependent upon peroxisome proliferator-activated receptor-alpha, but not NADPH oxidase. Free Radic Biol Med 42(3):335–342
Yang P, Zhou X (2013) MicroRNA: a new type of gene. Microrna 2(1):1
Young J, Giesbrecht H, Eskin M, Aliani M, Suh M (2014) Nutrition implications for fetal alcohol spectrum disorder. Adv Nutr 5:675–692
Zhu F et al (2009) Comparison of major phenolic constituents and in vitro antioxidant activity of diverse Kudingcha genotypes from Ilex kudingcha, Ilex cornuta, and Ligustrum robustum. J Agric Food Chem 57(14):6082–6089
Zoeller R et al (2012) Endocrine-disrupting chemicals and public health protection: a statement of principles from The Endocrine Society. Endocrinology 153(9):4097–4110

Chapter 5
Effects of Environmentally Acquired Heavy Metals and Nutrients on the Epigenome and Phenotype

David A. Skaar, Susan K. Murphy, and Cathrine Hoyo

Abstract Cadmium, arsenic, mercury and lead are ubiquitous environmental contaminants that tend to co-occur. Unlike organic compounds that are chemically, biologically, or photo-degraded, these metals persist in the environment for indefinite periods. Although protein disruption/misfolding, generation of oxidative stress, and endocrine disruption are known effects of toxic metal exposure, beyond the known toxic effects of high dose exposure, mechanisms causing these effects, especially at low chronic doses, are still largely unknown. Epigenetics is emerging as a viable mechanistic framework to explain how the environment interacts with the genome to alter disease risk. Alterations in DNA methylation, histone marks and chromatin structure have been proposed as useful exposure assessment biomarkers that can substantially improve assessment of risk in etiologic studies where exposure occurs early during the life course. If developed into exposure-specific biomarkers, these epigenetic marks can be a powerful tool to identify populations exposed to low doses where phenotypic response may not be immediately apparent, and also to evaluate the efficacy of therapeutic and public health interventions. This could be particularly important as exposed populations tend to be the socioeconomically disadvantaged who have limited contact with the health care system. In this review, we provide an overview of the current state of literature on heavy-metal-associated epigenetic alterations. We discuss the extent to which such epigenetic alterations alter susceptibility to common chronic diseases and how they might be mitigated by some nutrients, albeit within narrow margins. We conclude by

---

This work was supported by NIH grants R01DK085173, P30ES025128, P01ES022831 and US EPA grant RD-83543701. The content is solely the responsibility of the authors and does not necessarily represent the official views of the NIH or the USEPA.

D.A. Skaar • C. Hoyo (✉)
Department of Biological Sciences, and Center for Human Health and the Environment, North Carolina State University, Campus Box 7633, Raleigh, NC 27695-7633, USA
e-mail: choyo@ncsu.edu

S.K. Murphy
Department of Obstetrics and Gynecology, Duke University Medical Center, Durham, NC 27708, USA

C.L. Hughes, M.D. Waters (eds.), *Translational Toxicology*, Molecular and Integrative Toxicology, DOI 10.1007/978-3-319-27449-2_5

discussing key issues that must be resolved if human epigenetic data is to provide useful biomarkers and mechanistic insights into how low dose chronic exposure to these metals might alter the epigenome and increase disease susceptibility.

Keywords Cadmium • Lead • Arsenic • Mercury • Epigenetics • DNA methylation • Cancer • Neurodevelopment

5.1 Introduction

Cadmium, mercury, arsenic and lead are ubiquitous environmental contaminants that rank in the top ten metals of concern by environmental health agencies. Unlike organic compounds that are chemically, biologically, or photo-degraded, these metals persist in the environment for long periods. For example, while smoking is a major source of cadmium, the most common source of chronic exposure to cadmium for non-smokers is dietary intake of staples (e.g., potatoes, grains) and consumption of contaminated water or seafood (ATSDR 2015). Due to past industrial activity, however, these metals also co-contaminate soils of many urban and rural landscapes, are tracked indoors and become part of household dust that is ingested or inhaled. With no active biological mechanisms for elimination, these multiple metals bioaccumulate in soft tissue and bone and target multiple organ systems. Fetal exposure via transplacental transfer from concurrent sources or exposure mobilized from longer term physiological stores (e.g., soft tissue and bone) is well documented (Reynolds et al. 2006) directly or indirectly by comparing concentrations in parturition maternal blood with umbilical cord blood (Needham et al. 2011).

Reducing chronic exposure to these heavy metals and their effects on human health is a high priority for environmental health agencies (ATSDR 2015), as these metals readily enter the food supply. While low dose but chronic exposure to one or more of these heavy metals rarely elicits a phenotypic response, in cross-sectional studies, elevated urine and peripheral blood levels of these metals are associated with a broad range of dysfunctions in adult humans and animals (Satarug and Moore 2004, 2012). Elevated levels of these metals have long been implicated in cardiovascular disease (Agarwal et al. 2011), diabetes (Chen et al. 2006), skeletal disorders (Chen et al. 2011; Cho et al. 2012), renal disorders (Huang et al. 2013; Liang et al. 2012), some cancers (Christensen and Marsit 2011; Christensen et al. 2009; Gallagher et al. 2010; Benbrahim-Tallaa et al. 2009; Tokar et al. 2011), hypertension and elevated insulin and/or glucose (Gallagher and Meliker 2010; Wallia et al. 2014; Tellez-Plaza et al. 2013; Karim et al. 2013; Mendez et al. 2016; Kuo et al. 2013; Schober et al. 2006), with some suggestions of sex- or race-specific effects (Ferraro et al. 2012; Scinicariello et al. 2011), even if associations have not always been consistent. Although protein disruption/misfolding, generation of oxidative stress, and endocrine disruption are known effects of heavy metal exposure, mechanisms causing these effects are still largely unknown (Satarug and Moore 2012; Smeester et al. 2014; Sanders et al. 2014).

Epigenetics is emerging as a mechanistic framework that may explain how the environment interacts with the genome to alter disease risk. Alterations in DNA methylation, histone marks and chromatin structure have been proposed as useful exposure assessment biomarkers that can improve risk assessment (Ho et al. 2012; McKay and Mathers 2011; Heijmans et al. 2009; Hoyo et al. 2009). However, beyond a handful of epigenomic regions evaluated based on known gene function in relation to acute toxicity (Smeester et al. 2014; Li et al. 2015; Vidal et al. 2015), the epigenomic landscape targeted by these metals, especially at chronic low doses, is still unknown. Also unknown is the specificity of associations and stability over time of epigenomic marks identified thus far; these factors likely depend on the regulatory regions perturbed, the tissues they target—a function of the route of exposure—and the timing of exposure during the life course.

Here, we provide an overview of studies that have evaluated potential effects of these heavy metals on the epigenome of target tissues, depending on the route and source of exposure, and whether such epigenetic effects are associated with physiological effects. DNA methylation is the most intensively studied epigenetic modification in humans due to the stability of DNA (Hoyo et al. 2009) and the covalent nature of DNA methylation (Heijmans et al. 2009; Hoyo et al. 2009). In addition, collection of field or clinic-based specimens from otherwise healthy human can involve challenging storage or shipment conditions; thus, the majority of human studies discussed focus on DNA methylation. We conclude by discussing some issues that must be resolved if human epigenetic data are to yield useful biomarkers and mechanistic insights into how low-dose chronic exposure to these metals might alter the epigenome to increase disease susceptibility.

5.2 Lead Exposure

5.2.1 Sources of Lead Exposure

Lead toxicity is possibly the most studied of the four heavy metals we focus on humans, given the long history, pervasiveness, and severe impact of lead exposure and physiological response. Lead was one of the first metals worked by humans, and the ease and variety of uses resulted in its early implementation into cookware, pottery, cosmetics, pigments and plumbing, among others. More recently, lead release from sources including coal burning, gasoline additives, electronics, plastics, paint and batteries has resulted in continued widespread human exposure, environmental contamination, and the need for large amounts of hazardous waste to be disposed and contained. Because there is no apparent necessary or beneficial biological usage of lead, a “safe” level of exposure is unlikely; neither are there effective mechanisms of elimination once it enters the human body.

5.2.2 *Routes of Lead Exposure*

Lead exposure to humans is primarily through ingestion and inhalation. Absorption through the gastrointestinal tract is dependent on particle size, with smaller particles more readily absorbed, while nearly all lead inhaled as vapor is absorbed. Once absorbed, lead enters the blood, the means by which it is dispersed to other organs. In blood, 99 % of lead is contained in erythrocytes, with 1 % in plasma, the source of lead transfer to cells and organs. Long term storage of lead is in bone (Renner 2010; Silbergeld et al. 1993), holding 85–95 % of total lead in adults, and 70 % in children. These long term physiological deposits contribute to 40–70 % of lead released into peripheral blood in adults, with a bone surface lead pool that is easily exchangeable, and a deeper, cortical bone pool that can only exchange when bone is being re-absorbed (Patrick 2006). This slow turnover contributes to bone-storage of lead that can last decades, long after environmental lead exposure has ceased (ATSDR 2015; Mason et al. 2014).

5.2.3 *Tissues/Systems Targeted by Lead*

Depending on exposure route, the liver, lungs, and kidneys have the highest lead concentration after acute exposure, with the brain also a site of accumulation, particularly if exposure occurs during fetal development or early childhood. Although hematopoietic, hepatic, and renal systems (Kalia and Flora 2005) are also targeted, several lines of evidence suggest that the central nervous system may be the most sensitive and most strongly affected (Cory-Slechta 1996). Physiological uptake is likely due to Pb^{2+} substitution for Ca^{2+}, using Ca-ATPase pumps to enter cells; cellular accumulation of lead leads to depletion of calcium reserves.

5.2.4 *Physiological Effects of Lead Exposure*

The physiological effects of lead are widespread, primarily due to non-specific protein interactions and disruption of homeostasis of other cations, such as Ca^{2+}, or with specific cellular or physiological targets. The differences in symptoms between acute (>50–60μg/dl blood), and more common chronic (~10μg/dl) exposures is hypothesized to be the large scale disruption of multiple organ systems when exposure dose is acute, versus disruption of processes most sensitive to lead in low-level chronic exposures.

Mechanisms of lead toxicity include generation of oxidative stress, protein disruption, and disruption of Ca^{2+} homeostasis. Generation of oxidative stress is similar to mechanisms of other heavy metals: i.e., creation of free radicals, and reduction of cellular antioxidant pools, reducing the ability to inactivate free radicals. Both effects come from the affinity of divalent lead (Pb^{2+}) for sulfhydryl (–SH) groups, leading to reaction with and deactivation of antioxidant enzymes, including superoxide dismutase and catalase, as well as glutathione, the primary cellular antioxidant for removal of free radicals. Proteins affected by lead are widespread, and besides inhibition of antioxidants, enzymes involved in glutathione synthesis and regeneration, such as glutathione reductase, peroxidase, and S-transferase, are also affected, further depleting glutathione (Ahamed and Siddiqui 2007). In addition, the general protein inactivation caused by lead has known effects on cell signaling, cell adhesion, protein folding and processing, apoptosis, ion transport, enzyme regulation, and neurotransmitters, among others. Lead also affects protein activity through its primary ionic activity, as Pb^{2+} replaces other divalent protein cofactors, such as zinc, calcium, magnesium, and iron (Lidsky and Schneider 2003). Besides reductions in protein function due to substitution, the mobility of lead is increased when acting as a protein cofactor, allowing passage across the blood brain barrier, leading to neuronal accumulation. Finally, the disruption of Ca^{2+} homeostasis upon Pb^{2+} influx increases intracellular Ca^{2+} levels, a trigger for apoptosis (Wang et al. 2015).

Phenotypic effects including disability and death have been reported in response to lead exposure, likely due to widespread damage and system failure, with brain, kidney and liver as primary organ system targets. As lead effectively accumulates in the brain, and the high energy requirements of neurons generates reactive oxygen species, the nervous system is the most sensitive to lead toxicity (Cory-Slechta 1996). In adults, the peripheral nervous system is more strongly affected, while in children, effects on the central nervous system are more pronounced (Brent 2006; Bellinger 2004). Developing fetuses and young children are highly vulnerable, as a higher proportion of accumulated circulating lead is absorbed by immature brains than in adults (Needleman 2004). Lead exposure in children is associated with developmental delays, decreased growth, short-term memory defects, hearing loss and some suggestions of decreased IQ, as well as behavioral and attention defects and gray matter atrophy, with high levels causing permanent brain damage and death (Cleveland et al. 2008). Peripheral nervous system effects include reduced motor activity due to loss of myelin sheathing, leading to muscle weakness and lack of coordination (Sanders et al. 2009).

Effects of acute lead poisoning includes anemia (Guidotti et al. 2008; Cornelis 2005), of two possible types: frank anemia, caused by long term elevated lead levels resulting in reduced heme production (Vij 2009); and hemolytic anemia, a result of acute high-level exposure. Three enzymes in heme synthesis are most affected (Piomelli 2002), with δ-aminolevulinic acid dehydratase (ALAD) being the most sensitive, and useful as a clinical measure of the degree of lead poisoning, as ALAD inhibition is detectable at low (10–20μg/dl) lead levels. Heme synthesis is not affected until ALAD activity is reduced to 10–20 % of its normal levels (Ahamed et al. 2005). The mechanism for hemolytic anemia is not understood, but a biomarker for exposure is basophilic stippling of red blood cells, the result of accumulated dense aggregates of degraded ribonucleic acid (Patrick 2006). Lower doses of lead also may cause acute and chronic nephropathy, vascular damage leading to hypertension and cardiovascular disease, and reproductive health effects in men and women, in nearly all areas of gametogenesis, sexual function, and gestation (Wadi and Ahmad 1999; Ronis et al. 1998; Iavicoli et al. 2004, 2006; Said and Hernandez 2015; Batuman and Wedeen 2014; Rabito et al. 2014; Perkins et al. 2014; Anis et al. 2007)

5.3 Cadmium Exposure

5.3.1 Sources of Cadmium Exposure

As with lead, cadmium has no biological utility, and thus lacks mechanisms for elimination, resulting in bioaccumulation in soft tissues, primarily the kidney cortex. World Health Organization guidelines (2010) set a provisional tolerable weekly intake of 5.8μg/kg body weight, while the Consumer Product Safety Commission recommends a daily intake limit of 0.1μg/kg body weight for chronic exposure (Mead 2010). Most human exposure is cigarette smoking-related (Galazyn-Sidorczuk et al. 2008; Lampe et al. 2008), although occupational exposure in metal processing, battery production, and electroplating have also been reported (ATSDR 2015; Sahmoun et al. 2005; World Health Organization 1992). Agricultural exposure is increasing (World Health Organization 1992) with the usage of phosphate or sewage sludge fertilizers (Bandara et al. 2010).

5.3.2 Routes of Cadmium Exposure

Geographic-specific exposure occurs in populations living in regions with contaminated soil or water, the original sources of contamination are likely to be from past or present industrial activity. Inhalation and ingestion of household dust are the primary sources of intake, with 10–50 % absorbed, depending on particle size and solubility. Nearly all ingested cadmium passes through the gastrointestinal tract, with ~5 % of cadmium dissolved in water absorbed, and ~2.5 % of cadmium in food absorbed (Godt et al. 2006).

5.3.3 Tissues/Systems Targeted by Cadmium

The most severe effects of acute cadmium exposure are neurological, with inhaled cadmium from occupational exposures or smoking crossing into the central nervous system from nasal mucosa or olfactory pathways which easily bypass the blood-brain barrier (Lafuente et al. 1999; Lafuente and Esquifino 1999; Esquifino et al. 1999). In adults, the blood-brain barrier effectively blocks cadmium from the brain, but the barrier is not fully developed until adulthood, putting children at particular risk (Antonio et al. 1999; Gill et al. 1989). The fetal brain is particularly vulnerable during development, as cadmium crosses the placenta and is present in breast milk (Korpela et al. 1986). Other toxic effects include vertigo, decreased equilibrium, Parkinson's-like effects, peripheral neuropathy, decreased concentration, and learning disabilities. Elevated cadmium concentration measured in hair has also been associated with reduced IQ, dyslexia, and lower performance on visual-motor tasks (Pihl and Parkes 1977; Marlowe et al. 1983, 1985; Capel et al. 1981; Thatcher et al. 1982, 1984).

5.3.4 Physiological Effects of Cadmium Exposure

Animal models show effects on neuronal morphology and neuro-development in exposed fetuses, with smaller heads, unclear boundaries between brain subdivisions, and altered commitment of progenitor cells (Chow et al. 2008). Animal models have also indicated cadmium effects on a number of biochemical processes in the brain, affecting neurotransmitters including acetylcholine, serotonin and glutamate, and altering the balance of phospholipids in developing brains (Borisova et al. 2011; Borges et al. 2007; Herba et al. 2001; Desi et al. 1998; Gupta and Shukla 1996).

In adults with chronic exposure from cadmium inhalation, most damage occurs to the lungs (Hendrick 1996). When cadmium is ingested, damage is to the kidneys and liver, where ~50 % of cadmium accumulates (Johri et al. 2010). Cadmium is very slowly cleared from these organs, with half-lives estimated from 6 to nearly 40 years, such that damage could continue long after exposure has ceased. The damage causes symptoms of kidney or liver disease and decreased organ function, likely due to the accumulation of reactive oxygen species, substitution of other cations, and/or overall disruption of protein structure. Chronic cadmium exposure is also associated with generation of multinucleate osteoclasts, resulting in bone softening, osteoporosis, and increased fractures, as seen in Itai-Itai disease, even in the absence of detectable effects on renal function (Nawrot et al. 2010; Akesson et al. 2014; Kazantzis 2004).

While cadmium exposure generates reactive oxygen species, the mechanisms are unclear, as cadmium has no redox activity (existing solely as Cd^{2+}) and does not directly produce reactive oxygen species. However, its mimicry and replacement of other redox-active metals, such as Fe^{2+}, results in direct reaction of displaced cations to form free radicals (Casalino et al. 1997; Bonomi et al. 1994; Kehrer 2000).

The long term persistence of cadmium body loads and potential high bioaccumulation from chronic low exposures has raised public health concerns regarding risk of cardiometabolic disease and cancer. As a probable human carcinogen (EPA and IARC), it has been associated with lung and kidney cancer (Beyersmann and Hartwig 2008). Animal models have implicated cadmium in prostate cancer, and cadmium is being investigated as an agent in the development or progression of type 2 diabetes, and cancers of the breast, ovary, endometrium, and bladder (Feki-Tounsi and Hamza-Chaffai 2014; Cho et al. 2013). Human studies are inconsistent as there is likely residual confounding by dietary factors and population genetic variability.

5.4 Arsenic

5.4.1 Sources of Arsenic Exposure

The toxicity of arsenic has been known for centuries, although its natural presence in the environment has resulted in constant chronic exposure to humans, resulting in biological processing mechanisms and tolerance. It occurs in both organic and inorganic forms, with trivalent and pentavalent oxidation states. The trivalent state is more toxic (Watanabe and Hirano 2012). Organic arsenic compounds, such as pesticides, are less toxic than inorganic forms due to low biologic uptake and metabolism.

The most common source of human exposure is through food and contaminated groundwater, where concentrations are often as high as 100–1000μg/l, depending on the region. Populations depending on these water sources have the highest chronic exposures. Elevated arsenic levels in groundwater have been reported globally, with some of the highest levels in South America and Asia, where natural levels over 3000μg/l have been reported in some locations (Nordstrom 2002). Arsenic contamination is a particular problem in south and south-east Asia, where natural arsenic levels are high, and growing populations are increasingly dependent on wells for survival. Surface waters contain mostly the pentavalent form (As(V)/arsenate), while anoxic subsurface water contains mostly the more toxic trivalent form (As(III)/arsenite).

5.4.2 Routes of Arsenic Exposure

Over 90 % of inorganic As, both As(III) and As(V), is absorbed through the intestine, and excretion is mainly through urine. The half-life of absorbed arsenic is approximately 4 days, although after exposure ceases, significant amounts can be retained in skin, hair, nails, and muscle, with smaller amounts in teeth and bones.

Unlike other heavy metals, arsenic toxicity is more complex as it exists in multiple states, with biological processes capable of not only converting the inorganic forms, but also modifying inorganic arsenic into biological molecules. The most toxic form, As(III), present in anoxic groundwater, can also be produced by biological oxidation of absorbed As(V), with glutathione as the electron donor. Similar to cadmium and lead, depletion of glutathione by these reactions could limit cellular potential to eliminate reactive oxygen species, increasing oxidative stress. Further processing of inorganic arsenic is by methylation, primarily in the liver by the arsenic methyltransferase AS3MT, which uses single-carbon metabolism with S-adenosyl methionine (SAM) as the methyl donor, to methylate As(III) (Kojima et al. 2009). The final forms are monomethylarsenic(V) and dimethylarsenic(V) (MMA^{V} and DMA^{V}), with glutathione conjugates and methylated arsenic(III) forms (MMA^{III} and DMA^{III}) as intermediates(Sun et al. 2014). This further reduces glutathione, as well as SAM, which both require homocysteine generated through one-carbon metabolism for regeneration, putting multiple processing mechanisms in competition for resources.

5.4.3 *Tissue/Systems Targeted by Arsenic*

After ingestion of inorganic arsenic, levels peak in the kidney and liver after 1 h, with DMA^{V} being the primary form in liver after 4 h (Kenyon et al. 2005). MMA^{V} and DMA^{V} are both well absorbed through the intestines, and detectable in urine, along with both the tri- and pentavalent inorganic forms (Shraim et al. 2003). DMA^{V} is the most prevalent arsenic form observed in urine, but for reasons not understood, it is not detectable in bile (Kala et al. 2004; Suzuki et al. 2004). MMA^{III} and DMA^{III} are both highly reactive forms, and are responsible for arsenic toxicity, but only MMA^{III} is detectable in urine, likely due to rapid oxidation of DMA^{III} to DMA^{V}. Once modified in hepatic cells, it is hypothesized that arsenicals are conjugated with GSH, and transported by members of the families of ATP-binding cassette (ABC) transporters (Drobna et al. 2009). Depending on the transporter used, GSH-arsenical complexes can be released into bile or blood, with $MMAG^{III}$ (and possibly $DMAG^{III}$, which is difficult to detect due to instability) released into blood, where it complexes with hemoglobin as a thiol complex with cysteines (Shiobara et al. 2001).

Toxicity of arsenic is thought to operate through mechanisms similar to lead and cadmium, by oxidative stress, depletion of glutathione, protein disruption by binding to sulfhydryl groups, or replacement of cations such as zinc, being primary mechanisms of toxicity. As(III) binding to sulfhydryls is not as efficient as other heavy metals, with strong binding requiring two or three cysteine residues in proximity (Kitchin and Wallace 2005, 2006), such that arsenic affects a more specific set of proteins than other heavy metals. One of the most severe effects of trivalent arsenic, and a likely mechanism for acute toxicity damage, is as an inhibitor of pyruvate dehydrogenase, by a combination of thiol binding and reducing acetyl CoA production, and thus decreasing citric cycle activity and ATP production.

As(V) does not bind sulfhydryl groups, but can replace phosphate, interfering with ATP synthesis by reacting with ADP, consequently reducing ATP levels, as does AS(III), albeit by a different mechanism. As(V) can bind to zinc finger proteins, but is specific to motifs C3H1 or C4, not the C2H2 motifs (Zhou et al. 2011) common to transcription factors. While pentavalent arsenic alone cannot bind sulfhydryl groups, dimethyldithioarsinate ($DMDTA^V$), a product of complex conversions by gut microbes using hydrogen sulfide, has high reactivity with –SH, and toxicity comparable to MMA^{III} (Watanabe and Hirano 2012).

5.4.4 Physiological Effects of Arsenic Exposure

In acute arsenic poisoning, damage is primarily to the heart, resulting in cardiomyopathy and hypotension. Chronic arsenic poisoning affects all organs, with kidney being the most sensitive, and hypertension and cardiovascular disease often occurring. Unlike lead, arsenic does not strongly affect the nervous system, with peripheral neuropathy being the most common effect of long-term exposure. Gastrointestinal effects are toxic hepatitis and elevated liver enzymes. Arsenic is also a class I (confirmed) human carcinogen (Bustaffa et al. 2014), directly implicated in lung, skin, and bladder cancer. Arsenic is not a potent mutagen, but has genotoxic effects by induction of micronuclei, DNA strand breaks, sister chromatid exchanges, and aneuploidy. These may result from oxidative damage to DNA, combined with inhibition of DNA repair and DNA ligase, and interference with tubulin polymerization in the mitotic spindle. Given the low mutagenicity of arsenic, attention is currently being given to epigenetic effects of arsenic leading to carcinogenesis. The primary epigenetic effect of arsenic is to decrease the activity of DNA methyltransferase (DNMT), due to depletion of S-adenosylmethionine (SAM) pools from methylation of arsenic (Mass and Wang 1997; Zhao et al. 1997). This will be discussed further along with the epigenetic effects of other heavy metals.

Unlike lead and cadmium that require human activity to generate environmental exposures, arsenic has a significant environmental presence, such that mechanisms for tolerance and detoxification have evolved, resulting in the variety of methylated arsenic compounds produced after ingestion of inorganic arsenic. The arsenic methyltransferase *AS3MT* is an ancestral vertebrate gene, with homologs in xenopus and salmon (Kent et al. 2002). Some mammals are not able to methylate arsenic, including guinea pig, chimp, marmoset, tamarin, and squirrel monkey. It has been proposed that as trypanosomes are sensitive to As(III), animals that evolved in trypanosome endemic areas can prevent trypanosome infections by maintaining As(III) in the blood (Aposhian 1997; Zakharyan et al. 1996). In addition, transport of arsenic-glutathione complexes, both as inorganic and methylated arsenic, is also a possible mechanism of tolerance, with efflux from cells by the previously mentioned ABC transporters. These ABC transporters are also upregulated in response to inorganic As (Liu et al. 2001), with transporter inhibition or knockouts in cell culture and mouse models showing increased sensitivity and accumulation of arsenic, with multiple transporters involved (Liu et al. 2001, 2002; Rappa et al. 1997). These hypothesized mechanisms of tolerance should be investigated to determine their utility in preventing arsenic effects in humans.

5.5 Mercury

5.5.1 *Sources of Mercury Exposure*

Mercury is distinct from other common heavy metal toxins in that exposure can come from three different forms – elemental, inorganic, and organic – with different sources, transport within the organism, metabolism and toxicity (Syversen and Kaur 2012). These differences complicate determination of exposure risk and monitoring, requiring different detection assays. Similar to other heavy metals, much of mercury's biological activity is related to high reactivity with sulfhydryl (–SH) groups. However, in contrast to lead or cadmium, mercury does not replace other cations as enzymatic cofactors. Because mercury is part of the earth's crust, humans have been exposed via natural sources in air, water and food, prior to modern anthropogenic sources. The properties of mercury result in the three forms: elemental mercury (Hg^0), inorganic mercury as monovalent/mercurous and divalent/mercuric (Hg_2^{++} and Hg^{2+}, respectively), and organic mercury, primarily as alkyl compounds, such as methylmercury (MeHg).

5.5.2 *Routes of Exposure to Mercury*

Elemental mercury is distributed as vapor, resulting from evaporation from land and water. Volcanic activity is a primary natural source (Fitzgerald and Clarkson 1991), with burning of coal and municipal waste producing a major anthropogenic source of atmospheric mercury. Additionally, the use of mercury in gold extraction processes releases vapor, with estimates of Amazon basin gold extraction resulting in 130 tons of mercury released annually (Pfeiffer et al. 1989). Monoatomic mercury gas is chemically stable in the atmosphere, with residence time of approximately 1 year, as it is oxidized in the upper atmosphere to water soluble ionic mercury which is returned to the surface in rainwater (Clarkson 2002). This cycling process results in global distribution of mercury from point sources. Inorganic mercury in fresh and ocean water sediments is methylated by microorganisms, enters the food chain, resulting in the bioaccumulation of mercury in fish, and in fish-consuming animals, with sharks, marine mammals, and humans as apex consumers. The toxicology of each of the forms of mercury, and the current state of known long term effects of mercury exposure via epigenetic mechanisms will be described with a focus on MeHg, as it is a primary source of concern.

Inorganic mercury exposure is oral and topical, through compounds used medically or cosmetically: antiseptics, teething powders, and skin-lightening creams. Uptake is low, with 1–16 % of Hg^{2+} taken orally absorbed (Hattula and Rahola 1975), and ~8 % of $HgCl_2$ applied to the skin absorbed (Friberg et al. 1961). Accumulation is primarily in kidney, followed by liver, with uneven absorption in brain, depending on cell type (Moller-Madsen and Danscher 1986). Motor neurons accumulate more than sensory neurons, neurons accumulate more than glial cells, and while present in the cerebellum, mercury is not accumulated in Purkinje cells. Toxicity after acute poisoning is rapid, with corrosive effects on the epithelium, and kidney failure from necrosis of the tubular epithelium occurring within 24 h (Pollard and Hultman 1997).

Methylmercury is of interest for human health, due to its pervasiveness in the food chain, difficulty in measuring exposures, latency of effect after exposure has ceased, retention in the body, and unknown mechanisms of action. As MeHg contaminates water, food or air, primary routes of exposure include inhalation or dermal absorption with ~80 % retention (Berlin et al. 2007). If ingested, gastrointestinal tract absorption is nearly 100 %, with distribution throughout the body occurring within 30 h; equilibrium between blood and tissue is reached after approximately 4 days, with ~5 % in blood, and ~10 % in brain (Syversen and Kaur 2012). Measuring MeHg is blood is complicated by the fact that concentration in red blood cells is ~20 times that in plasma, which also affects bioavailability to the rest of the body. Hair provides a useful means of measuring exposure level, as MeHg accumulates proportionally to blood levels, but ~250 times higher, and exposure timing is possible, as longitudinal analysis of single hairs can show positions of MeHg deposition (Amin-Zaki et al. 1974).

5.5.3 Tissues/Systems Targeted by Mercury

The rapid and widespread distribution of MeHg is due to its reactivity with thiol groups, placing it in water soluble molecules (Syversen and Kaur 2012). MeHg binds the thiol of reduced glutathione, which is then pumped out of mammalian cells, broken down in the bile ducts and gall bladder to L-cysteine, then reabsorbed into the bloodstream (Ballatori and Clarkson 1985; Dutczak and Ballatori 1992, 1994). Cellular uptake of this MeHg-L-cysteine complex is efficient, as it closely mimics L-methionine, and is transported by the large neutral amino acid (or L-methionine) transporter, which also allows crossing the blood-brain barrier (Kerper et al. 1992; Yin et al. 2008). The cycling through the biliary and digestive systems is also the mechanism for very gradual elimination of MeHg, as ~1 % of this MeHg is reduced to Hg^0 in the intestines, which is excreted, rather than reabsorbed. This reduction is likely dependent on the gut biome, providing an explanation for variability in MeHg toxicity among individuals, due to differences in bacterial populations.

5.5.4 Physiological Effects of Mercury

The primary toxic effect of MeHg is on the central nervous system, with highest susceptibility in cerebellar granule cells and the visual cortex. The actual mechanism by which MeHg causes CNS damage is still unclear. It has been suggested that the production of Hg^{2+} in the brain could be a defensive response by glial cells to eliminate MeHg (Clarkson and Magos 2006), and that the MeHg radical is the true toxic form (Magos et al. 1985). This defensive hypothesis is consistent with organic mercury detoxification occurring by conversion to Hg^{2+}, with ethylmercury more readily converted than MeHg; this would explain the relative toxicities. Methylmercury is particularly hazardous to brain tissues of developing embryos, which results in a different syndrome than seen in exposure of developed nervous systems. The threshold for damage to prenatal CNS is significantly lower than for adults, and fetal brains can show widespread damage (Castoldi et al. 2003), instead of the focal lesions seen in adults. Milder effects from prenatal exposure have also been observed, in which delayed developmental milestones and neurological abnormalities have been reported (Marsh et al. 1987; Choi et al. 1978).

5.6 Epigenetic Effects of Toxic Metal Exposure

Beyond acute toxic effects of heavy metal exposure on individuals, chronic longer term effects are less studied. Of particular interest are changes in gene expression and the underlying chromatin structure affecting gene regulation that result from alterations generated during short-term exposure that then have persistent effects, or alternatively, accumulation of alterations from long term, low-level exposure, below the threshold for acute toxicity. Epigenetics has the potential to explain observed associations between heavy metal exposure and phenotypic outcomes such as cardiometabolic diseases, neurodevelopmental abnormalities and malignancies. Consequently, understanding the behavior of these epigenetic marks throughout the life course can be useful in improving our knowledge of the etiology underlying the associations between these heavy metals and chronic disease risk. Also, while disease risks are often assessed in cross-sectional studies, stable epigenetic marks could be cost-effective screeners to identify past exposure, while plastic ones can be used to monitor therapeutic and public health interventions. Here we provide an overview of known epigenetic effects associated with exposure to cadmium and arsenic as well as to lead and methyl mercury.

5.6.1 Zinc Finger Proteins and Cadmium or Arsenic Substitution

The low observed mutagenicity of most heavy metals in mammalian cells and the lack of mutagenicity of heavy metals in bacterial assays indicates that long-term health outcomes are due to indirect effects on gene regulation and expression. One key factor for direct, indirect, short-, and long-term effects may be the susceptibility of zinc-finger proteins to heavy metal substitution. As DNA-binding factors, zinc-finger proteins have roles in DNA repair, and as transcription factors, functions that could have widespread effects if disrupted. Zinc-finger proteins are dependent on zinc as a cofactor for proper folding (Chang et al. 2010; Li et al. 2008; Miura et al. 1998), with the observed replacement of zinc by other metals, including arsenic and cadmiums, proposed as a mechanism for toxic and carcinogenic effects of heavy metals (Beyersmann and Hartwig 2008; Hartwig 1998, 2001; Hartwig et al. 2002, 2010; Razmiafshari et al. 2001; Sarkar 1995).

Many studies have investigated substitution of other metals into zinc finger proteins, with particular attention to cadmium in relation to DNA binding, repair and maintenance activity, due to the similar chemical properties of cadmium and zinc, as cadmium inhibits nucleotide excision repair (Asmuss et al. 2000), and diminishes the DNA-protein interactions necessary for recognition of DNA damage (Hartwig 1998). *In vitro* studies with mouse xeroderma pigmentosum group A protein (XPA), a zinc-finger protein that binds damaged DNA, including lesions caused by benzo[a] pyrene (B[a]P) and UV radiation, showed that exposure to cadmium inhibited DNA-binding, and that co-exposure with zinc could prevent inhibition (Asmuss et al. 2000); however, re-exposure to zinc could not reverse the inhibition (Asmuss et al. 2000; Hartwig et al. 2003).

General inhibition of nucleotide excision repair (NER) by cadmium was demonstrated in cell culture, with removal of DNA adducts and UVC-induced lesions, and reduced nuclear protein levels of XPC, a probable initiator of global genome NER (Schwerdtle et al. 2010). These widespread effects may be due, in part, to effects of cadmium on the tumor suppressor protein p53 with observations of altered p53 conformation, reduced DNA binding, and lower transcription of a reporter gene (Méplan et al. 1999). As p53 activates transcription of NER genes *XPC* and *p48*, this could be the source of cadmium inhibition of DNA repair, and health effects at low but chronic, non-cytotoxic doses. The fact that cigarette smoke is a major source of both environmental cadmium and B[a]P implies that this one exposure can both induce DNA damage and inhibit repair of the damage.

Similarly, arsenite and its methylated metabolites have shown inhibited repair of B[a]P induced DNA damage. *In vitro* studies with human XPA have shown displacement of zinc by As(III), MMA(III), and DMA(III) (Schwerdtle et al. 2003). Tests with other heavy metals have found no inhibition of XPA DNA binding by lead or mercury, but other DNA repair proteins have variable inhibition. In human whole cell extracts, the apurinic/apyrimidinic endonuclease 1 (Ape1), which has magnesium-dependent repair function, is inhibited by lead and cadmium, but not by As(III) (McNeill et al. 2004). In cultured HeLa cells, activity of PARP-1 is inhibited by non-cytotoxic (10nM) amounts of As(III); PARP-1 stimulates repair of DNA breaks by inducing dissociation of nuclear proteins, including histones and topoisomerases (Hartwig et al. 2002). Another key role of DNA-binding zinc finger proteins is as transcription factors, especially as key elements of regulatory networks, such that disruption of these proteins would affect gene expression, with far-reaching and long-term consequences. This interaction of zinc-finger proteins with heavy metals is a path, that once started, can self-perpetuate in the absence of the initial exposure.

As with some of the mechanisms of toxicity, while the effects can be measured, the means by which heavy metals induce epigenetic change are mostly unknown.

Effects that have been observed include both widespread DNA hypo- and hypermethylation, changes in histone modifications, overall histone levels, and altered gene expression, possibly due to these altered regulatory epigenetic features (Smeester et al. 2011, 2014; Sanders et al. 2014; Vidal et al. 2015; Bailey and Fry 2014; Cheng et al. 2012; Gadhia et al. 2012; Koedrith et al. 2013). Potential means include the generation of reactive oxygen species which cause DNA damage, increasing cellular stress, and activating repair pathways; depletion of cellular S-adenosylmethionine (SAM), the methyl donor for DNA methyltransferases; and cascading effects from initial disruption of key epigenetic regulators.

Cadmium has also been shown to inhibit DNMT activity. Cell-culture model systems have shown that acute doses orders of magnitude lower than the IC50 for cell viability (1uM vs. ~200μM) decrease overall DNMT activity by 25 % or more (Takiguchi et al. 2003). Carcinogenic properties of cadmium are also hypothesized to be due to widespread hypomethylation, leading to increased proliferation of cultured cells stimulated by cadmium exposure (Jiang et al. 2008). While cadmium-induced hypomethylation may be a short-term response, chronic exposure can cause widespread hypermethylation, presumably by compensatory increases in methylation, seen by increased methyltransferase activity in chronically exposed cultured cells (Takiguchi et al. 2003; Jiang et al. 2008; Benbrahim-Tallaa et al. 2007). Increased expression and activity of DNMTs has been reported in cells chronically exposed, along with increased cellular proliferation (Takiguchi et al. 2003; Jiang et al. 2008). In addition, activation of oncogenes and promoters of proliferation, such as *c-myc*, *c-jun*, or *c-fos*, has been reported *in vivo* and *in vitro* after cadmium exposure (Joseph et al. 2001; Spruill et al. 2002). Cadmium-induced malignant transformation in cell culture has been observed, with associated increased DNMT activity due to increased expression of the *de novo* methyltransferase DNMT3b (Benbrahim-Tallaa et al. 2007). Thus, while mechanisms resulting in epigenetic response are still unknown, they are likely due to zinc substitution, loss of protein function, and generation of reactive oxygen species from reactions with the released Zn^{2+}.

Arsenic exhibits similar effects to cadmium, but with both hypo- and hypermethylation of different genes (Zhong and Mass 2001), including silencing of tumor suppressor genes by promoter hypermethylation, and altered histone modifications, including phosphorylation, methylation and acetylation (Li et al. 2003; Ramirez et al. 2008; Chu et al. 2011; Somji et al. 2011; Ren et al. 2011; Arita et al. 2012). In a human population with arsenic exposure through drinking water, global increases in the silencing modification H3K9me2 and the activating modification H3K9Ac have been reported (Arita et al. 2012). Altered global histone methylation has been observed in cell culture as well, also with opposing increased and decreased methylation of silencing marks H3K9me2 and H3K27me3, respectively, and increased global methylation of the activating mark H3K4me3, for both lower and higher arsenic levels (Zhou et al. 2008). In cell culture, the apparent contradiction of increases in both activating and silencing marks was resolved, at least for H3K9me2 (silencing), and H3K4Me3 (activating) by determination that different nuclear regions are specifically affected, with increased silencing marks primarily in heterochromatin, and increased activating marks primarily in euchromatin (Zhou et al. 2008). Additionally, H3K9 methylation is a marker for recruiting DNMTs (Liu et al. 2013), tying together these two processes correlated to loss of gene expression. Whether other observed opposing activation/silencing has gene or nuclear region specificity remains to be seen.

Widespread hypomethylation may be related to depletion of SAM, as methyl donors are required to process arsenic for excretion. This depletion has been observed in combination with DNA hypomethylation and malignant transformation in cultured cells exposed to arsenic, including upregulation of multiple oncogenes (Chen et al. 2001). However, the simultaneous increases and decreases in histone and DNA methylation observed in response to arsenic appear paradoxical. Investigation of a potential non-linear relationship between arsenic levels and DNA methylation in a population of healthy adults with varying chronic exposures from drinking water (Niedzwiecki et al. 2013) found a positive association in peripheral blood mononuclear cells, with the highest DNA methylation associated with the highest arsenic exposures. However, with the limited knowledge of the sequence-specific targets of DNA methylation, determining the loci whose methylation is altered in response to arsenic and/or cadmium has been challenging.

5.6.2 *Epigenetic Effects of Lead and Methylmercury*

Epigenetic effects of lead and mercury are less understood when compared to effects of cadmium and arsenic. Lead exposure studies in primates have shown epigenetic effects in adult and aging animals after lead exposure in development and infancy (Basha et al. 2005; Bihaqi et al. 2011). DNMT1 activity in the brain is reduced, and in gene expression analysis, 95 % of genes with altered expression are CpG-rich (CpGs are the target of DNMT methylation activity). In adult primates that had been exposed to lead during the first year of life, DNMT1 and DNMT3A expression was reduced, as were multiple histone modifications, including both methylation and acetylation. Among affected CpG-rich brain transcripts are amyloid precursor protein and beta-amyloid, evidence for a hypothesized link between infant lead exposure and adult-onset Alzheimer's disease (Basha et al. 2005; Wu et al. 2008).

MeHg decreases expression of DNA methyltransferases in rats exposed *in utero*, with decreased DNA methylation at specific promoters examined, particularly brain-derived neurotrophic factor (*BDNF*) (Desaulniers et al. 2009). The *BDNF* promoter showed increases in multiple histone repressive signatures, and increased DNA methylation, consistent with decreased *BDNF* expression and depression-like behavior in the exposed rats. A genome-wide methylation analysis in polar bear brains showed that increased mercury levels correlated to a decrease in global DNA methylation, specifically in males (Pilsner et al. 2010). A human study that examined methylation at candidate regions including repetitive sequences, *DNMT1*, and mercury protective genes (*SEPW1* and *SEPP1*), identified a correlation between increased hair mercury levels and *SEPP1* hypomethylation, also specific to males (Goodrich et al. 2013). As with cadmium, MeHg exposure also correlates with hypermethylation at other promoters, along with associated shifts of histone modifications to more repressive markings.

One particular difference between metals has been determined by studies with mouse embryonic stem cells exposed to multiple heavy metal toxins, and the effects on total protein and histone protein levels determined (Gadhia et al. 2012). While both arsenic and mercury decreased total protein levels and histone protein levels, cadmium exposure resulted in a ~200 % increase in histone proteins, suggesting a repair response to metal insult that requires histone proteins.

In addition to direct effects to the exposed individual, epigenetic effects can potentially affect offspring by altering epigenetic marks in the early embryo. Such effects have been observed due to endocrine disruptors such as fungicides, pesticides, herbicides and plastics (Skinner 2014). Thus, the disruption of epigenetic regulatory systems by heavy metals, combined with the indications that heavy metals, including cadmium, arsenic, mercury and lead have endocrine disruption activity (Iavicoli et al. 2009), suggests a pathway through which these metals can have similar effects, also potentially multigenerational. Further work into the endocrine effects of heavy metals, for which our current understanding is limited, will determine how metals act on and through this system, which may be a key element in long term health effects of these toxicants.

5.6.3 Regulatory Regions of Genomically Imprinted Genes and Heavy Metals

While epigenetic mechanisms have the potential to explain observed associations between heavy metals and chronic diseases, the regions of the epigenome most responsive to cadmium and/or lead exposure are still unknown. Consequently, epigenetic marks evaluated in humans and animals most often reside in regions of the genome with known function. Alternatively, agnostic approaches are employed with limited coverage of the epigenome using commercially-available platforms. Imprinted genes, which are expressed monoallelically due to parent of origin specific epigenetic regulation, are a group of special interest. These genes may form a particularly susceptible class for heavy metal sensitivity at different times during the life course. Imprint marks are reprogrammed during gametogenesis, resist reprogramming post-fertilization, and are faithfully maintained in somatic tissues throughout life. The appropriate imprinting and expression of these genes depend on allele-specific CpG methylation at differentially methylated regions (DMRs). The monoallelic expression is a means of controlling gene dose, which is critical for proper development, as the set of imprinted genes is enriched for growth promoting and regulating factors that have severe effects if over- or under-expressed.

In adults with inorganic arsenic exposure, the known imprinted gene *ANO1* (a calcium activated chloride channel), and the predicted imprinted gene *FOXF1* (a forkhead transcription factor with developmental functions) show increased promoter methylation in leukocytes (Smeester et al. 2011). In pregnant women who smoke, and thus have potential increased cadmium exposure, ten imprinted genes, including *GFI1* (a transcription repressor), *CYP1A1* (a cytochrome P450 family member), and *AHRR* (aryl-hydrocarbon receptor repressor) are hypomethylated in newborn umbilical cord blood (Joubert et al. 2012). In the Newborn Epigenetics STudy, imprinted *IGF2* (insulin-like growth factor 2) is hypermethylated in male newborns of smokers (Murphy et al. 2012) and prenatal cadmium exposure is associated with lower methylation of multiple imprinted genes including *PLAGL1* in newborns (Vidal et al. 2015).

The Cincinnati Lead Study (Sivaprasad et al. 2004) has been a valuable resource for studying the effects of lead exposure in early life on later life outcomes (Wright et al. 2008; Brubaker et al. 2010; Yuan et al. 2006; Dietrich et al. 1993). Peripheral whole blood was obtained from participants and lead levels were measured at age 10 days, then every 3 months for the first 5 years of life, and then every 6 months until age ~7 years (Wang et al. 2015). Genomic DNA isolated from blood collected from these participants when aged 28–32 years was examined at 22 imprinted regulatory regions and showed lower sex-specific methylation of *PEG3 and IGF2/H19* DMRs, and higher methylation of the *PLAGL1/ HYMA1* DMR that was not sex-specific. Similar DNA methylation patterns in regulatory regions of these genes have been associated with increased gene expression, loss of imprinting (Nye et al. 2013; Cui et al. 2003) and lower birth weight (Hoyo et al. 2012). These data together suggest that DNA methylation changes at regulatory DMRs are associated with changes in expression, but the precise molecular interactions that lead to these changes are unknown. The PEG3 protein is a CD8 T cell antigen, allowing mononuclear cells to infiltrate the insulin producing beta cells, resulting in beta cell destruction and insulin dysregulation (Mukherjee et al. 2014). In murine models, disrupted *Peg3* is associated with low birth weight (Chiavegatto et al. 2012). *Plagl1* knockout disrupts a network of coordinately regulated genes containing a large number that are imprinted (Varrault et al. 2006). Multiple imprinted genes have altered expression when *PLAGL1* is overexpressed (Iglesias-Platas et al. 2014), supporting that this regulatory DMR has the potential to alter network-wide imprinted gene expression. *PLAGL1* impairs glucose-stimulated insulin translation and secretion (Du et al. 2012), effects that may be mechanisms by which environmental exposures contribute to obesity and insulin resistance, outcomes also tied to exposure to endocrine disrupting chemicals. Therefore, the apparent endocrine disrupting activity of heavy metals is a possible mechanism for their potential effects on cardiometabolic risk.

In considering genes with altered mRNA or protein expression, or DNA methylation, that is correlated with exposure to arsenic, cadmium, lead, and mercury, two particular pathways were identified (Smeester et al. 2014). Multiple components of each pathway show effects correlated to metal exposure, with some affected genes belonging to both pathways. The first pathway is the aryl-hydrocarbon receptor (AhR) signaling pathway, which has components that respond to xenobiotics and that regulate cellular proliferation and differentiation, as well as having members that are known oncogenes. The second is the TP53 signaling pathway, a master regulatory system that responds to cellular stresses and DNA damage, with control over DNA repair, cell cycle arrest and apoptosis. Taken together, these data suggest that regulatory sequences of imprinted genes may be one of the most important groups of genomic loci to evaluate when examining a persistent epigenetic response to heavy metals.

5.6.4 *Possible Nutritional Mitigation of Heavy Metal Exposure and Its Effects*

While use of chelators of lead and arsenic is possible for treatment of acute doses, removal of mercury in its multiple forms is difficult. The effectiveness of cadmium chelation is unclear (Smith 2013). Some even contend that the re-mobilization of sequestered metals may actually increase health risks. Even when chelation may be effective, the windows for treatment are narrow. For chronic low-dose exposures, there are no active mechanisms for effective elimination, and the health benefits of chelation are uncertain (Smith 2013; Kosnett 2010, 2013). Thus, identifying ways to mitigate exposure are urgently needed.

5.6.4.1 Essential Metals

Single nutrients such as zinc, magnesium, selenium, copper and iron can competitively displace heavy metals via trans-metallation reactions. These essential metals are key in many metabolic processes (Katzen-Luchenta 2007), participate in gene expression (Burdge and Lillycrop 2010), and are critical in development (Katzen-Luchenta 2007). Human data suggest that, within very narrow concentration ranges, blood lead and cadmium levels are highest in individuals with low iron and calcium levels, presumably due to reduction in intestinal absorption of cadmium in the presence of the essential metals (Gallagher et al. 2011). Weak inverse relationships between zinc, copper and magnesium with cadmium have also been reported in some populations (Djukić-Cosić et al. 2006). While these observations have not been confirmed by others (Cheng et al. 2013), deficiency of copper (Wildman and Mao 2001), iron (Bourque et al. 2008; Komolova et al. 2008; Menzie et al. 2008; Yanoff et al. 2007), manganese (Ganeshan et al. 2011), magnesium (Guerrero-Romero and Rodríguez-Morán 2002, 2006, 2011) and zinc (Sanchez et al. 2015; Azab et al. 2014) have been associated with obesity risk and higher insulin levels. It is still unknown whether these elevated insulin levels are due to insulin sensitivity, insulin secretion or insulin clearance.

5.6.4.2 One Carbon Cycle Nutrients

Elevated levels of one carbon cycle nutrients, including folate, choline and vitamins B12 and B6, which are critical in generating S-adenosyl-methionine and key in the biotransformation and clearance of arsenic, have been associated with lower concentrations of inorganic arsenic in mice (Tsang et al. 2012) and humans (Sanders et al. 2014; Lambrou et al. 2012; Gruber et al. 2012; Heck et al. 2007). Depletion of one carbon metabolism co-enzymes reduces arsenic excretion and methylation (Rossman and Klein 2011; Vahter and Marafante 1987), whereas their repletion lowers blood arsenic levels (Heck et al. 2007). In 6-year old arsenic-exposed children, supplementation with one-carbon cycle nutrients was associated with re-methylation of interspersed repeat elements and lower arsenic levels (Hall et al. 2009). Inverse but weak associations between individual one carbon cycle nutrients and lead or cadmium level have also been shown (Suarez-Ortegón et al. 2013). Together, these data support a cautious repletion of some nutrients to mitigate toxic metal levels and their effects, within very narrow ranges. Understandably, such data have not been used to recommend nutrient-based interventions. Epigenetic responses with high specificity could be useful in determining nutrient ranges that may restore normal gene function. Some unpublished data suggest that inverse associations between nutrients and toxic metals are robust and persistent only when some but not all nutrient mixtures are examined. These findings are consistent with nutrients being beneficial when delivered as combinations in a diet, rather than single supplements.

5.6.5 Effects of Metal Mixtures

Similar to nutrients being potent in specific combinations, increased attention is now being paid to epigenetic effects of heavy metals in combinations. This is in part due to the recognition of the significance of low-dose exposures on development and lifetime disease risk. Such low exposures coming from environmental sources, such as household dust or well water, tend to contain multiple toxic metals, with the common combination of lead, arsenic and cadmium. As these toxic metals have different target tissues and proteins, metal doses that would individually be within current limits may have additive or synergistic effects.

A few studies have examined the toxicology of metal mixtures, and some identifications of biomarkers of exposure to mixtures have been made (reviewed in Wang and Fowler 2008). Several lines of evidence suggest that enhanced nephrotoxicity due to combined cadmium and arsenic and co-exposure to lead and arsenic increased metallothionein expression in kidney over that observed with individual metal exposures. Exposure to lead, cadmium, and arsenic have more complex effects, both *in vivo* and *in vitro*. In cell culture, increasing levels from very low to high produces results from hormesis to increased damage and then antagonistic effects. The antagonistic effects may be the result of competition for interactions at high doses, with similar observations in coexposures of lead and cadmium, in which cadmium apparently blocks lead absorption (Wang and Fowler 2008).

Two studies with particular relevance evaluated developmental and neurological effects of lead/arsenic/cadmium mixtures at levels comparable to those found in drinking water (Paul and Giri 2015; Rooney 2013). Developmental exposure to this metal mixture causes demyelination in white matter of rat brains and optic nerves, with a potential mechanism being reduction in glutamine synthase (Rooney 2013). The mixture had a synergistic effect compared to the individual metals, and the myelin loss weakened the blood brain barrier, allowing increased metal accumulation in the brain. Similarly, lead, arsenic, and cadmium effects on the brain identified synergistic effects on increased beta-amyloid and amyloid precursor protein levels, markers for Alzheimer's Disease (AD) (Paul and Giri 2015).

Imprinted gene regulation is also an area with great potential for examining the epigenetic effects of heavy metal mixtures, with implications for early development and lifelong health outcomes. In the NEST cohort, peripheral blood leukocytes of newborns showed DNA hypermethylation of the regulatory sequences at the imprinted *DLK1/MEG3* domain in newborns of women with elevated prenatal lead (Nye et al. 2015) and socioeconomic disadvantage (King et al. 2015). However, because co-exposure with cadmium was also common in this sub-cohort, it remains to be seen whether the synergistic effects of cadmium and lead together with socioeconomic conditions altered the methylation profile of this regulatory region. Larger studies are required to clarify these relationships if the DNA methylation patterns in these regulatory regions are to be developed as biomarkers of early exposure. Nonetheless, these results are all intriguing, calling for further studies of mixtures of heavy metals, particularly in looking at additive, synergistic, or antagonistic epigenetic effects. The observed interactive effects involving stress response elements, including glutathione and oxidative stress, are highly indicative that metal mixtures at environmentally relevant levels will have significant developmental and epigenetic effects.

5.7 Conclusion and Future Research

Whereas toxic heavy metals have significant effects on health and these effects operate at least in part through epigenetic mechanisms, a number of unresolved questions remain regarding the mechanisms of action, differences in response depending on dose, timing during the life course and specificity of regulatory region-specific epigenetic modifications reported thus far. Answers to some of these issues, particularly the question of opposing epigenetic activating and silencing marks in response to metal exposure, can be determined by full genome methyl-sequencing to identify specific regions of hypo- and hypermethylation. Likewise, chromatin immmunoprecipitation and sequencing (ChIP-seq), examining relevant activating and suppressing histone marks have the potential to identify region-specific alterations in response to heavy metal exposure. However, resources for such comprehensive strategies have thus far been limited.

Possibly the most important work to be done to understand the effects of heavy metals is the examination of heavy metal mixtures that occur at low doses. While some exposures, such as arsenic in drinking water, mercury in fish, or cadmium and lead in soil and house dust, are primarily single exposures, many populations are exposed to multiple heavy metals, particularly when the source comes from undegraded past industrial activity or waste. While most current studies have focused on responses to single exposures, it has been difficult to account for potential synergistic and antagonistic effects. Finally, existing work has found intriguing interactions regarding mitigating effects of select nutrients in populations exposed to these heavy metals, interactions that could be followed up with sequence-specific analysis, and particularly exposure to multiple metals. Understanding gene-specific effects of metals, singly and in combination, will establish how these exposures have short and long term health effects, including carcinogenesis, life-long disease risk, and multi-generational effects.

References

Agarwal S et al (2011) Heavy metals and cardiovascular disease: results from the National Health and Nutrition Examination Survey (NHANES) 1999–2006. Angiology 62(5):422–429

Agency for Toxic Substances and Disease Registry (ATSDR) (2015) http://www.atsdr.cdc.gov/

Ahamed M, Siddiqui MK (2007) Low level lead exposure and oxidative stress: current opinions. Clin Chim Acta 383(1–2):57–64

Ahamed M et al (2005) Environmental exposure to lead and its correlation with biochemical indices in children. Sci Total Environ 346(1–3):48–55

Akesson A et al (2014) Non-renal effects and the risk assessment of environmental cadmium exposure. Environ Health Perspect 122(5):431–438

Amin-Zaki L et al (1974) Intra-uterine methylmercury poisoning in Iraq. Pediatrics 54(5):587–595

Anis TH et al (2007) Chronic lead exposure may be associated with erectile dysfunction. J Sex Med 4(5):1428–1434; discussion 1434–6

Antonio MT, Corpas I, Leret ML (1999) Neurochemical changes in newborn rat's brain after gestational cadmium and lead exposure. Toxicol Lett 104(1–2):1–9

Aposhian HV (1997) Enzymatic methylation of arsenic species and other new approaches to arsenic toxicity. Annu Rev Pharmacol Toxicol 37:397–419

Arita A et al (2012) The effect of exposure to carcinogenic metals on histone tail modifications and gene expression in human subjects. J Trace Elem Med Biol 26(2–3):174–178

Asmuss M et al (2000) Differential effects of toxic metal compounds on the activities of Fpg and XPA, two zinc finger proteins involved in DNA repair. Carcinogenesis 21(11):2097–2104

Azab SF et al (2014) Serum trace elements in obese Egyptian children: a case-control study. Ital J Pediatr 40:20

Bailey KA, Fry RC (2014) Arsenic-associated changes to the epigenome: what are the functional consequences? Curr Environ Health Rep 1:22–34

Ballatori N, Clarkson TW (1985) Biliary secretion of glutathione and of glutathione-metal complexes. Fundam Appl Toxicol 5(5):816–831

Bandara JM et al (2010) Chronic renal failure in Sri Lanka caused by elevated dietary cadmium: Trojan horse of the green revolution. Toxicol Lett 198(1):33–39
Basha MR et al (2005) The fetal basis of amyloidogenesis: exposure to lead and latent overexpression of amyloid precursor protein and beta-amyloid in the aging brain. J Neurosci 25(4):823–829
Batuman V, Wedeen RP (2014) The persistence of chronic lead nephropathy. Am J Kidney Dis 64(1):1–3
Bellinger DC (2004) Lead. Pediatrics 113(4 Suppl):1016–1022
Benbrahim-Tallaa L et al (2007) Tumor suppressor gene inactivation during cadmium-induced malignant transformation of human prostate cells correlates with overexpression of de novo DNA methyltransferase. Environ Health Perspect 115(10):1454–1459
Benbrahim-Tallaa L et al (2009) Cadmium malignantly transforms normal human breast epithelial cells into a basal-like phenotype. Environ Health Perspect 117(12):1847–1852
Berlin M, Zalups RK, Fowler BA (2007) Mercury. In: Nordberg GF, Fowler BA, Nordberg M, Friberg LT (eds) Handbook on the toxicology of metals. Elsevier, New York
Beyersmann D, Hartwig A (2008) Carcinogenic metal compounds: recent insight into molecular and cellular mechanisms. Arch Toxicol 82(8):493–512
Bihaqi SW et al (2011) Infant exposure to lead (Pb) and epigenetic modifications in the aging primate brain: implications for Alzheimer's disease. J Alzheimers Dis 27(4):819–833
Bonomi F et al (1994) Reversible and non-denaturing replacement of iron by cadmium in Clostridium pasteurianum ferredoxin. Eur J Biochem 222(2):639–644
Borges VC et al (2007) Heavy metals modulate glutamatergic system in human platelets. Neurochem Res 32(6):953–958
Borisova T et al (2011) Presynaptic malfunction: the neurotoxic effects of cadmium and lead on the proton gradient of synaptic vesicles and glutamate transport. Neurochem Int 59(2):272–279
Bourque SL et al (2008) Long-term circulatory consequences of perinatal iron deficiency in male Wistar rats. Hypertension 51(1):154–159
Brent J (2006) Review of: "medical toxicology". Clin Toxicol 44:355
Brubaker CJ et al (2010) The influence of age of lead exposure on adult gray matter volume. Neurotoxicology 31(3):259–266
Burdge GC, Lillycrop KA (2010) Nutrition, epigenetics, and developmental plasticity: implications for understanding human disease. Annu Rev Nutr 30:315–339
Bustaffa E et al (2014) Genotoxic and epigenetic mechanisms in arsenic carcinogenicity. Arch Toxicol 88(5):1043–1067
Capel ID et al (1981) Comparison of concentrations of some trace, bulk, and toxic metals in the hair of normal and dyslexic children. Clin Chem 27(6):879–881
Casalino E, Sblano C, Landriscina C (1997) Enzyme activity alteration by cadmium administration to rats: the possibility of iron involvement in lipid peroxidation. Arch Biochem Biophys 346(2):171–179
Castoldi AF, Coccini T, Manzo L (2003) Neurotoxic and molecular effects of methylmercury in humans. Rev Environ Health 18(1):19–31
Chang S et al (2010) Stability and folding behavior analysis of zinc-finger using simple models. Int J Mol Sci 11(10):4014–4034
Chen H et al (2001) Genetic events associated with arsenic-induced malignant transformation: applications of cDNA microarray technology. Mol Carcinog 30(2):79–87
Chen L et al (2006) Plasma metallothionein antibody, urinary cadmium, and renal dysfunction in a Chinese type 2 diabetic population. Diabetes Care 29(12):2682–2687
Chen X et al (2011) Effects of cadmium on bone microstructure and serum tartrate-resistant acid phosphatase 5b in male rats. Exp Biol Med (Maywood) 236(11):1298–1305
Cheng TF, Choudhuri S, Muldoon-Jacobs K (2012) Epigenetic targets of some toxicologically relevant metals: a review of the literature. J Appl Toxicol 32(9):643–653
Cheng Z et al (2013) Trophic relationships and health risk assessments of trace metals in the aquaculture pond ecosystem of Pearl River Delta, China. Chemosphere 90(7):2142–2148

Chiavegatto S et al (2012) Hypothalamic expression of Peg3 gene is associated with maternal care differences between SM/J and LG/J mouse strains. Brain Behav 2(4):365–376

Cho GJ et al (2012) The relationship between blood mercury level and osteoporosis in postmenopausal women. Menopause 19(5):576–581

Cho YA et al (2013) Dietary cadmium intake and the risk of cancer: a meta-analysis. PLoS One 8(9):e75087

Choi BH et al (1978) Abnormal neuronal migration, deranged cerebral cortical organization, and diffuse white matter astrocytosis of human fetal brain: a major effect of methylmercury poisoning in utero. J Neuropathol Exp Neurol 37(6):719–733

Chow ES et al (2008) Cadmium inhibits neurogenesis in zebrafish embryonic brain development. Aquat Toxicol 87(3):157–169

Christensen BC, Marsit CJ (2011) Epigenomics in environmental health. Front Genet 2:84

Christensen BC et al (2009) Differentiation of lung adenocarcinoma, pleural mesothelioma, and nonmalignant pulmonary tissues using DNA methylation profiles. Cancer Res 69(15):6315–6321

Chu F et al (2011) Quantitative mass spectrometry reveals the epigenome as a target of arsenic. Chem Biol Interact 192(1–2):113–117

Clarkson TW (2002) The three modern faces of mercury. Environ Health Perspect 110(Suppl 1):11–23

Clarkson TW, Magos L (2006) The toxicology of mercury and its chemical compounds. Crit Rev Toxicol 36(8):609–662

Cleveland LM et al (2008) Lead hazards for pregnant women and children: part 1: immigrants and the poor shoulder most of the burden of lead exposure in this country. Part 1 of a two-part article details how exposure happens, whom it affects, and the harm it can do. Am J Nurs 108(10):40–49; quiz 50

Cornelis R (2005) Handbook of elemental speciation II: species in the environment, food, medicine & occupational health. Wiley, Hoboken. ISBN: 978-0-470-01465-3

Cory-Slechta DA (1996) Legacy of lead exposure: consequences for the central nervous system. Otolaryngol Head Neck Surg 114(2):224–226

Cui H et al (2003) Loss of IGF2 imprinting: a potential marker of colorectal cancer risk. Science 299(5613):1753–1755

Desaulniers D et al (2009) Effects of mixtures of polychlorinated biphenyls, methylmercury, and organochlorine pesticides on hepatic DNA methylation in prepubertal female Sprague-Dawley rats. Int J Toxicol 28(4):294–307

Desi I, Nagymajtenyi L, Schulz H (1998) Behavioural and neurotoxicological changes caused by cadmium treatment of rats during development. J Appl Toxicol 18(1):63–70

Dietrich KN et al (1993) The developmental consequences of low to moderate prenatal and postnatal lead exposure: intellectual attainment in the Cincinnati Lead Study Cohort following school entry. Neurotoxicol Teratol 15(1):37–44

Djukić-Cosić D et al (2006) Effect of supplemental magnesium on the kidney levels of cadmium, zinc, and copper of mice exposed to toxic levels of cadmium. Biol Trace Elem Res 114(1–3):281–291

Drobna Z et al (2009) Metabolism of arsenic in human liver: the role of membrane transporters. Arch Toxicol 84(1):3–16

Du X et al (2012) Overexpression of ZAC impairs glucose-stimulated insulin translation and secretion in clonal pancreatic beta-cells. Diabetes Metab Res Rev 28(8):645–653

Dutczak WJ, Ballatori N (1992) gamma-Glutamyltransferase-dependent biliary-hepatic recycling of methyl mercury in the guinea pig. J Pharmacol Exp Ther 262(2):619–623

Dutczak WJ, Ballatori N (1994) Transport of the glutathione-methylmercury complex across liver canalicular membranes on reduced glutathione carriers. J Biol Chem 269(13):9746–9751

Esquifino AI et al (1999) Effects of chronic alternating cadmium exposure on the episodic secretion of prolactin in male rats. J Trace Elem Med Biol 12(4):205–210

Feki-Tounsi M, Hamza-Chaffai A (2014) Cadmium as a possible cause of bladder cancer: a review of accumulated evidence. Environ Sci Pollut Res Int 21(18):10561–10573

Ferraro PM et al (2012) Temporal trend of cadmium exposure in the United States population suggests gender specificities. Intern Med J 42(6):691–697

Fitzgerald WF, Clarkson TW (1991) Mercury and monomethylmercury: present and future concerns. Environ Health Perspect 96:159–166

Friberg L, Skog E, Wahlberg JE (1961) Resorption of mercuric chloride and methyl mercury dicyandiamide in guinea-pigs through normal skin and through skin pretreated with acetone, alkylaryl-sulphonate and soap. Acta Derm Venereol 41:40–52

Gadhia SR, Calabro AR, Barile FA (2012) Trace metals alter DNA repair and histone modification pathways concurrently in mouse embryonic stem cells. Toxicol Lett 212(2):169–179

Galazyn-Sidorczuk M, Brzóska MM, Moniuszko-Jakoniuk J (2008) Estimation of Polish cigarettes contamination with cadmium and lead, and exposure to these metals via smoking. Environ Monit Assess 137(1–3):481–493

Gallagher CM, Meliker JR (2010) Blood and urine cadmium, blood pressure, and hypertension: a systematic review and meta-analysis. Environ Health Perspect 118(12):1676–1684

Gallagher CM, Chen JJ, Kovach JS (2010) Environmental cadmium and breast cancer risk. Aging (Albany NY) 2(11):804–814

Gallagher CM, Chen JJ, Kovach JS (2011) The relationship between body iron stores and blood and urine cadmium concentrations in US never-smoking, non-pregnant women aged 20–49 years. Environ Res 111(5):702–707

Ganeshan M et al (2011) Maternal manganese restriction increases susceptibility to high-fat diet-induced dyslipidemia and altered adipose function in WNIN male rat offspring. Exp Diabetes Res 2011:486316

Gill KD, Pal R, Nath R (1989) Effect of cadmium on lipid peroxidation and antioxidant enzymes in undernourished weanling rat brain. Pharmacol Toxicol 65(1):73–77

Godt J et al (2006) The toxicity of cadmium and resulting hazards for human health. J Occup Med Toxicol 1:22

Goodrich JM et al (2013) Mercury biomarkers and DNA methylation among Michigan dental professionals. Environ Mol Mutagen 54(3):195–203

Gruber JF et al (2012) Associations between toenail arsenic concentration and dietary factors in a New Hampshire population. Nutr J 11:45

Guerrero-Romero F, Rodríguez-Morán M (2002) Low serum magnesium levels and metabolic syndrome. Acta Diabetol 39(4):209–213

Guerrero-Romero F, Rodríguez-Morán M (2006) Hypomagnesemia, oxidative stress, inflammation, and metabolic syndrome. Diabetes Metab Res Rev 22(6):471–476

Guerrero-Romero F, Rodríguez-Morán M (2011) Magnesium improves the beta-cell function to compensate variation of insulin sensitivity: double-blind, randomized clinical trial. Eur J Clin Invest 41(4):405–410

Guidotti TL, McNamara J, Moses MS (2008) The interpretation of trace element analysis in body fluids. Indian J Med Res 128(4):524–532

Gupta A, Shukla GS (1996) Ontogenic profile of brain lipids following perinatal exposure to cadmium. J Appl Toxicol 16(3):227–233

Hall MN et al (2009) Folate, cobalamin, cysteine, homocysteine, and arsenic metabolism among children in Bangladesh. Environ Health Perspect 117(5):825–831

Hartwig A (1998) Carcinogenicity of metal compounds: possible role of DNA repair inhibition. Toxicol Lett 102–103:235–239

Hartwig A (2001) Zinc finger proteins as potential targets for toxic metal ions: differential effects on structure and function. Antioxid Redox Signal 3(4):625–634

Hartwig A et al (2002) Interference by toxic metal ions with zinc-dependent proteins involved in maintaining genomic stability. Food Chem Toxicol 40(8):1179–1184

Hartwig A et al (2003) Very low concentrations of arsenite suppress poly(ADP-ribosyl)ation in mammalian cells. Int J Cancer 104(1):1–6

Hartwig A, Schwerdtle T, Bal W (2010) Biophysical analysis of the interaction of toxic metal ions and oxidants with the zinc finger domain of XPA. Methods Mol Biol 649:399–410
Hattula T, Rahola T (1975) The distribution and biological half-time of 203Hg in the human body according to a modified whole-body counting technique. Environ Physiol Biochem 5(4):252–257
Heck JE et al (2007) Consumption of folate-related nutrients and metabolism of arsenic in Bangladesh. Am J Clin Nutr 85(5):1367–1374
Heijmans BT et al (2009) The epigenome: archive of the prenatal environment. Epigenetics 4(8):526–531
Hendrick DJ (1996) Occupational and chronic obstructive pulmonary disease (COPD). Thorax 51(9):947–955
Herba E et al (2001) The effect of serotonin on flash visual evoked potential in the rat prenatally exposed to cadmium. Klin Oczna 103(2–3):81–84
Ho SM et al (2012) Environmental epigenetics and its implication on disease risk and health outcomes. ILAR J 53(3–4):289–305
Hoyo C, Murphy SK, Jirtle RL (2009) Imprint regulatory elements as epigenetic biosensors of exposure in epidemiological studies. J Epidemiol Community Health 63(9):683–684
Hoyo C et al (2012) Association of cord blood methylation fractions at imprinted insulin-like growth factor 2 (IGF2), plasma IGF2, and birth weight. Cancer Causes Control 23(4):635–645
Huang M et al (2013) Evaluation of factors associated with cadmium exposure and kidney function in the general population. Environ Toxicol 28(10):563–570
Iavicoli I et al (2004) Effects of low doses of dietary lead on puberty onset in female mice. Reprod Toxicol 19(1):35–41
Iavicoli I et al (2006) Low doses of dietary lead are associated with a profound reduction in the time to the onset of puberty in female mice. Reprod Toxicol 22(4):586–590
Iavicoli I, Fontana L, Bergamaschi A (2009) The effects of metals as endocrine disruptors. J Toxicol Environ Health B Crit Rev 12(3):206–223
Iglesias-Platas I et al (2014) Altered expression of the imprinted transcription factor PLAGL1 deregulates a network of genes in the human IUGR placenta. Hum Mol Genet 23(23):6275–6285
Jiang G et al (2008) Effects of long-term low-dose cadmium exposure on genomic DNA methylation in human embryo lung fibroblast cells. Toxicology 244(1):49–55
Johri N, Jacquillet G, Unwin R (2010) Heavy metal poisoning: the effects of cadmium on the kidney. Biometals 23(5):783–792
Joseph P et al (2001) Cadmium-induced cell transformation and tumorigenesis are associated with transcriptional activation of c-fos, c-jun, and c-myc proto-oncogenes: role of cellular calcium and reactive oxygen species. Toxicol Sci 61(2):295–303
Joubert BR et al (2012) 450 K epigenome-wide scan identifies differential DNA methylation in newborns related to maternal smoking during pregnancy. Environ Health Perspect 120(10):1425–1431
Kala SV et al (2004) Formation and urinary excretion of arsenic triglutathione and methylarsenic diglutathione. Chem Res Toxicol 17(2):243–249
Kalia K, Flora SJ (2005) Strategies for safe and effective therapeutic measures for chronic arsenic and lead poisoning. J Occup Health 47(1):1–21
Karim MR et al (2013) Increases in oxidized low-density lipoprotein and other inflammatory and adhesion molecules with a concomitant decrease in high-density lipoprotein in the individuals exposed to arsenic in Bangladesh. Toxicol Sci 135(1):17–25
Katzen-Luchenta J (2007) The declaration of nutrition, health, and intelligence for the child-to-be. Nutr Health 19(1–2):85–102
Kazantzis G (2004) Cadmium, osteoporosis and calcium metabolism. Biometals 17(5):493–498
Kehrer JP (2000) The Haber-Weiss reaction and mechanisms of toxicity. Toxicology 149(1):43–50

Kent WJ et al (2002) The human genome browser at UCSC. Genome Res 12(6):996–1006
Kenyon EM, Del Razo LM, Hughes MF (2005) Tissue distribution and urinary excretion of inorganic arsenic and its methylated metabolites in mice following acute oral administration of arsenate. Toxicol Sci 85(1):468–475
Kerper LE, Ballatori N, Clarkson TW (1992) Methylmercury transport across the blood-brain barrier by an amino acid carrier. Am J Physiol 262(5 Pt 2):R761–R765
King K, Murphy S, Hoyo C (2015) Epigenetic regulation of Newborns' imprinted genes related to gestational growth: patterning by parental race/ethnicity and maternal socioeconomic status. J Epidemiol Community Health 69(7):639–647
Kitchin KT, Wallace K (2005) Arsenite binding to synthetic peptides based on the Zn finger region and the estrogen binding region of the human estrogen receptor-alpha. Toxicol Appl Pharmacol 206(1):66–72
Kitchin KT, Wallace K (2006) Arsenite binding to synthetic peptides: the effect of increasing length between two cysteines. J Biochem Mol Toxicol 20(1):35–38
Koedrith P et al (2013) Toxicogenomic approaches for understanding molecular mechanisms of heavy metal mutagenicity and carcinogenicity. Int J Hyg Environ Health 216(5):587–598
Kojima C et al (2009) Requirement of arsenic biomethylation for oxidative DNA damage. J Natl Cancer Inst 101(24):1670–1681
Komolova M et al (2008) Sedentariness and increased visceral adiposity in adult perinatally iron-deficient rats. Int J Obes (Lond) 32(9):1441–1444
Korpela H et al (1986) Lead and cadmium concentrations in maternal and umbilical cord blood, amniotic fluid, placenta, and amniotic membranes. Am J Obstet Gynecol 155(5):1086–1089
Kosnett MJ (2010) Chelation for heavy metals (arsenic, lead, and mercury): protective or perilous? Clin Pharmacol Ther 88(3):412–415
Kosnett MJ (2013) The role of chelation in the treatment of arsenic and mercury poisoning. J Med Toxicol 9(4):347–354
Kuo CC et al (2013) Environmental chemicals and type 2 diabetes: an updated systematic review of the epidemiologic evidence. Curr Diab Rep 13(6):831–849
Lafuente A, Esquifino AI (1999) Cadmium effects on hypothalamic activity and pituitary hormone secretion in the male. Toxicol Lett 110(3):209–218
Lafuente A et al (1999) Cadmium affects the episodic luteinizing hormone secretion in male rats: possible age-dependent effects. Toxicol Lett 104(1–2):27–33
Lambrou A et al (2012) Arsenic exposure and DNA methylation among elderly men. Epidemiology 23(5):668–676
Lampe BJ et al (2008) Association between 24-hour urinary cadmium and pulmonary function among community-exposed men: the VA Normative Aging Study. Environ Health Perspect 116(9):1226–1230
Li J et al (2003) Tumor promoter arsenite stimulates histone H3 phosphoacetylation of proto-oncogenes c-fos and c-jun chromatin in human diploid fibroblasts. J Biol Chem 278(15):13183–13191
Li W et al (2008) Metal-coupled folding of Cys2His2 zinc-finger. J Am Chem Soc 130(3):892–900
Li Y et al (2015) Lead exposure during early human development and DNA methylation of imprinted gene regulatory elements in adulthood. Environ Health Perspect
Liang Y et al (2012) Renal function after reduction in cadmium exposure: an 8-year follow-up of residents in cadmium-polluted areas. Environ Health Perspect 120(2):223–228
Lidsky TI, Schneider JS (2003) Lead neurotoxicity in children: basic mechanisms and clinical correlates. Brain 126(Pt 1):5–19
Liu J et al (2001) Overexpression of glutathione S-transferase II and multidrug resistance transport proteins is associated with acquired tolerance to inorganic arsenic. Mol Pharmacol 60(2):302–309
Liu J et al (2002) Multidrug-resistance mdr1a/1b double knockout mice are more sensitive than wild type mice to acute arsenic toxicity, with higher arsenic accumulation in tissues. Toxicology 170(1–2):55–62

Liu X et al (2013) UHRF1 targets DNMT1 for DNA methylation through cooperative binding of hemi-methylated DNA and methylated H3K9. Nat Commun 4:1563

Magos L et al (1985) The comparative toxicology of ethyl- and methylmercury. Arch Toxicol 57(4):260–267

Marlowe M, Errera J, Jacobs J (1983) Increased lead and cadmium burdens among mentally retarded children and children with borderline intelligence. Am J Ment Defic 87(5):477–483

Marlowe M et al (1985) Main and interaction effects of metallic toxins on classroom behavior. J Abnorm Child Psychol 13(2):185–198

Marsh DO et al (1987) Fetal methylmercury poisoning. Relationship between concentration in single strands of maternal hair and child effects. Arch Neurol 44(10):1017–1022

Mason LH, Harp JP, Han DY (2014) Pb neurotoxicity: neuropsychological effects of lead toxicity. Biomed Res Int 2014:840547

Mass MJ, Wang L (1997) Arsenic alters cytosine methylation patterns of the promoter of the tumor suppressor gene p53 in human lung cells: a model for a mechanism of carcinogenesis. Mutat Res 386(3):263–277

McKay JA, Mathers JC (2011) Diet induced epigenetic changes and their implications for health. Acta Physiol (Oxf) 202(2):103–118

McNeill DR et al (2004) Inhibition of Ape1 nuclease activity by lead, iron, and cadmium. Environ Health Perspect 112(7):799–804

Mead MN (2010) Cadmium confusion: do consumers need protection? Environ Health Perspect 118(12):a528–a534

Mendez MA, González-Horta C, Sánchez-Ramírez B, Ballinas-Casarrubias L, Hernández Cerón R, Viniegra Morales D, Baeza Terrazas FA, Ishida MC, Gutiérrez-Torres DS, Saunders RJ, Drobná Z, Fry RC, Buse JB, Loomis D, García-Vargas GG, Del Razo LM, Stýblo M (2016) Chronic exposure to arsenic and cardiometabolic risk – a cross-sectional study in Chihuahua, Mexico. Environ Health Perspect 124(1):104–111

Menzie CM et al (2008) Obesity-related hypoferremia is not explained by differences in reported intake of heme and nonheme iron or intake of dietary factors that can affect iron absorption. J Am Diet Assoc 108(1):145–148

Méplan C, Mann K, Hainaut P (1999) Cadmium induces conformational modifications of wild-type p53 and suppresses p53 response to DNA damage in cultured cells. J Biol Chem 274(44):31663–31670

Miura T, Satoh T, Takeuchi H (1998) Role of metal-ligand coordination in the folding pathway of zinc finger peptides. Biochim Biophys Acta 1384(1):171–179

Moller-Madsen B, Danscher G (1986) Localization of mercury in CNS of the rat. I. Mercuric chloride (HgCl2) per os. Environ Res 41(1):29–43

Mukherjee G et al (2014) Glucagon-reactive Islet-infiltrating CD8 T Cells in NOD mice. Immunology 144:631–640

Murphy SK et al (2012) Gender-specific methylation differences in relation to prenatal exposure to cigarette smoke. Gene 494(1):36–43

Nawrot T et al (2010) Occupational cadmium exposure and calcium excretion, bone density, and osteoporosis in men. J Bone Miner Res 25(6):1441–1445

Needham LL et al (2011) Partition of environmental chemicals between maternal and fetal blood and tissues. Environ Sci Technol 45(3):1121–1126

Needleman HL (2004) Low level lead exposure and the development of children. Southeast Asian J Trop Med Public Health 35(2):252–254

Niedzwiecki MM et al (2013) A dose-response study of arsenic exposure and global methylation of peripheral blood mononuclear cell DNA in Bangladeshi adults. Environ Health Perspect 121(11–12):1306–1312

Nordstrom DK (2002) Public health. Worldwide occurrences of arsenic in ground water. Science 296(5576):2143–2145

Nye MD et al (2013) Associations between methylation of paternally expressed gene 3 (PEG3), cervical intraepithelial neoplasia and invasive cervical cancer. PLoS One 8(2):e56325

Nye MD et al (2015) Maternal blood lead concentrations, DNA methylation of DLK1/MEG3 imprinted domain and early growth in a multiethnic cohort. Under Rev
Patrick L (2006) Lead toxicity, a review of the literature. Part 1: Exposure, evaluation, and treatment. Altern Med Rev 11(1):2–22
Paul S, Giri AK (2015) Epimutagenesis: a prospective mechanism to remediate arsenic-induced toxicity. Environ Int 81:8–17
Perkins M et al (2014) Very low maternal lead level in pregnancy and birth outcomes in an eastern Massachusetts population. Ann Epidemiol 24(12):915–919
Pfeiffer WC et al (1989) Mercury concentrations in inland waters of gold-mining areas in Rondonia, Brazil. Sci Total Environ 87–88:233–240
Pihl RO, Parkes M (1977) Hair element content in learning disabled children. Science 198(4313):204–206
Pilsner JR et al (2010) Mercury-associated DNA hypomethylation in polar bear brains via the LUminometric Methylation Assay: a sensitive method to study epigenetics in wildlife. Mol Ecol 19(2):307–314
Piomelli S (2002) Childhood lead poisoning. Pediatr Clin North Am 49(6):1285–1304, vii
Pollard KM, Hultman P (1997) Effects of mercury on the immune system. Met Ions Biol Syst 34:421–440
Rabito FA et al (2014) Changes in low levels of lead over the course of pregnancy and the association with birth outcomes. Reprod Toxicol 50:138–144
Ramirez T et al (2008) Sodium arsenite modulates histone acetylation, histone deacetylase activity and HMGN protein dynamics in human cells. Chromosoma 117(2):147–157
Rappa G et al (1997) Evidence that the multidrug resistance protein (MRP) functions as a co-transporter of glutathione and natural product toxins. Cancer Res 57(23):5232–5237
Razmiafshari M et al (2001) NMR identification of heavy metal-binding sites in a synthetic zinc finger peptide: toxicological implications for the interactions of xenobiotic metals with zinc finger proteins. Toxicol Appl Pharmacol 172(1):1–10
Ren X et al (2011) An emerging role for epigenetic dysregulation in arsenic toxicity and carcinogenesis. Environ Health Perspect 119(1):11–19
Renner R (2010) Exposure on tap: drinking water as an overlooked source of lead. Environ Health Perspect 118(2):A68–A72
Reynolds LP et al (2006) Evidence for altered placental blood flow and vascularity in compromised pregnancies. J Physiol 572(Pt 1):51–58
Ronis MJ, Gandy J, Badger T (1998) Endocrine mechanisms underlying reproductive toxicity in the developing rat chronically exposed to dietary lead. J Toxicol Environ Health A 54(2):77–99
Rooney JP (2013) The retention time of inorganic mercury in the brain – a systematic review of the evidence. Toxicol Appl Pharmacol 274(3):425–435
Rossman TG, Klein CB (2011) Genetic and epigenetic effects of environmental arsenicals. Metallomics 3(11):1135–1141
Sahmoun AE et al (2005) Cadmium and prostate cancer: a critical epidemiologic analysis. Cancer Invest 23(3):256–263
Said S, Hernandez GT (2015) Environmental exposures, socioeconomics, disparities, and the kidneys. Adv Chronic Kidney Dis 22(1):39–45
Sanchez A et al (2015) Micronutrient deficiencies in morbidly obese women prior to bariatric surgery. Obes Surg
Satarug S, Moore MR (2004) Adverse health effects of chronic exposure to low-level cadmium in foodstuffs and cigarette smoke. Environ Health Perspect 112(10):1099–1103
Sanders T et al (2009) Neurotoxic effects and biomarkers of lead exposure: a review. Rev Environ Health 24(1):15–45
Sanders AP et al (2014) Cadmium exposure and the epigenome: exposure-associated patterns of DNA methylation in leukocytes from mother-baby pairs. Epigenetics 9(2):212–221

Sarkar B (1995) Metal replacement in DNA-binding zinc finger proteins and its relevance to mutagenicity and carcinogenicity through free radical generation. Nutrition 11(5 Suppl):646–649

Satarug S, Moore MR (2012) Emerging roles of cadmium and heme oxygenase in type-2 diabetes and cancer susceptibility. Tohoku J Exp Med 228(4):267–288

Schober SE et al (2006) Blood lead levels and death from all causes, cardiovascular disease, and cancer: results from the NHANES III mortality study. Environ Health Perspect 114(10):1538–1541

Schwerdtle T, Walter I, Hartwig A (2003) Arsenite and its biomethylated metabolites interfere with the formation and repair of stable BPDE-induced DNA adducts in human cells and impair XPAzf and Fpg. DNA Repair (Amst) 2(12):1449–1463

Schwerdtle T et al (2010) Genotoxicity of soluble and particulate cadmium compounds: impact on oxidative DNA damage and nucleotide excision repair. Chem Res Toxicol 23(2):432–442

Scinicariello F, Abadin HG, Murray HE (2011) Association of low-level blood lead and blood pressure in NHANES 1999–2006. Environ Res 111(8):1249–1257

Shiobara Y, Ogra Y, Suzuki KT (2001) Animal species difference in the uptake of dimethylarsinous acid (DMA(III)) by red blood cells. Chem Res Toxicol 14(10):1446–1452

Shraim A et al (2003) Arsenic speciation in the urine and hair of individuals exposed to airborne arsenic through coal-burning in Guizhou, PR China. Toxicol Lett 137(1–2):35–48

Silbergeld EK et al (1993) Lead in bone: storage site, exposure source, and target organ. Neurotoxicology 14(2–3):225–236

Sivaprasad TR, Malarkodi SP, Varalakshmi P (2004) Therapeutic efficacy of lipoic acid in combination with dimercaptosuccinic acid against lead-induced renal tubular defects and on isolated brush-border enzyme activities. Chem Biol Interact 147(3):259–271

Skinner MK (2014) Endocrine disruptor induction of epigenetic transgenerational inheritance of disease. Mol Cell Endocrinol 398:4–12

Smeester L et al (2011) Epigenetic changes in individuals with arsenicosis. Chem Res Toxicol 24(2):165–167

Smeester L et al (2014) Imprinted genes and the environment: links to the toxic metals arsenic, cadmium, lead and mercury. Genes (Basel) 5(2):477–496

Smith SW (2013) The role of chelation in the treatment of other metal poisonings. J Med Toxicol 9(4):355–369

Somji S et al (2011) Differences in the epigenetic regulation of MT-3 gene expression between parental and Cd+2 or As+3 transformed human urothelial cells. Cancer Cell Int 11(1):2

Spruill MD et al (2002) Proto-oncogene amplification and overexpression in cadmium-induced cell transformation. J Toxicol Environ Health A 65(24):2131–2144

Suarez-Ortegón MF et al (2013) Nutrients intake as determinants of blood lead and cadmium levels in Colombian pregnant women. Am J Hum Biol 25(3):344–350

Sun HJ et al (2014) Arsenic and selenium toxicity and their interactive effects in humans. Environ Int 69:148–158

Suzuki KT et al (2004) Distributions and chemical forms of arsenic after intravenous administration of dimethylarsinic and monomethylarsonic acids to rats. Toxicol Appl Pharmacol 198(3):336–344

Syversen T, Kaur P (2012) The toxicology of mercury and its compounds. J Trace Elem Med Biol 26(4):215–226

Takiguchi M et al (2003) Effects of cadmium on DNA-(Cytosine-5) methyltransferase activity and DNA methylation status during cadmium-induced cellular transformation. Exp Cell Res 286(2):355–365

Tellez-Plaza M et al (2013) Cadmium exposure and clinical cardiovascular disease: a systematic review. Curr Atheroscler Rep 15(10):356

Thatcher RW et al (1982) Effects of low levels of cadmium and lead on cognitive functioning in children. Arch Environ Health 37(3):159–166

Thatcher RW, McAlaster R, Lester ML (1984) Evoked potentials related to hair cadmium and lead in children. Ann N Y Acad Sci 425:384–390

Tokar EJ, Benbrahim-Tallaa L, Waalkes MP (2011) Metal ions in human cancer development. Met Ions Life Sci 8:375–401
Tsang V et al (2012) The epigenetic effects of a high prenatal folate intake in male mouse fetuses exposed in utero to arsenic. Toxicol Appl Pharmacol 264(3):439–450
Vahter M, Marafante E (1987) Effects of low dietary intake of methionine, choline or proteins on the biotransformation of arsenite in the rabbit. Toxicol Lett 37(1):41–46
Varrault A et al (2006) Zac1 regulates an imprinted gene network critically involved in the control of embryonic growth. Dev Cell 11(5):711–722
Vidal AC et al (2015) Maternal cadmium, iron and zinc levels DNA methylation and birth weight. BMC Pharmacol Toxicol 16:20
Vij A (2009) Hemopoietic, hemostatic and mutagenic effects of lead and possible prevention by zinc and vitamin C. Al Ameen J Med Sci 2:27–36
Wadi SA, Ahmad G (1999) Effects of lead on the male reproductive system in mice. J Toxicol Environ Health A 56(7):513–521
Wallia A et al (2014) Association between urinary cadmium levels and prediabetes in the NHANES 2005–2010 population. Int J Hyg Environ Health 217(8):854–860
Wang G, Fowler BA (2008) Roles of biomarkers in evaluating interactions among mixtures of lead, cadmium and arsenic. Toxicol Appl Pharmacol 233(1):92–99
Wang H et al (2015) Redistribution of subcellular calcium and its effect on apoptosis in primary cultures of rat proximal tubular cells exposed to lead. Toxicology 333:137–146
Watanabe T, Hirano S (2012) Metabolism of arsenic and its toxicological relevance. Arch Toxicol 87(6):969–979
Wildman RE, Mao S (2001) Tissue-specific alterations in lipoprotein lipase activity in copper-deficient rats. Biol Trace Elem Res 80(3):221–229
World Health Organization (1992) Environmental Health Criteria 134 and 135. Cadmium – environmental health aspects. World Health Organization, Geneva. ISBN 92 1571392, p 156
Wright JP et al (2008) Association of prenatal and childhood blood lead concentrations with criminal arrests in early adulthood. PLoS Med 5(5):e101
Wu J et al (2008) Alzheimer's disease (AD)-like pathology in aged monkeys after infantile exposure to environmental metal lead (Pb): evidence for a developmental origin and environmental link for AD. J Neurosci 28(1):3–9
Yanoff LB et al (2007) Inflammation and iron deficiency in the hypoferremia of obesity. Int J Obes (Lond) 31(9):1412–1419
Yin Z et al (2008) The methylmercury-L-cysteine conjugate is a substrate for the L-type large neutral amino acid transporter. J Neurochem 107(4):1083–1090
Yuan W et al (2006) The impact of early childhood lead exposure on brain organization: a functional magnetic resonance imaging study of language function. Pediatrics 118(3):971–977
Zakharyan RA, Wildfang E, Aposhian HV (1996) Enzymatic methylation of arsenic compounds. III. The marmoset and tamarin, but not the rhesus, monkeys are deficient in methyltransferases that methylate inorganic arsenic. Toxicol Appl Pharmacol 140(1):77–84
Zhao CQ et al (1997) Association of arsenic-induced malignant transformation with DNA hypomethylation and aberrant gene expression. Proc Natl Acad Sci U S A 94(20):10907–10912
Zhong CX, Mass MJ (2001) Both hypomethylation and hypermethylation of DNA associated with arsenite exposure in cultures of human cells identified by methylation-sensitive arbitrarily-primed PCR. Toxicol Lett 122(3):223–234
Zhou X et al (2008) Arsenite alters global histone H3 methylation. Carcinogenesis 29(9):1831–1836
Zhou X et al (2011) Arsenite interacts selectively with zinc finger proteins containing C3H1 or C4 motifs. J Biol Chem 286(26):22855–22863

Chapter 6
Fetal Imaging and Effects of Exposures on Growth and Function

Elena Demicheva and Fatima Crispi

Abstract Fetal growth and function are regulated and influenced by a multitude of complex factors including genetic profile of the embryo, maternal predispositions, placental state, fetal and maternal hormonal environment, and adequate nutrient and oxygen supply to the developing fetus. Every abnormal change that occurs during this sensitive developmental period programs health condition of the individual in postnatal life – the phenomenon known as "fetal programming." One of the most prevalent disorders affecting fetal development rate in almost 10 % of all pregnancies is fetal growth restriction (FGR) – a very complex and multifactorial disorder that often results in multiple adverse perinatal and postnatal complications including death. A growing number of studies over the last few decades confirms that the complex interaction between genetic constitution, prenatal and early postnatal environment determines the growth and development of the fetus and defines the susceptibility to certain disorders in adult life, like cardiovascular disease, neurobehavioral and metabolic disorders. Considering the high prevalence of FGR and the progressing availability of intervention strategies, it is of the highest clinical relevance to detect potential health risks as early as possible, to introduce timely preventive interventions and to adapt the life style in order to improve the long-term outcome of FGR cases.

Keywords Fetal programming • Fetal growth restriction • Fetal imaging Echocardiography • Cardiovascular remodeling

List of Abbreviations

ACTH	Adrenocorticotrophic Hormone
11βHSD2	11β-Hydroxysteroid Dehydrogenase-2
ANP	Atrial Natriuretic Peptide
BNP	B-type Natriuretic Peptide
DV	Ductus Venosus

E. Demicheva, PhD • F. Crispi, MD, PhD (✉)
BCNatal | Barcelona Center for Maternal Fetal and Neonatal Medicine, Hospital Clínic and Hospital Sant Joan de Déu, Universitat de Barcelona, Barcelona, Spain
e-mail: demincheva@clinic.ub.es; fcrispi@clinic.ub.es

C.L. Hughes, M.D. Waters (eds.), *Translational Toxicology*, Molecular and Integrative Toxicology, DOI 10.1007/978-3-319-27449-2_6

ECG	Electrocardiography
EFW	Estimated Fetal Weight
ELBW	Extremely Low Birth Weight
FGR	Fetal Growth Restriction
HPA	Hypothalamic Pituitary Axis
IRT	Isovolumetric Relaxation Time
IUGR	Intrauterine Growth Restriction
LGA	Large for Gestational Age
MPI	Myocardial Performance Index
SGA	Small for Gestational Age
STIC	4D Spatiotemporal Image Correlation
TDI	Tissue Doppler Imaging
TTTS	Twin-to-Twin Transfusion Syndrome
VLBW	Very Low Birth Weight

6.1 Introduction

6.1.1 Fetal Programming

6.1.1.1 Fetal Programming and Fetal Growth Restriction (FGR)

Normal fetal development is regulated by complex mechanisms involving various critical factors, such as genetic profile of the embryo, maternal predispositions, placental state, fetal and maternal hormonal environment and adequate nutrient and oxygen supply to the developing fetus. If one or more of these developmental factors is abnormal, the growth of the fetus might be impaired resulting in low birth weight and possible maldevelopment of the key organs.

Accumulating evidence from a large number of epidemiological and animal studies performed over the last few decades confirm the existence of the direct link between the conditions of embryonic/fetal development and long-term consequences in adult life. Taken together, the data obtained suggests that the complex interaction between genetic constitution and prenatal and early postnatal environment determines the growth and development of the fetus and defines the susceptibility to certain disorders in adult life, like hypertension, diabetes, dyslipidemia, coagulation and neurobehavioral disorders (Gluckman 2004). This phenomenon is known as "fetal programming", when an insult *in utero* leads to functional changes in key organs persisting into postnatal life and leading to a greater risk of various diseases in adulthood (Palinski and Napoli 2008; Phillips 1996; Gluckman et al. 2008; Tintu et al. 2009). Rapid cellular proliferation and differentiation during fetal growth are very sensitive to even small environmental changes that can lead to permanent alterations in structural and functional constitution, which then may persist into the ensuing adult life (Gluckman 2004).

Besides premature birth, low birth weight in the majority of the cases is caused by fetal growth restriction (FGR), also known as intrauterine growth restriction (IUGR) (Diderholm 2009). FGR is defined as a failure of a fetus to achieve its genetic growth potential, characterized by the birth weight and body mass lower than normal with respect to the number of gestational weeks – below percentile 10 (Committee on Practice Bulletins-Gynecology, American College of Obstetricians and Gynecologists 2001). This functional definition, however, does not include fetuses that are small for gestation age (SGA), but at the same time are not pathologically small. Not all SGA fetuses are growth restricted and, similarly, not all fetuses that have not met their genetic growth potential are in less than the 10th percentile for estimated fetal weight (EFW).

FGR is a major cause of perinatal morbidity and mortality (Alberry and Soothill 2007) and may complicate 7–10 % of all pregnancies, depending on the population, geographical location and the standard growth curves used as reference in different countries.

The gestational age at which the fetal growth restriction develops is of the major significance for the outcome of the pregnancy (Baschat 2004). Therefore, FGR is also classified by early- and late-onset. As most FGR cases are associated with placental insufficiency, early FGR is diagnosed by the classical Doppler flow measurements before 34 weeks of gestation, recognized as an increase in impedance to blood-flow at the umbilical artery and subsequent fetal vasodilation mechanism to maintain cerebral oxygen supply (Hecher et al. 2001). Accordingly, late-onset of FGR after 34 weeks differs by the clinical manifestations, patterns of deterioration, and severity of placental dysfunction, which normally is milder and with normal umbilical artery blood flow (Baschat 2010, 2011). Unfortunately, the lack of information in the literature makes it very difficult to differentiate between early- and late-FGR, and most of the studies fail to identify the origins of FGR in order to select accordingly the correct patient cohorts.

6.1.1.2 Long-Term Consequences of FGR

Cardiovascular Remodeling

In the modern world 23 % of mortality is caused by cardiovascular disorders, counting over four million deaths each year in Europe. For many years the cardiovascular condition was thought to be determined only by the genetic factors and the lifestyle of individual, particularly by the amount of physical activity and quality of nutrition. However, in most cases, cardiovascular diseases seem to be triggered at the very beginning of life and then undergo long subclinical phase that can last decades before the first clinical symptoms appear (Berenson 2002).

As early as 1989 the group of David Barker in Southampton, UK has established a direct correlation between low birth weight and cardiovascular disease in adulthood, including hypertension and cardiovascular mortality (Barker et al. 1989a, b). Later on, multiple studies revealed that fetuses, which have been diagnosed with

FGR, often show symptoms of cardiac systolic and diastolic dysfunction (Hecher et al. 1995), such as increased E/A-ratios and myocardial performance index (Crispi et al. 2008), together with reduced myocardial tissue velocities (Comas et al. 2010). Moreover, the placental dysfunction leads to hypoxia and insufficient nutrition supply for the FGR fetus, which then has to adapt accordingly by hemodynamic redistribution, resulting in hypertension and hypervolemia (Kiserud et al. 2006a). Further follow-up evaluation of the children who suffered FGR *in utero*, as well as experimental research, have confirmed that the triggered cardiovascular remodeling persist as a permanent feature in their postnatal life (Tintu et al. 2009). The cardiovascular system of these children showed pronounced dilated cardiomyopathy-like heart remodeling, vascular dysfunction, increased blood pressure and carotid intima-media thickness.

In the affected hearts of FGR fetuses significant changes of the cardiac myocyte contractile machinery has been found on the molecular level, similar to those described in dilated cardiomyopathy, diastolic heart failure and other cardiac diseases driven by high pressure and/or volume (Bijnens et al. 2012). As such, altered gene expression of the sarcomere regulatory proteins and decreased sarcomere length discovered in the FGR hearts result in a less efficient contraction (Iruretagoyena et al. 2014) and are likely to persist into the adulthood.

In addition to the obvious importance of the intrauterine environment during critical developmental stages, the conditions of early childhood are likewise of a great significance. The accelerated postnatal growth, so-called "catch-up growth", is one of the most important triggers of hypertension in SGA children and increased susceptibility to cardiovascular complications in adult life (Huxley et al. 2000).

Metabolic Syndrome

Apart from the cardiovascular disorders, FGR can lead to obesity, hypercholesterolemia, impaired glucose tolerance, and/or diabetes mellitus type 2 (Curhan et al. 1996; Harder et al. 2007), the phenomenon known as "metabolic syndrome." Likewise, the early studies of Barker and colleagues revealed that metabolic syndrome is ten times more prevalent in the subjects with low birth weight (Barker et al. 1993).

As a central factor in metabolic syndrome, obesity is a major public health problem that is prevalent in over 33 % of the world population (Kelly et al. 2008). Moreover, currently worldwide over 20 million children under 5 years old are overweight and this tendency is alarmingly increasing each year (Han et al. 2010). Moreover, obesity is clearly associated with hyperlipidemia and hyperglycemia, inevitably leading to a decrease in β-cell mass and its dysfunction – parameters that characterize the onset of type 2 diabetes (Lingohr et al. 2002).

A growing number of recent studies further confirm the evident link between restricted fetal growth and development of the metabolic syndrome or its components later in life, such as hypertension (Hult et al. 2010), dyslipidemia (Lussana et al. 2008; Kajantie et al. 2008), obesity (Kensara et al. 2005; Parsons et al. 2001), impaired glucose tolerance and insulin resistance (Phillips et al. 1994; Jaquet et al. 2000).

Neurodevelopmental Impairment

Additionally, impaired fetal growth might provoke critical changes in neurodevelopmental processes, such as motor, behavioral and cognitive deficiencies.

The earliest symptom of neurodevelopmental impairment is motor dysfunction that becomes evident at birth and certain by 2 years of age. At this stage other neurological abnormalities might start to manifest themselves and can be then diagnosed (Eixarch et al. 2008).

One of the major risk factor in the development of FGR children is a decrease of head growth resulting in smaller brain dimensions, which in turn is associated with psychomotor retardation, cognitive delay and abnormal behavior rating index in infancy, followed by persistent cognitive and speech delays, motor dysfunction and lower scholastic performance from childhood all the way to adolescence (Harvey et al. 1982; Parkinson et al. 1986; Padilla et al. 2010; McCowan et al. 2002; Kutschera et al. 2002; Ley et al. 1996).

The abnormal neurodevelopment effect seems to occur in children who suffered early- as well as late-onset FGR, and the severity of the growth restriction correlates predominantly with parameters of motor development in infancy and early childhood (Padilla et al. 2010; Wienerroither et al. 2001; Leppänen et al. 2010; Baschat 2011; Torrance et al. 2010).

6.1.1.3 Fetal Programming in Other Fetal Conditions

Prematurity

Very low birth weight (VLBW, <1.5 kg) and extremely low birth weight (ELBW, <1 kg) infants are of major concern to clinicians when dealing with premature birth. Over the last 60 years the mortality rate of VLBW and ELBW babies has been dramatically improved due to the available technical equipment and constant research on intensive care and nutrition. However, the consequences of premature birth remain severe and include a wide range of complications in early as well as in adult life of the individual.

Despite intensive care and early parenteral and enteral nutrition programs, a majority of premature infants suffer postnatal growth restriction throughout childhood and delays in neurodevelopment (Dusick et al. 2003; Neubauer et al. 2008; Ehrenkranz et al. 1999). Follow-up studies of premature VLBW infants have revealed critically elevated levels of glucose and insulin concentrations, as well as significantly increased systolic blood pressure (Hovi et al. 2007).

Furthermore, premature infants are likely to have impaired cognitive development and to exhibit anxiety, depression, and attention deficit and hyperactivity disorder

(Hack 2006; Dahl et al. 2006; Hack et al. 2004). Due to perinatal and postnatal stress, the hypothalamic pituitary axis (HPA) becomes activated, which in turn leads to elevated levels of adrenocorticotrophic hormone (ACTH) and adrenal hypertrophy (Ward et al. 2000). These alterations are not only linked to neurobehavioral programming, but also serve as a predisposition for atherosclerosis, dyslipidemia, type 2 diabetes mellitus and immunosuppression later in life.

Maternal Diabetes

Maternal obesity and/or diabetes present another adverse form of fetal programming. Mothers with gestational diabetes are at very high risk of delivering a large for gestational age (LGA) infant with hyperinsulinemia as a consequence of maternal hyperglycemia. For every 1 kg increase of birth weight, full-term neonates have a 50 % increase in adolescent obesity (Gillman et al. 2003).

In mothers with gestational diabetes, fetal growth is accelerated by increased production and delivery of glucose and other macronutrients, resulting in increased fetal insulin production, the main fetal growth hormone. Therefore, LGA infants have increased adiposity and elevated insulin and leptin levels which have been programmed *in utero* (Catalano et al. 2003; Vela-Huerta et al. 2008).

In addition to obesity, LGA children are at risk to develop hypertension, hypertriglyceridemia (Gillman et al. 2003; Eriksson et al. 1999; Taveras et al. 2009; Wang et al. 2009) and type 2 diabetes mellitus not only themselves, but to pass it to their future offspring as well (McLean et al. 2006).

6.1.2 Fetal Cardiac Function and Pathophysiology

The primary function of the heart is to provide adequate perfusion of organs by ejecting blood from the ventricle through the aortic/pulmonary valve and into the aorta/pulmonary artery (systole) by contracting its muscular walls (Guyton and Hall 2006). To maintain normal cardiac function, both blood ejection in systole and filling of the ventricle from the atria (diastole) must be adequate and occur in a synchronized manner (Guyton and Hall 2006; Bijnens et al. 2012).

6.1.2.1 Basics of Fetal Cardiac Function and Physiology

The normal cardiac cycle consists of five major phases. The diastolic part of the cycle includes: (1) *isovolumetric relaxation phase*, when after aortic/pulmonary valve closure with an isovolumetric relaxation period no blood enters or ejects from the ventricles while the myocardium starts to relax and the intraventricular pressure drops; (2) *early diastolic phase*, when ventricular pressure lowers the atrial pressure, the mitral/tricuspid valve opens and blood from the atria starts to fill the ventricle in a passive manner; (3) *atrial contraction phase*, when the atria contract and complete the filling of the ventricle (late diastole).

The ejection process for systole includes: (4) *isovolumetric contraction phase*, when cardiomyocytes start to contract, increasing intraventricular pressure which in turn opens the aortic/pulmonary valve (isovolumetric contraction time) without any change in volume; (5) *ejection phase*, when due to the increased ventricular pressure, the aortic/pulmonary valves opens and the myocardium deforms ejecting blood from the ventricle.

These main components of the cardiac cycle define the main features of cardiac blood flow movement and myocardial motion and deformation (Guyton and Hall 2006; Bijnens et al. 2012; Sutherland et al. 2006; Kasper et al. 2005).

6.1.2.2 Fetal Cardiac Remodeling and Dysfunction

Heart failure is defined as the inability of the heart to supply sufficient blood flow to meet the needs of the organism (Jessup et al. 2009). This is usually a late event that can be easily recognized by cardiomegaly, atrioventricular insufficiency, and fetal hydrops, which can be quantified by measuring a significant decrease in cardiac output or ejection fraction (Huhta 2004; Jessup et al. 2009).

In the initial stages of an insult, the heart usually manages to adapt, undergoing a long subclinical period of cardiac dysfunction before the actual heart failure occurs (Huhta 2004; Rychik et al. 2007; Crispi et al. 2008). The adaptational changes in cardiac function, shape and size, can be measured and define the process of "cardiac remodeling" (Opie et al. 2006).

Changes in cardiac function and shape are determined not only by the causal insult, but also by myocardial contractility, fiber orientation, tissue elasticity, heart geometry, segment interaction, loading conditions, electrical activation, and myocardial perfusion (Bijnens et al. 2009). In the fetal heart, cardiac remodeling also depends on myocardial maturation and blood circulation (Kiserud and Acharya 2004).

Myocardial Contractility

Myocardial contractility is defined by the intrinsic ability of cardiac muscle to develop force for a given muscle length (Guyton and Hall 2006; Bijnens et al. 2012) It may be affected by genetic disposition to cardiac diseases or by hypoxia and essentially conditions myocardial motion and deformation during systole (Bijnens et al. 2009, 2012).

Myocardial *motion* is defined as the distance covered by one point over a certain period of time and is determined by displacement (distance) and velocity (distance divided by time). Accordingly, myocardial *deformation* is defined as the change in the length/thickness of a segment (two points) and is determined by strain (percentage of change) and strain rate (velocity of segment change) (Bijnens et al. 2009).

When myocardial fibers contract, all segments deform moving the heart's base toward the apex to eject blood (Bijnens et al. 2009). *Global* longitudinal myocardial motion reflects the motion of all myocardial segments and is usually measured at the mitral/tricuspid annulus as this is a fibrose area with no intrinsic capacity for deformation. Conversely, myocardial deformation should be assessed in a specific myocardial segment reflecting *regional* function.

Fiber Orientation

The complex three-directional motion of myocardial contraction is determined by the complex geometry of myocardial fibers and muscle band orientation (Fig. 6.1) (Bijnens et al. 2012) and involves longitudinal and radial contractions, and rotation (circumferential axis) (Anderson et al. 2009; Sengupta et al. 2008).

Longitudinal motion from the apex to the base of the heart is mainly determined by endocardial longitudinal fibers – farthest from the epicardial blood supply and consequently the most sensitive under milder degrees of hypoxia. Therefore, longitudinal motion usually becomes abnormal in the very early stages of cardiac dysfunction (Bijnens et al. 2009, 2012; Gardiner et al. 2006a, b).

Radial motion, perpendicular to the epicardium, is determined by radial fibers located mainly in the mid part of the ventricular wall and usually becomes abnormal in the late stages of fetal deterioration (Bijnens et al. 2009, 2012; Jessup et al. 2009).

Rotational motion of the left ventricle myocardial wall is executed on the circumferential axis, which is perpendicular to both the longitudinal and the radial axes. The geometry of the myofibers changes smoothly from a right-handed helix in the subendocardium to a left-handed helix in the subepicardium, so that the angle of

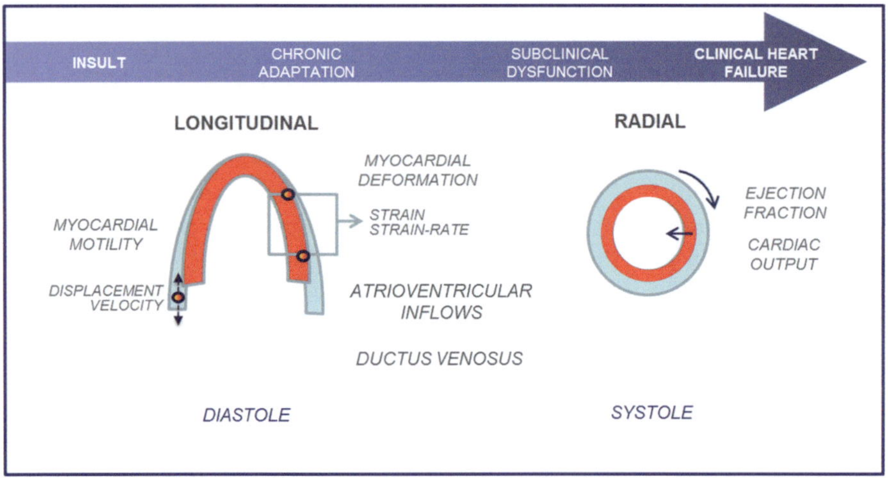

Fig. 6.1 Different stages of fetal cardiovascular adaptation, subclinical dysfunction and clinical failure

the helix varies continuously from positive at the endocardium to negative at the epicardium (Anderson et al. 2009; Sengupta et al. 2008). Thus, the base and apex of the left ventricle rotate in opposite directions, leading to the twisting motion described as "the wringing of a wet linen cloth."

Rotation and twisting have been shown to become abnormal in the very early stages of cardiac dysfunction (Burns et al. 2008), but very few studies have successfully assessed rotation *in utero* and, therefore, it's utility in fetal life remains to be elucidated (Barker et al. 2009).

Changes in Myocardial Maturation During in Utero Development

During gestation, maturational changes occur within the myocardium leading to changes in elasticity and contractility throughout the pregnancy, which should be taken into account when evaluating and interpreting fetal cardiac function (Sedmera 2011). The density and compaction of myofibrils increases particularly in early pregnancy, but contractility and elasticity continue to improve during the second half of pregnancy (Tobita et al. 2005). Once the structural details have been organized during the embryonic period, the fetal heart continues to grow by cell division until birth, followed by the continued growth due to cell enlargement (Sedmera 2011).

Ventricular Loading

Cardiac function is determined mainly by the volume and pressure loading conditions (Bijnens et al. 2009). In pathological conditions, volume overload [e.g. due to fetal anemia, twin-to-twin transfusion syndrome (TTTS), valve leakage, etc.] will mainly lead to heart dilatation to help the heart manage the increased blood volume more efficiently (Opie et al. 2006). The mechanism is based on *preload*, defined as a muscle length prior to contractility, depending on ventricular filling or blood volume in end-diastole (Guyton and Hall 2006; Bijnens et al. 2012). The most important determining factor for preload is venous return.

On the other hand, pressure overload (e.g. due to valvular stenosis or TTTS) will mainly lead to myocardial hypertrophy in order to increase the contractile mass to overcome the elevated afterload (Opie et al. 2006). *Afterload* – the tension (or the arterial pressure) against which the ventricle must contract – depends on the maximum tension of the myocardial muscle mass in end-systole (Guyton and Hall 2006; Bijnens et al. 2012) and is determined by aorta pressure for the left ventricle by pulmonary artery pressure for the right.

Particularities of Fetal Circulation

In contrast to postnatal life, the fetal systemic circulation is fed from the left and right ventricles in parallel, with a small proportion of the right output being spared for the lungs (Kiserud and Acharya 2004). The well-oxygenated blood is directed from the umbilical vein through the *ductus venosus* (DV) across the inferior vena cava, through the foramen ovale, left atrium, and ventricle and up the ascending aorta to join the low oxygenated blood in the descending aorta. Deoxygenated blood from the superior and inferior vena cava is directed through the right atrium and ventricle, pulmonary trunk, and ductus arteriosus.

Additionally, the three shunts – DV, ductus arteriosus, and foramen ovale – are essential distributional arrangements, making fetal circulation a flexible and adaptable system for intrauterine life (Kiserud and Acharya 2004). The hemodynamic properties and functional ranges of these shunts are important determinants of the fetal heart development and circulation during the second and third trimesters.

6.2 Effects of Exposures on Growth and Function

6.2.1 Epidemiological and Clinical Studies

6.2.1.1 FGR and Cardiovascular Mortality

The correlation between low birth weight and hypertension was reported as early as 1988 in the Swedish study analyzing a cohort of male army recruits who had been small at birth (Gennser et al. 1988), which was then confirmed later in several larger studies in men and women (Leon et al. 2000; Martyn et al. 1995; Hales et al. 1991; Curhan et al. 1996).

Shortly after, Barker and colleagues discovered the correlation between low birth weight and increased risk of death from cardiovascular disease (Barker et al. 1989a) and stroke. Moreover, men with the lowest birth weights at 1 year of age had highest death rates from ischemic heart disease (Barker et al. 1989b).

One large study of almost 15,000 Swedish men and women with a 97 % follow-up over a period of more than 50 years (Leon et al. 1998) showed a strong correlation between lower birth weight and death rate from ischemic heart disease. Another large cohort study performed by the group of Rich-Edwards in 1997 in more than 120,000 American women revealed strong negative trends of recorded birth weight and incidence of non-fatal coronary heart disease and stroke (Rich-Edwards et al. 1997). An interesting observation was that despite the similar increase in general death rate in men and women inversely related to their lower birth weight, only men had developed cardiovascular disease in adult life (Osmond et al. 1993). Therefore, a hypothesis was proposed, that the promotion of the growth during prenatal and early postnatal life could improve cardiovascular condition in adulthood, especially in boys who weighed less than 3.5 kg (Barker et al. 1989a, b). Later, however, it was proved to be wrong, as in the cohort study in Finnish men the cardiovascular mortality was also increased as the result of the rapid weight gain in the first 3 years of life (Eriksson et al. 2001). The increased mortality rate was also confirmed in the obese men who had the BMI above average between 7 and 15 years of age (Eriksson et al. 1999). Moreover, this effect was greater in males born to obese mothers (Forsén et al. 1997).

Another study in women born short in length showed that an increase in height later in life was associated with the higher risk of mortality from coronary heart disease (Forsén et al. 1999). Since most of these women had tall mothers, it was suggested that their prenatal growth was constrained.

Despite the certainty of the observations described above, one needs to take into consideration other environmental factors that may have great influence on the cardiovascular development. For example, a cohort study in South Africa showed the link between low birth weight and adult glucose intolerance and blood pressure elevation that occurs in young adults despite the lack of full catch-up growth as a consequence of high-risk, disadvantaged environment of the population (Levitt et al. 2000).

Therefore, due to the obvious correlation between the early life environment and susceptibility to cardiovascular diseases, the proper follow-up of the FGR patients is critical for their health in adult life.

6.2.1.2 FGR Effect on Fetal and Postnatal Cardiovascular System

One large cohort study of almost 150,000 adolescents in Sweden showed that the systolic blood pressure was significantly higher in the young men who had lowest birth weight, thus supporting the notion of a programming effect of fetal growth retardation *in utero* on hemodynamic regulation in early adult life (Nilsson et al. 1997; Law and Shiell 1996).

The well-known Generation R study has shown that fetal hemodynamic patterns change in the presence of reduced fetal growth while the fetus is still within the normal estimated fetal weight range. The condition of decreased fetal growth triggers cardiac remodeling and output that are consistent with a gradual increase in afterload and compromised arterial compliance (Verburg et al. 2008).

Recently, the group of Gratacós has obtained confirmatory data in a prospective cohort study in children 2–6 years of age, suggesting that the fetal cardiovascular programming is not exclusive for premature and severe forms of growth restriction, but it also occurs in mild late-onset of FGR (Comas et al. 2011; Cruz-Martinez et al. 2011), which often is referred to as SGA (Figueras et al. 2008; Soothill et al. 1999). In both, early- and late-onset FGR children, more globular hearts, reduced longitudinal motion and impaired relaxation were observed. Late-onset FGR could compensate by increased radial function, while the more severe early-onset cases showed decreased radial function leading to lower stroke volume and increased heart rate in order to maintain cardiac output (Crispi et al. 2012a). Additionally, both groups showed signs of vascular remodeling including increased blood pressure, as well as carotid intima-media thickness (Crispi et al. 2012a). Significantly progressive increased aortic intima-media thickness and elevated blood pressure were also observed in another recent study in FGR fetuses and infants at 18 months of age, suggesting predisposition of the infants to hypertension early in life and cardiovascular risk in adulthood (Zanardo et al. 2013).

6.2.2 Animal Studies on Fetal Programming

Despite the performance of many studies investigating different aspects of FGR in human pregnancies, the pathophysiological processes underlying this disorder remain very complex and not completely understood. Therefore, various animal models have been developed, not only to clarify processes regulating fetal growth in normal condition and FGR but also to develop potential strategies of intervention.

6.2.2.1 Possible Mechanistic Pathways of FGR

Maternal Nutrition

The strong correlation between maternal nutritional status (diet) and fetal development was described decades ago (Villar and Belizan 1982), confirming fetal nutrition as the most critical factor that determines pregnancy outcome. Many animal studies have subsequently confirmed that the limitation of general food intake to pregnant animals, besides significantly lower birth weight, also programs increased blood pressure, decreased angiogenesis, and increased risk of endothelial dysfunction in the adult who was an undernourished offspring (Ergaz et al. 2005; Ozaki et al. 2001; Vickers et al. 2000). The programmed outcome, however, depends on the severity and the gestational period of the restriction (Barnes and Ozanne 2011; Tarry-Adkins and Ozanne 2011; Garofano et al. 1998).

Protein intake seems to play a major role in the fetal programming, since compared to the balanced reduction in maternal nutrient intake, the specific restriction of maternal protein in rats programs a more consistent increase in blood pressure in adult life (McMillen and Robinson 2005; Langley and Jackson 1994; Snoeck et al. 1990). Since the early 1990s, many studies have shown that a maternal low-protein diet during the gestational period leads to lower birth weight, hypertension, vascular dysfunction, increased angiotensin-converting enzyme activity, decreased nephron number, increased oxidative stress in adulthood and shorter life span (Tarry-Adkins and Ozanne 2011; Nuyt and Alexander 2009; Watkins et al. 2010). In non-litter-bearing species, maternal low protein diet has also been shown to program endothelial dysfunction (Nishina et al. 2003). Similar to the general caloric restriction, if dams are fed the low-protein diet only for a single week, the magnitude of the effect of the diet on postnatal blood pressure is greatest when consumed during the last week of gestation (Langley-Evans et al. 1996a, b).

Impairment of Uteroplacental Function

Maternal nutritional manipulation provides an easy and reproducible tool to study fetal programming and mechanisms of FGR, but it fails to reproduce the restriction of oxygen supply (Huizinga et al. 2004), which is considered to be among the most critical factors in fetal development. Alternatively, placental insufficiency can be induced by direct restriction of vascular perfusion.

One of the oldest surgical models is permanent ligation of both uterine arteries, which leads to hypoxia, decreased growth factor availability and hypoglycemia (Wigglesworth 1964). The high degree of the blood flow obstruction results in the significantly lower birth weight and programs reduced nephron numbers, hypertension, type 2 diabetes and proteinurea in the adult offspring (Barnes and Ozanne 2011). Unfortunately, the model has a major disadvantage due to the high mortality rate as a consequence of invasive and complicated surgery.

Likewise, permanent ligation of only one uterine artery performed at mid-gestation in guinea pig causes alterations of the placenta, heart, aorta and kidneys in the offspring (Briscoe et al. 2004). Moreover, growth restricted fetuses and neonates are chronically hypoxic (Lafeber et al. 1984), hyperglycemic (Jones et al. 1984) and have altered brain development (Rehn et al. 2004). Further development of the model is selective ligation of 30–50 % of uteroplacental vessels late in gestation. This technique programs pronounced cardiovascular Doppler changes in fetuses, particularly in the DV, which partially reproduces the hemodynamic features of FGR condition in human fetuses (Eixarch et al. 2011).

Other interventions mimicking placental insufficiency such as utero-placental embolization, carunclectomy or maternal hyperthermia (Anthony et al. 2003) result in a similar phenotype and cardiovascular dysfunction as in the described surgical models (Morrison 2008), suggesting a common underlying mechanisms of cardiovascular programming.

Prenatal Exposure to Exogenous Factors

In 1997 the group of Langley-Evans discovered that hypertension, induced by fetal exposure to a maternal low-protein diet in the rat, can be prevented by treatment with metyrapone – an inhibitor of maternal glucocorticoid synthesis (Langley-Evans 1997). In humans, the association of maternal exposure to glucocorticoids with reduced birth weight and adult disease was suspected long ago (Nyirenda et al. 1998; Newhham 2001; Edwards et al. 1993). Consequent animal studies not only confirmed that synthetic glucocorticoids induce lower birth weight but also program elevation of arterial blood pressure that persists into adulthood (Benediktsson et al. 1993; Levitt et al. 1996).

The programming mechanism seems to depend on the function of the specific placental enzyme, 11β-hydroxysteroid dehydrogenase-2 (11βHSD2), which normally metabolizes corticosterone to the inert 11-dehydrocorticosterone (Lindsay et al. 1996) but is unable to process synthetic glucocorticoids. Interestingly, maternal nutritional restriction has also been associated with decreased activity of 11βHSD2 in placenta, which would in turn increase access of endogenous maternal glucocorticoids to the fetus (Langley-Evans et al. 1996a, b; Lesage et al. 2001; Dwyer and Stickland 1992).

Other exogenous environmental factors like high altitude can also act independently to reduce birth weight. Apparently, increased maternal ventilation and ventilator response to hypoxia during pregnancy raises arterial oxygen saturation and correlates with the offspring's birth weight (Moore 1987). Similarly, the exposure of pregnant rats to chronic hypoxia in the last trimester of gestation leads to an increase in the percentage and size of binucleated myocytes, apoptotic cells in the fetal heart (Bae et al. 2003) and susceptibility of the adult heart to ischemic-reperfusion injury (Li et al. 2003). In turn, massive cell death in fetal heart could lead to the cardiac hypertrophy, resulting in asymmetric enlargement of the heart chambers and increased cardiovascular risk in adulthood.

Among other risk factors, maternal smoking has been clearly identified to have a strong association with fetal growth restriction, dramatically affecting the placental vessels due to nicotine and hypoxia (Bernstein et al. 2005; Jaddoe et al. 2007; Larsen et al. 2002). Moreover, fetal exposure to maternal smoking might also lead to adverse cardiovascular effects, like hypertension persisting into adulthood (Blake et al. 2000; Brion et al. 2008). Accordingly, it has been shown in an animal experimental model that daily inhalation of tobacco throughout gestation causes significantly reduced birth weight in rat offspring (Leichter 1995; Younoszai et al. 1969).

Similar adverse effects of growth restriction are observed after fetal exposure to alcohol (Carter et al. 2013; Ponnappa and Rubin 2000) or harmful drugs like cocaine (Bateman et al. 1993), confirming that cardiac overload due to the changes in placental function or maternal nutrition may result in a change in the developmental profile of cardiomyocytes and cardiac function in adult life.

6.2.2.2 Clinical Relevance

Possible Therapeutic Targets

Quite a few studies have already shown that different nutritional interventions in perinatal and early postnatal life can revert or alleviate the cardiovascular consequences of FGR programming. For example, supplementation with vitamin C, E and NO donors has been shown to reduce blood pressure and prevents proteinuria in rats (Racasan et al. 2005). On the other hand, inhibition of certain factors could also improve the cardiovascular outcome, as shown for the NF-κB inhibitor Pyrrolidine dithiocarbamate and soluble epoxide hydrolase inhibitor 12-(3-adamantan-1-yl-ureido)-dodecanoic acid, which significantly reduce blood pressure in perinatal treated offspring of spontaneously hypertensive rats (Koeners et al. 2011a, b).

Furthermore, FGR fetal programming induced by low-protein diet in rats has been shown to be reversed or even prevented by co-treatments with urea, which normalized body weight, or with glycine, preventing increased blood pressure (Jackson et al. 2002). These findings emphasize the importance of nitrogen-containing compounds for healthy development and growth of the fetus.

Other promising results have been obtained in the rat model of maternal low-protein diet, where sconsequences of induced FGR has been improved with various treatments like perinatal administration of folate (Torrens et al. 2006), metyrapone (11β-hydroxylase inhibitor) (Bogdarina et al. 2010) or lazaroid (an inhibitor of lipid peroxidation) (Cambonie et al. 2007), as well as postnatal treatment with fish oil (n-3 polyunsaturated fatty acids) (Catta-Preta et al. 2006) or atorvastatin (cholesterol-lowering and anti-atherosclerotic drug) (Bezerra et al. 2008).

It is clear that further intensive research is still needed to fully understand the mechanisms of fetal programming and its precise timing. However, the progress that has been made so far seems very promising for successful development of early intervention therapies to prevent and maybe even reverse the consequences of FGR in adult life.

Clinical Implications

Although current clinical guidelines still do not include FGR as a risk factor for chronic diseases in adult life, intensive research over the last few decades points towards strong necessity of thorough monitoring and follow-up studies in children as well as adults who suffered FGR in fetal life. Hypertension and pre-hypertension in the child has been associated with substantial long-term health risks and considered an indication for lifestyle modifications. Accordingly, it has been shown that promoting physical activity and avoiding exposure to secondary smoking or obesity together with dietary interventions significantly improve cardiovascular health in hypertensive children (Williams et al. 2002). Particularly, high intake of dietary long-chain ω-3 fatty acids has been associated with lower blood pressure and may prevent progression of subclinical atherosclerosis in children born with low birth weight (Skilton et al. 2013a, b). A recent randomized trial in a large cohort of children showed that inverse association of fetal growth with arterial wall thickness in childhood can be prevented by dietary ω-3 fatty acid supplementation over the first 5 years of life (Skilton et al. 2012).

It is critical to adequately select the high-risk population that may benefit from these therapeutic interventions. Currently the severity of FGR is determined mainly by the gestational age and birth centile, which are inversely related to the perinatal and neurodevelopmental outcomes (Rossi et al. 2011; Savchev et al. 2012; Cruz-Lemini et al. 2012; Figueras and Gardosi 2011). For further antepartum surveillance of the viable FGR fetus, the umbilical artery Doppler analysis has been proposed; thus significantly decreasing the necessity of labor induction, delivery via cesarean section, and perinatal death (Berkley et al. 2012). However, cardiovascular programming occurs not only in premature and severe FGR, but also in mild late-onset FGR cases with normal umbilical artery parameters (Crispi et al. 2008, 2010, 2012a; Comas et al. 2011). Therefore, standard perinatal parameters such as gestational age, birth weight centile or umbilical Doppler show poor performance for predicting long-term cardiovascular outcome. To improve diagnostic accuracy, fetal echocardiography has been recently pointed out as a very promising tool for the early identification of those FGR cases with postnatal hypertension and arterial remodeling (Cruz-Lemini et al. 2013; Crispi and Gratacós 2012).

Considering the strong prevalence of FGR and the progressing availability of intervention strategies, it is of the highest clinical relevance to detect cardiovascular risks as early as possible, to introduce timely preventive interventions and to adopt protective life style changes in order to improve the long-term cardiovascular health outcome of the FGR patients.

6.3 Fetal Imaging

6.3.1 Techniques of Fetal Imaging to Access Cardiac Function

Traditionally, fetal cardiac function was assessed by measuring blood flow through conventional Doppler or cardiac morphometry in 2D or M-mode. Recently, the diagnostic evaluation has been improved by employing tissue Doppler imaging (TDI) and 2D speckle tracking imaging (Bijnens et al. 2009; Comas et al. 2011a, b; Germanakis and Gardiner 2012) for the direct assessment of myocardial motion and deformation. Additionally, 4D spatiotemporal image correlation (STIC) has been proposed to evaluate more accurately cardiac dimensions and volumes (Godfrey et al. 2012a, b).

The most suitable parameters for assessing fetal cardiac function are mainly determined by the cause of the dysfunction. Abnormal values of ejection fraction or cardiac output are usually found in the late stages of deterioration, and, therefore, more sensitive parameters have been proposed for earlier diagnosis and monitoring of fetal cardiac dysfunction. In most cases of cardiac dysfunction, diastolic parameters (such as DV or IRT) are the first to be altered, reflecting impaired relaxation and compliance due to a stiffer or less effective heart. Similarly, parameters reflecting longitudinal function (such as annular displacement or velocities) are typically affected in the early stages as compared to radial function (such as ejection fraction) (Fig. 6.2).

6.3.1.1 Current Available Techniques

Conventional Doppler

One of the first parameters assessed in order to determine cardiac function are blood outflow (systole) and inflow (diastole) in the heart, as well as appropriate timing (Lee et al. 2008; Hernandez-Andrade et al. 2012).

Conventional Doppler allows assessment of blood flow through the outflow tracts, which reflects *systolic* function. This measurement can be multiplied by the area of the outflow tracts to calculate the stroke volume – the amount of blood ejected per heart beat (Guyton and Hall 2006). Combining this information with the fetal heart rate allows estimation of cardiac output (volume per minute), which should normally be expressed as the cardiac index (cardiac output adjusted by fetal weight) (Guyton

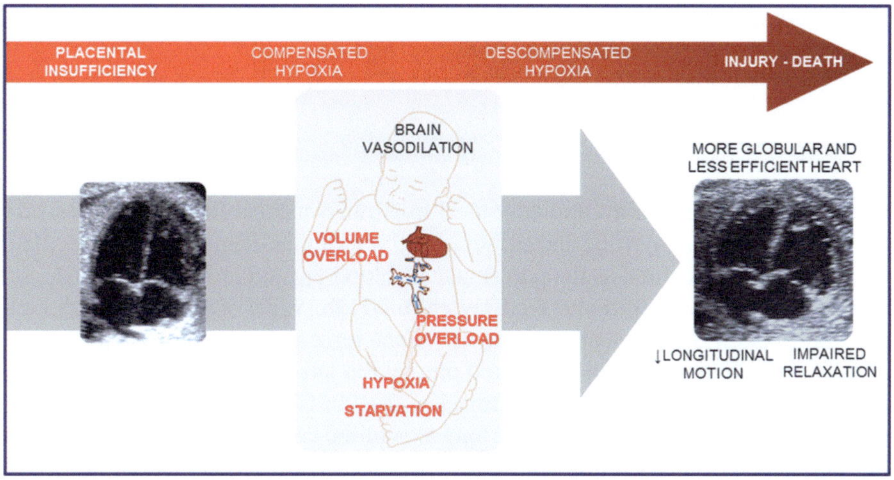

Fig. 6.2 Fetal cardiovascular adaptation to placental insufficiency with more globular and less efficient hearts

and Hall 2006; Hernandez-Andrade et al. 2012). Cardiac output is one of the classical parameters that directly reflects cardiac function, but it becomes abnormal only in the very late stages of deterioration (Hernandez-Andrade et al. 2012).

The main Doppler indices used to evaluate *diastolic* function are the early diastolic filling (atrial contraction), E/A ratios and precordial vein pulsatility indices (Lee et al. 2008; Kasper et al. 2005; Hernandez-Andrade et al. 2012). Ventricle filling typically has a biphasic pattern, reflecting picks E and A, and calculation of the E/A ratio essentially reflects ventricular relaxation (Jessup et al. 2009). However, this parameter is of little use in fetal life as it is strongly affected by respiratory and corporal movements and a high fetal heart rate usually leads to temporarily fused E/A waves (Van Mieghem et al. 2009a, b, c; Godfrey et al. 2012a, b). Another important limitation of this ratio is that impaired relaxation can be reflected by an increased, decreased, or pseudo-normal value, hampering interpretation (Jessup et al. 2009). Diastolic function can also be indirectly evaluated with Doppler assessment of the precordial veins, which reflect pressure changes in the right atrium (Van Mieghem et al. 2009a, b, c; Godfrey et al. 2012a, b). Additionally, assessment of DV is used in fetal medicine to reflect impaired relaxation and as an early marker of disease (Baschat et al. 2007; Timmerman et al. 2010).

Furthermore, Doppler can also be used to calculate *time periods* (Hernandez-Andrade et al. 2012) such as isovolumetric contraction and relaxation times – time elapsed from the start of contraction/relaxation and closure/opening of the outflow valve, respectively. In the very early stages of dysfunction these parameters become abnormal, particularly the isovolumetric relaxation time (IRT), when the time required to properly relax the myocardium increases. Time events can be evaluated individually or as a composite parameter, such as the myocardial performance index (MPI), which is considered a marker of global cardiac function (Cruz-Martinez et al. 2011; Van Mieghem et al. 2009a, b, c) and takes into account several systolic and diastolic time events (Tei et al. 1997).

M-mode

M-mode techniques are traditionally used in a transverse cardiac view to measure the difference in end-systolic and end-diastolic ventricular diameter and to calculate ejection fraction, defined as the percentage of blood ejected in each heart cycle, by applying the Teicholz formula (Godfrey et al. 2012b; Yagel et al. 2009). Although ejection fraction is the essential parameter characterizing heart failure in adulthood (Jessup et al. 2009), it is usually altered only in the late stages of deterioration as it mainly reflects radial function (Godfrey et al. 2012b; Cikes et al. 2010).

M-mode can be also applied in the long axis of the heart to evaluate tricuspid and mitral annular displacement – sensitive markers of cardiac dysfunction reflecting global longitudinal function (Gardiner et al. 2006a, b; Carvalho et al. 2001).

Tissue Doppler Imaging (TDI)

While conventional echocardiographic techniques are based on blood flow, TDI uses frequency shifts in ultrasound waves to calculate myocardial velocity characterized by a lower velocity and higher amplitude (Sutherland et al. 2006; Ho and Solomon 2006). TDI can be applied online to evaluate annular or myocardial velocities, as well as offline to assess deformation parameters (strain and strain rate).

Peak systolic strain and strain rate assessed at each myocardial segment provide information on myocardial deformation and interaction with neighboring segments (Sutherland et al. 2006). Accordingly, peak velocities evaluated at the mitral or tricuspid annulus reflect global systolic (S’) or diastolic (E’ and A’) myocardial motion and have been demonstrated to be an early and sensitive marker of cardiac dysfunction (Yu et al. 2007; Comas et al. 2010).

Although TDI may provide valuable information on global and regional myocardial motion and deformation, this technique should be used with caution in the fetal heart as it has several major limitations (Comas et al. 2012).

2D Speckle Tracking

A non-Doppler technology, 2D speckle tracking techniques, allows myocardial deformation to be quantified by using frame-by-frame tracking of bright myocardial areas (speckles) (Bijnens et al. 2009). 2D speckle tracking requires post-processing and off-line analysis of 2D images and allows myocardial strain and strain rate to be measured. Despite its potential advantages, this is a recent technique that still requires validation for use in the fetal heart (Germanakis and Gardiner 2012; Van Mieghem et al. 2009a, b, c; Crispi et al. 2012b).

4D Spatiotemporal Image Correlation

4D STIC technique is based on a sweep (volume data set) of the fetal heart containing a complete reconstructed cardiac cycle and permits 3D reconstruction of the fetal heart over time, so that any target region of interest can be obtained at any stage of the cardiac cycle. 4D STIC has been proposed to measure ventricular volumes that allow more accurate estimation of the cardiac output and ejection fraction, and the off-line analysis also allows mitral/tricuspid annulus displacement to be assessed. 4D STIC is a promising technique that requires further studies to improve its applicability in fetal cardiac function assessment (Godfrey et al. 2012b).

6.3.1.2 Technical Challenges of Fetal Imaging

Most fetal echocardiographic techniques are derived from parameters previously developed and validated in the adult heart. However, one has to take into account the specific functional anatomy of the fetal heart when interpreting the results of echocardiography.

Unlike in the adult heart, changes in fetal cardiomyocyte maturation (myocardium stiffness and intrinsic contractility) and loading occur in myocardium during gestation (Sedmera 2011; Tobita et al. 2005). Additionally, the fetal circulation pattern not only differs from that in the adult, with a predominant right heart and both circulations being connected (Kiserud and Acharya 2004), but may also change during pregnancy thus hampering the understanding of cardiac adaptation due to different insults (volume or pressure overload, hypoxia, cardiac compression, etc.) *in utero*.

Fetal heart evaluation is challenging due to the small heart size, high heart rate, and limited access to the fetus. Fetal echocardiography requires specific training and expertise to acquire images and interpret the results (Lee et al. 2008). Furthermore, several limitations should be taken into account when assessing fetal cardiac function, which are particularly important in techniques requiring offline analysis (4D STIC, color TDI, and 2D speckle tracking).

Fetal Position, Movement, Size

Since a fetus lies deep in the maternal pelvis/abdomen, factors such as maternal adiposity, oligoamnios, or an anterior placenta may interfere with image quality. In addition, fetal position changes constantly, requiring different angles to view the fetal heart, and evaluation of longitudinal or radial motion at an apical/basal or transverse view might be impossible if the fetal spine is persistently in an anterior position.

Both conventional and tissue Doppler are critically affected by the angle of acquisition, which should be as close to 0 as possible (Sutherland et al. 2006; Yagel et al. 2009). Other techniques such as 4D STIC or M-mode are less angle-dependent, but a good angle is still required to obtain reliable results. Fetal corporal and respiratory movements may also interfere with the quality of acquisition.

Moreover, the fetal heart is much smaller than the adult heart and changes with gestational age, requiring recalculation of z-scores and standardization of the measurements. Some fetal conditions may also affect heart size (e.g. cardiomegaly) and, therefore, reference values adjusted by heart size or specific fetal biometrics may be necessary to correct parameters that depend on myocardial size (such as annular displacement or myocardial velocities) (Comas et al. 2012). The small size of the fetal heart also reduces the accuracy of cardiac or vessel dimensions estimation and of the complex parameters based on several measurements (e.g. cardiac output), resulting in relatively wide variability as the error induced by one inaccurate dimension is multiplied in the final calculation (Hernandez-Andrade et al. 2012). Furthermore, heart size strongly limits any attempt to differentially evaluate endocardial and epicardial layers within the thin myocardium.

Fetal Heart Rate and Frame Rate Requirements

Proper acquisition, processing, and interpretation are even more critical in techniques requiring offline analysis, such as TDI or 2D speckle tracking, since software tools for offline analysis of deformation were initially designed for the adult heart with a low heart rate, fixed position, and electrocardiographic (ECG) co-registration (Comas et al. 2012; Germanakis and Gardiner 2012; D'hooge et al. 2000). The intrinsic characteristics of the fetal heart may hamper correct performance of the algorithms and, therefore, limit their reliability.

ECG co-registration, impossible in fetus, is not only critical to identify time events (D'hooge et al. 2000), but is also mandatory for the correct functioning of the offline cardiac software tools. Alternatively, manual indication of time events based on the underlying M-mode (Willruth et al. 2011) or 2D images (Crispi et al. 2012b) has been proposed in order to improve offline analysis of both TDI and 2D speckle tracking. Additionally, as the heart rate is about 2–3 times faster in fetuses than in adults, a higher frame rate would probably be necessary for proper offline analysis compared to adults (Sutherland et al. 2006), but optimal values remain to be defined.

Poor quality acquisitions with a low frame rate or lack of ECG co-registration may lead to incorrect results. A clear example of inconsistent data is the disagreement in reports on longitudinal strain changes throughout gestation, which first had been described as increasing in the studies performed with low frame rate acquisitions, but then have been shown to decrease when using more appropriate methodology (Matsui et al. 2011; Willruth et al. 2011; Simpson 2011).

Lack of Validation of Techniques in the Fetal Heart

Because invasive study of the fetal circulation is not feasible, most of the techniques used in fetal functional echocardiography have not been validated. Additionally, there are discrepancies in the literature on many fetal cardiac function parameters regarding methodology, normal values, and interpretation.

For example, measurement of MPI using either blood flow or valve clicks as landmarks leads to different normality values (Meriki and Welsh 2012). Another example is the Teichholz formula for ejection fraction, which assumes only normal adult heart geometry, but not fetal (Yagel et al. 2009). Moreover, the E/E' ratio has been demonstrated to correlate with intracavitary pressure at end-diastole (Nagueh et al. 1997) but its significance in fetal life is unknown. Finally, TDI and 2D speckle tracking techniques have been validated for deformation analysis in the adult heart by experimental settings including sonomicrometry (Sutherland et al. 2006). However, no validation studies using invasive procedures can be performed to ascertain the real strain and strain rate values in the fetal heart during the maturation process.

Despite these limitations, recent reports have demonstrated that deformation can be assessed in a reproducible manner when the appropriate methodology is employed (Matsui et al. 2011; Willruth et al. 2011; Crispi et al. 2012b).

As discussed above, fetal cardiac function assessment may have major limitations and therefore any technique or parameter proposed for its assessment should follow several steps for validation before being incorporated into clinical practice.

The first phase is to demonstrate *feasibility* and *reproducibility* in well-designed and conducted studies. Use of the proposed parameter following strict *methodological criteria* is also critical to ensure proper applicability. Then, the behavior of the parameter in normal fetal conditions (physiology), as well as in each clinical disease (pathophysiology), must be described before the technique or parameter can applied in clinical conditions.

6.3.2 Fetal Conditions Affecting Cardiac Function

6.3.2.1 Placental Insufficiency

Placental insufficiency is one of the main causes of FGR, affecting up to 10 % of all pregnancies with severe consequences of perinatal mortality and morbidity (Alberry and Soothill 2007). Early- and late-onset FGR are increasingly being recognized as distinct forms of the disease in view of their differences in incidence, pathophysiology, and natural history (Comas et al. 2010). Particularly, cardiac dysfunction is recognized as being among the central pathophysiologic features of both early- and late-onset FGR as fetal adaptive mechanisms to placental insufficiency (Van Mieghem et al. 2009a, b, c; Comas et al. 2011; D'hooge et al. 2000; Simpson 2011).

Early-Onset FGR

Early FGR (<34 gestational weeks) affects less than 1 % of deliveries, but has extremely high perinatal mortality and morbidity (Alberry and Soothill 2007), prediction of which is critical for the clinical management of these fetuses (Baschat et al. 2007). However, cardiac dysfunction in the fetus is largely subclinical and requires sensitive methods for its identification (Crispi et al. 2008). Previously, systolic dysfunction has been detected only in severely affected fetuses, suggesting that ejection fraction deteriorates only in the very late stages (Figueras et al. 2003) and the cardiac output adjusted by fetal weight remains normal throughout disease progression (Kiserud et al. 2006a). Advancing further in myocardial imaging techniques, recent studies have demonstrated that systolic annular peak velocities are decreased from the very early stages of FGR, long before atrial flow in the DV becomes abnormal (Comas et al. 2010; Larsen et al. 2009). Additionally, fetuses with early FGR show signs of impaired relaxation (diastolic dysfunction) from the early stages of deterioration (Crispi et al. 2008; Comas et al. 2010), as measured by increased pulsatility in precordial veins (particularly DV) (Baschat et al. 2007; Bilardo et al. 2004), higher E/A ratios (Baschat et al. 2007; Larsen et al. 2009; Mäkikallio et al. 2002; Watanabe et al. 2009; Naujorks et al. 2009), increased IRT (Niewiadomska-Jarosik et al. 2005), reduced diastolic annular peak velocities (Larsen et al. 2011), and increased cord blood levels of atrial and B-type natriuretic peptide (ANP/BNP) (Girsen et al. 2007). This decrease in longitudinal motion and impaired relaxation may be a fetal adaptive mechanism to the chronic hypoxia and volume/pressure overload of placental insufficiency.

Over the years, several cardiac parameters have been identified as main predictors of acidemia and adverse perinatal outcome in early FGR (Baschat et al. 2007; Bilardo et al. 2004; Girsen et al. 2007). In a recent systematic review, DV emerged as a strong predictor of mortality, with a sensitivity ranging from 40 to 60 % (Baschat 2010). Although preliminary data suggested that MPI could also help to predict perinatal mortality in early FGR (Hernandez-Andrade et al. 2009), a recent multicenter study has not confirmed these results (Cruz-Lemini et al. 2012).

Late-Onset FGR

Recently discovered evidence showed that in a proportion of small fetuses near term that have been considered to have good prognosis (Soothill et al. 1999), in reality represent true forms of late FGR with a mild degree of placental insufficiency (not reflected by umbilical artery Doppler). This population shows poorer perinatal (Illa et al. 2009) and long-term outcomes, including suboptimal neurodevelopment (Figueras et al. 2009; Eixarch et al. 2008) and a higher postnatal cardiovascular risk (Barker et al. 1989a, b; Crispi et al. 2010; Tintu et al. 2009).

A few recent studies have demonstrated that late-FGR fetuses might also exhibit features of cardiac dysfunction (Chaiworapongsa et al. 2002; Comas et al. 2011a, b; Cruz-Martinez et al. 2011). Moreover, increased cord blood levels of troponins have been detected in some late-SGA newborns (Chaiworapongsa et al. 2002) and more than 30 % of late-onset SGA fetuses even with normal umbilical artery Doppler show increased values of MPI (Comas et al. 2011; Cruz-Martinez et al. 2011) and decreased annular peak velocities (Comas et al. 2011a, b).

Therefore, it is critical to thoroughly evaluate the aforementioned sensitive cardiovascular parameters for timely identification and follow-up of this subgroup of small fetuses which might be at higher perinatal and long-term risk.

FGR and Postnatal Cardiovascular Risk

Strong postnatal persistence of cardiovascular remodeling has been recently demonstrated in a cohort of 5-year-old children who suffered early or late FGR (Crispi et al. 2010). It seems that FGR insult *in utero* induces critical primary cardiac changes, which not only remain in postnatal life, but might also develop into severe cardiovascular disease if left unattended.

The cardiovascular remodeling in FGR children manifests itself mainly in altered cardiac shape (more globular morphology), subclinical cardiac dysfunction (increased heart rate and reduced stroke volume and myocardial peak velocities), and vascular remodeling (increased blood pressure and carotid intima media thickness). Both cardiac and vascular changes are present in early and late FGR, with a tendency toward worse results in early-onset cases.

The conclusion can be drawn that a proper fetal cardiac function assessment might significantly improve identification of FGR cases at high cardiovascular risk later in life, therefore allowing timely preventive interventions.

6.3.2.2 Maternal Diabetes

Maternal diabetes is the most common cause of fetal cardiac structural defects (Lisowski et al. 2010) and hypertrophic cardiomyopathy, which is characterized by an enlarged heart, increased shortening fraction and systolic longitudinal motion (Gardiner et al. 2006a, b), thickening of the interventricular septum, and ventricular outflow obstruction. These changes occur in an otherwise structurally normal fetal heart in about a quarter of affected pregnancies (Gandhi et al. 1995).

Furthermore, the fetuses of diabetic mothers show signs of impaired relaxation (diastolic dysfunction), as measured by increased pulsatility in precordial veins (Zielinsky et al. 2003; Rizzo et al. 1992; Zielinsky et al. 2004; Stuart et al. 2010), a lower E/A ratio, an increased IRT (Turan et al. 2011), increased diastolic annular peak velocities and E/E' ratio (Hatém et al. 2008), and increased cord blood levels of ANP, BNP and troponin (Oran et al. 2003). The described cardiac hypertrophy and dysfunction persist postnatally in a proportion of these fetuses, leading to worse neonatal outcomes (Kozák-Bárány et al. 2004) and increased cardiovascular risk in adulthood (Manderson et al. 2002).

Since it is quite challenging to detect cardiac hypertrophy by ultrasonography early enough (Hatém et al. 2008; Turan et al. 2011), cardiac function parameters might serve as useful early predictors of outcomes in such complicated pregnancies.

6.3.2.3 Congenital Heart Disease

First-Trimester Screening of Congenital Heart Disease

Cardiac function assessment in the first trimester may help in the early detection of congenital heart disease. DV flow (Van Mieghem et al. 2009a, b, c; Martínez et al. 2010) and tricuspid regurgitation (Pereira et al. 2011) are associated with an increased prevalence of cardiac defects. Fetal nuchal translucency above the 95th percentile, DV atrial reversed flow, or tricuspid regurgitation are present in 35, 28, and 33 % of fetuses with cardiac defects, respectively, and in 5, 2, and 1 % of those without cardiac defects. A combination of these three markers may permit detection of about 50 % of cardiac defects, with a false-positive rate of 8 % (Pereira et al. 2011).

Fetal Cardiac Function in Congenital Heart Disease

Evaluation of cardiac function in fetuses with congenital heart disease has been proposed to better understand the pathophysiology of these diseases, predict the perinatal outcome, and guide and monitor any *in utero* therapy.

In 2004 Huhta et al. proposed a cardiovascular score, including the presence of hydrops, cardiomegaly, abnormal myocardial function (measured as a monophasic inflow pattern, tricuspid/mitral regurgitation, or reduced shortening fraction), abnormal DV or umbilical artery or vein Doppler (Huhta 2004). This score showed a high sensitivity and specificity in identifying fetuses with cardiac defects at high risk of perinatal mortality (Wieczorek et al. 2008). These data paved the way for future studies incorporating newer techniques for the evaluation of cardiac function.

However, various types of congenital heart disease show specific differences in volume and/or pressure load, as well as distinct degrees of cardiac remodeling and adaptation and therefore, TDI and 2D speckle tracking techniques have been proposed as additional diagnostic tools. Particularly peak annular velocities, ventricular strain, and the left/right ventricular strain ratio have been proposed as prognostic markers in these entities. It has been shown that fetuses with left heart obstruction (hypoplastic left heart syndrome and critical stenosis of the aorta) show reduced left ventricular strain and left/right ventricular strain ratio values, while right ventricular values are maintained within the normal range (Larsen et al. 2006; Willruth et al. 2011; Germanakis et al. 2012). In contrast, right heart defects such as Ebstein's anomaly, lead to reduced longitudinal right ventricular strain with increased left/right ventricular strain ratio values (Germanakis et al. 2012). Additionally, deformation parameters within normal ranges with a similar left/right ventricular strain ratio have been reported in other anomalies such as tetralogy of Fallot, double outlet right ventricle and atrial/ventricular septal defects (Willruth et al. 2011; Germanakis et al. 2012). The aforementioned findings suggest that specific cardiovascular profile scores adjusted for the type of cardiac defect might provide clinically useful information.

In Utero Cardiac Intervention

Fetal aortic balloon valvuloplasty improves the perinatal and postnatal outcome of critical aortic stenosis. Monophasic mitral inflow together with qualitative left ventricular dysfunction are included in the selection criteria to predict progression of hypoplastic left heart syndrome and eligibility for prenatal intervention (McElhinney et al. 2010). Additionally, left ventricle diastolic dysfunction including TDI peak velocity assessment has been suggested as a prognostic marker of poor long-term outcome in these cases (Friedman et al. 2011).

Differential Diagnosis of Cardiomyopathies

Hypertrophic cardiomyopathy is characterized by increased ventricular wall thickness that leads to impaired relaxation together with systolic dysfunction (demonstrated by a decreased shortening fraction), present in almost half of all cases of cardiomyopathy. In addition, most fetuses with dilated cardiomyopathy will have systolic dysfunction with or without chamber enlargement but without increased wall thickness. The presence of diastolic dysfunction is similar in both types of disease (about 60 % of cases) including abnormal venous flows (umbilical vein pulsations and biphasic inferior vena cava flow), a decreased E/A ratio, and an increased IRT.

Therefore, assessment of fetal cardiac function has been proposed as a diagnostic and prognostic tool in fetal cardiomyopathies. Moreover, the presence of systolic dysfunction and significant atrioventricular valve regurgitation are risk factors for mortality, when diastolic dysfunction (particularly abnormal systemic venous flow patterns and umbilical venous pulsations) is considered the best predictor of perinatal mortality (Pedra et al. 2002, 2005).

6.3.2.4 Twin-to-Twin Transfusion Syndrome (TTTS)

TTTS complicates 10–15 % of monochorionic twin pregnancies and is characterized by unbalanced chronic blood transfer from one twin, defined as the donor twin, to the other, defined as the recipient, through placental anastomoses (Quintero et al. 1999). Significant hemodynamic changes are commonly observed in both fetuses in TTTS including hypervolemia in the recipient and hypovolemia in the donor, resulting in chronic activation of the renin-angiotensin system and release of vasoactive factors, which leads to pressure overload in both fetuses (Karatza et al. 2002).

Prenatal Cardiac Function in TTTS

Despite the normal heart structure, cardiac function seems to be altered in TTTS fetuses. Most changes that have been discovered are related to volume and pressure overload, which lead to cardiomegaly and hypertrophy, usually with impaired relaxation but preserved systolic function (Stirnemann et al. 2010).

Volume overload is reflected by increased umbilical blood flow (Gratacós et al. 2002; Gungor et al. 2008) and DV pulsatility (Van Mieghem et al. 2009a, b, c; Karatza et al. 2002; Stirnemann et al. 2010), tricuspid and mitral insufficiency, cardiomegaly (Zosmer et al. 1994; Szwast et al. 2007), and increased levels of ANP and BNP (Van Mieghem et al. 2009a, b, c; Bajoria et al. 2002).

Pressure overload leads to cardiac hypertrophy in more than half of recipients (Karatza et al. 2002; Stirnemann et al. 2010; Szwast et al. 2007; Michelfelder et al. 2007), usually with an increase in ejection fraction (Zosmer et al. 1994). This explains the impaired relaxation and increased ventricular filling pressures that lead to reduced E/A ratios or a monophasic Doppler inflow profile together with increased E/E' ratios and MPI (mainly due to an increased IRT), with 10–20 % of recipients also showing some degree of right tract obstruction – pulmonary stenosis, dysplasia, insufficiency, or atresia (Van Mieghem et al. 2009a, b, c; Stirnemann et al. 2010; Divanović et al. 2011; Raboisson et al. 2004).

Another recent study reported a reduced right ventricular systolic strain and strain rate in recipients, which supports the concept of subclinical systolic dysfunction from the early stages of TTTS (Van Mieghem et al. 2009a, b, c). At the same time, most studies did not discover any significant changes in the donor's cardiac function apart from arterial and venous (umbilical artery and vein and DV) Doppler changes, reflecting hypovolemic status (Stirnemann et al. 2010; Michelfelder et al. 2007).

Several studies have evaluated the correlation between cardiac function and TTTS severity, demonstrating that cardiomegaly, systolic dysfunction, and right outflow tract obstruction are more prevalent in advanced Quintero stages (Gungor et al. 2008; Michelfelder et al. 2007). However, 20–60 % of recipients in stages I–II may also have cardiac hypertrophy, tricuspid regurgitation, increased MPI, BNP or troponin T blood levels (Van Mieghem et al. 2010; Stirnemann et al. 2010; Raboisson et al. 2004).

Postnatal Cardiac Function in TTTS

Most studies report normal postnatal cardiac evaluation in TTTS pregnancies treated by fetoscopy (Fesslova et al. 1998), with the exception of some recipients with pulmonary stenosis requiring postnatal valvuloplasty (Karatza et al. 2002; Nizard et al. 2001; Gray et al. 2006). However, a recent study evaluating TTTS survivors at school age has demonstrated reduced diastolic function in recipients and abnormal cardiac dimensions in some donors (Halvorsen et al. 2009). Increased blood pressure levels in the neonatal period (Mercanti et al. 2011) and persistent pulmonary hypertension after birth has also been described in some recipient twins (Takahashi et al. 2012).

6.3.2.5 Congenital Diaphragmatic Hernia

Left heart underdevelopment is commonly observed in fetuses with congenital diaphragmatic hernia, due to compression of the left atrium by herniated abdominal organs (mainly the liver), redistribution of fetal cardiac output, and/or low pulmonary venous return (Van Mieghem et al. 2009a, b, c; Stressig et al. 2010; Messing et al. 2011). However, cardiac dimensions seem not to be directly related to significant changes in cardiac function (as measured by ejection fraction or MPI) (Van Mieghem et al. 2009a, b, c) and the heart generally even catches up after surgical correction in the neonatal period, resulting in normal cardiac size in long-term survivors of isolated congenital diaphragmatic hernia (Stefanutti et al. 2004).

However, recent data in children has shown that MPI, systolic/diastolic time, and plasma levels of BNP are able to predict neonatal mortality and the need for extracorporeal membrane oxygenation (Aggarwal et al. 2011; Cua et al. 2009; Baptista et al. 2008). Additionally, tracheal occlusion improves MPI by shortening the isovolumic contraction time interval (Van Mieghem et al. 2009a, b, c).

6.3.2.6 Other Fetal Conditions

Several other fetal anomalies have been reported to be associated with cardiac dysfunction due to heart compression, volume overload or a direct insult to the myocardium.

Large intrathoracic masses such as a congenital cystic adenomatoid malformation with mediastinal deviation have been described as compressing the heart and affecting cardiac volume loading, sometimes leading to heart failure and hydrops. Increased E/A ratio, right MPI and poor filling secondary to cardiac compression and a tamponade effect (cardiomegaly, reduced cardiac output, preserved ejection fraction, and increased reversal atrial contraction in the inferior vena cava) usually appear before hydrops is observed (Mahle et al. 2000; Szwast et al. 2007).

Similarly, sacrococcygeal teratoma or arteriovenous malformations, characterized by high blood flow through the tumor (volume overload), can also lead to heart failure (Lee et al. 2011; Heling et al. 2000). The inferior vena cava and the cardiac ventricles dilate due to increased venous return from the tumor, even though these fetuses typically maintain a normal shortening fraction as intrinsic contractility is not affected until the final stages. Absence of the DV with extrahepatic shunt that directly connects umbilical vein flow with the heart causes volume overload that may lead to cardiomegaly, tricuspid regurgitation, heart failure, and hydrops in half of these cases (Berg et al. 2006; Acherman et al. 2010).

Lower urinary tract obstruction with massive bladder distension may cause vascular compression, leading to an increased cardiothoracic ratio, small pericardial effusion, ventricular hypertrophy, and diastolic dysfunction (altered E/A ratio and increased DV-PI) (Rychik et al. 2010).

Fetal anemia leads to hyperdynamia, relative volume overload, and hypoxia, which may cause cardiomegaly, heart failure, and hydrops. Surviving children who received intrauterine transfusions show a reduction in left ventricular mass and atrial area with preserved ventricular function in childhood (Dickinson et al. 2010).

Acute parvovirus B19 infection in pregnancy may lead to severe anemia caused by the destruction of red blood cell precursors but may also be a result of hypoalbuminemia, hepatitis, myocarditis, and placentitis, which can culminate in cardiac failure and subsequent hydrops fetalis or fetal death. Cardiac failure may be a result of severe anemia but may also be associated with myocarditis, which can cause arrhythmias or even cardiac arrest without evidence of anemia, cardiac failure, or hydrops (Lamont et al. 2011; Fishman et al. 2011).

Preterm rupture of membranes, particularly in cases with intra-amniotic infection, is associated with changes in fetal cardiac function consistent with increased left ventricular compliance and systolic dysfunction measured as an increased E/A ratio, shorter ejection time, and ventricular strain (Romero et al. 2004; Di Naro et al. 2010; Letti Müller et al. 2010).

Cardiac function assessment in the first trimester may help in the early detection of trisomy 21, as the presence of tricuspid regurgitation, atrial reversed flow in the DV, increased right E/A ratio and MPI, and shortening fraction and stroke volume are associated with Down syndrome even in the absence of structural heart defects (Calda et al. 2010). Signs of cardiac dysfunction persist in the second and third trimester in cases of trisomy 21 (Clur et al. 2011).

6.4 Conclusions and Future Perspectives

Fetal growth and function are regulated and influenced by a multitude of complex factors of the environment *in utero*, maternal status and genetic predisposition. Every abnormal change that occurs during this sensitive developmental period programs the health condition of the individual in postnatal life – thus, the phenomenon of "fetal programming" is one of the major research focuses of clinicians and scientist worldwide.

One of the most prevalent disorders affecting fetal development rate is fetal growth restriction (FGR) – a very complex and multifactorial disorder that often results in multiple adverse perinatal and postnatal complications including death. A growing number of studies over the last few decades continue to add evidence for the consistent long-term persistence of various severe FGR consequences, such as cardiovascular disease, neurobehavioral and metabolic disorders. Considering the high prevalence of FGR and the progressing availability of intervention strategies, it is of the highest clinical relevance to detect potential health risks as early as possible, to introduce timely preventive interventions and to adapt the life style in order to improve the long-term outcome of FGR cases.

In this context, significant progress has been made recently in recognizing the potential value of monitoring cardiac function for clinical management. However, most parameters are still in the research phase with the exception of DV, which is already being used in clinical practice for staging TTTS or monitoring early FGR. Some cardiac parameters with high sensitivity such as MPI or annular peak velocities have shown promising results in monitoring and predicting outcomes in FGR or congenital diaphragmatic hernia. Another example is the integration of some cardiac function parameters (mitral inflow and myocardial contractility) into the inclusion criteria for fetal valvuloplasty in critical aortic stenosis. Additionally, cardiac function assessment has proven utility in the differential diagnosis of cardiomyopathies or prediction of perinatal mortality in congenital heart disease. Similarly, first-trimester DV and tricuspid regurgitation assessment has an additional predictive value for Down syndrome and congenital heart disease. Other promising parameters that have recently been proposed remain to be validated in fetal heart assessment. Conversely, cardiac function assessment does not have demonstrated utility in the staging or prognosis of TTTS.

Complete assessment and integration of the distinct components of systolic and diastolic function are crucial as no single simple test can fully evaluate fetal cardiac function. Therefore, some scoring systems have been proposed to standardize the current practice of multimodal assessment, as a prognostic marker or to define heart failure. Although scoring may be superior to single measurements in defining compromised cardio-circulatory function, there are still many limitations for the incorporation of the proposed scores in clinical practice. Scores need to be validated not only to predict perinatal mortality but also to predict morbidity and long-term outcomes. Furthermore, many definitions of abnormal cardiac function need to be standardized, since cardiac function can be affected by various fetal abnormalities through volume or pressure overload, hypoxia, hyperglycemia, heart compression, direct myocardial damage, etc. Although disease-specific cardiovascular profile scores would most probably improve the predictive value for adverse outcomes, constructing such a profile would be highly complex. Despite these limitations, cardiac function can be adequately evaluated in most fetuses when appropriate expertise, equipment, and time are available. Many cardiac function parameters are sufficiently sensitive for selecting high-risk populations and predicting outcomes.

Although, there are still no general pediatric guidelines used in clinical practice, the implementation of correct follow-up and management could help tremendously the high-risk patients who suffered from fetal conditions affecting the cardiovascular system such as FGR. Therefore, based on the collective research data known today, such guidelines ideally should be developed and introduced in the clinical practice as soon as possible.

References

Acherman R, Rollins R, Castillo W, Evans W (2010) Stenosis of alternative umbilical venous pathways in absence of the ductus venosus. J Ultrasound Med 29(8):1227–1231

Aggarwal S, Stockman P, Klein M, Natarajan G (2011) The right ventricular systolic to diastolic duration ratio: a simple prognostic marker in congenital diaphragmatic hernia? Acta Paediatr 100(10):1315–1318

Alberry M, Soothill P (2007) Management of fetal growth restriction. Arch Dis Child Fetal Neonatal Ed 2007:62–67

Anderson R, Smerup M, Sanchez-Quintana D, Loukas M, Lunkenheimer P (2009) The three-dimensional arrangement of the myocytes in the ventricular walls. Clin Anat 22(1):64–76

Anthony R, Scheaffer A, Wright C, Regnault T (2003) Ruminant models of prenatal growth restriction. Reprod Suppl 2003:183–194

Bae S, Xiao Y, Li G, Casiano C, Zhang L (2003) Effect of maternal chronic hypoxic exposure during gestation in apoptosis in fetal heart. Am J Physiol Heart Circ Physiol 2003:H983–H990

Bajoria R, Ward S, Chatterjee R (2002) Natriuretic peptides in the pathogenesis of cardiac dysfunction in the recipient fetus of twin-twin transfusion syndrome. Am J Obstet Gynecol 186(1):121–127

Baptista M, Rocha G, Clemente F, Azevedo L, Tibboel D, Leite-Moreira A, Guimarães H, Areias J, Correia-Pinto J (2008) N-terminal-pro-B type natriuretic peptide as a useful tool to evaluate pulmonary hypertension and cardiac function in CDH infants. Neonatology 94(1):22–30

Barker D, Osmond C, Golding J, Kuh D, Wadsworth M (1989a) Growth in utero, blood pressure in childhood and adult life, and mortality from cardiovascular disease. BMJ 1989:564–567

Barker D, Winter P, Osmond C, Margetts B, Simmonds S (1989b) Weight in infancy and death from ischemic heart disease. Lancet 1989:577–580

Barker D, Hales C, Fall C, Osmond C, Phipps K, Clark P (1993) Type 2 diabetes (non insulin dependent) diabetes millitus, hypertension and hyperlipidaemia (syndrome X) relation to reduced fetal growth. Diabetologia 36:62–67

Barker P, Houle H, Li J, Miller S, Herlong J, Camitta M (2009) Global longitudinal cardiac strain and strain rate for assessment of fetal cardiac function: novel experience with velocity vector imaging. Echocardiography 26(1):28–36

Barnes S, Ozanne S (2011) Pathways linking the early environment to long-term health and lifespan. Prog Biophys Mol Biol 2011:323–336

Baschat AA (2004) Fetal responses to placental insufficiency: an update. BJOG 2004:1031–1041

Baschat AA (2010) Fetal growth restriction – from observation to intervention. J Perinat Med 2010:239–246

Baschat AA (2011) Neurodevelopment following fetal growth restriction and its relationship with antepartum parameters of placental dysfunction. Ultrasound Obstet Gynecol 2011:501–514

Baschat A, Cosmi E, Bilardo C, Wolf H, Berg C, Rigano S, Germer U, Moyano D, Turan S, Hartung J, Bhide A, Müller T, Bower S, Nicolaides K, Thilaganathan B, Gembruch U, Ferrazzi E, Hecher K, Galan H, Harman C (2007) Predictors of neonatal outcome in early-onset placental dysfunction. Obstet Gynecol 109(2 Pt 1):253–261

Bateman D, Ng S, Hansen C, Heagarty M (1993) The effects of intrauterine cocaine exposure in newborns. Am J Public Health 1993:190–193

Benediktsson R, Lindsay R, Noble J, Seckl J, Edwards C (1993) Glucocorticoid exposure in utero: new model for adult hypertension. Lancet 1993:339–341

Berenson G (2002) Childhood risk factors predict adult risk associated with subclinical cardiovascular disease. The Bogalusa Heart Study. Am J Cardiol 2002:3L–7L

Berg C, Kamil D, Geipel A, Kohl T, Knöpfle G, Hansmann M, Gembruch U (2006) Absence of ductus venosus-importance of umbilical venous drainage site. Ultrasound Obstet Gynecol 28(3):275–281

Berkley E, Chauhan S, Abuhamad A (2012) Doppler assessment of the fetus with intrauterine growth restriction. Am J Obstet Gynecol 2012:300–308

Bernstein I, Mongeon J, Badger G, Solomon L, Heil S, Higgins S (2005) Maternal smoking and its association with birth weight. Obstet Gynecol 2005:986–991

Bezerra D, Lacerda Andrade L, Pinto da Cruz F, Mandarim-de-Lacerda C (2008) Atorvastatin attenuates cardiomyocyte loss in adult rats from protein-restricted dams. J Card Fail 2008:151–160

Bijnens B, Cikes M, Claus P, Sutherland G (2009) Velocity and deformation imaging for the assessment of myocardial dysfunction. Eur J Echocardiogr 10(2):216–226

Bijnens B, Cikes M, Butakoff C, Sitges M, Crispi F (2012) Myocardial motion and deformation: what does it tell us and how does it relate to function? Fetal Diagn Ther 2012:5–16

Bilardo C, Wolf H, Stigter R, Ville Y, Baez E, Visser G, Hecher K (2004) Relationship between monitoring parameters and perinatal outcome in severe, early intrauterine growth restriction. Ultrasound Obstet Gynecol 23(2):119–125

Blake K, Gurrin L, Evans S, Beilin L, Landau L, Stanley F, Newnham J (2000) Maternal cigarette smoking during pregnancy, low birth weight and subsequent blood pressure in early childhood. Early Hum Dev 2000:137–147

Bogdarina I, Haase A, Langley-Evans S, Clark A (2010) Glucocorticoid effects on the programming of AT1b angiotensin receptor gene methylation and expression in the rat. PLoS One 2010, e9237

Brion M, Leary S, Lawlor D, Smith G, Ness A (2008) Modifiable maternal exposures and offspring blood pressure: a review of epidemiological studies of maternal age, diet, and smoking. Pediatr Res 2008:593–598

Briscoe T, Rehn A, Dieni S, Duncan J, Wlodek M, Owens J, Rees S (2004) Cardiovascular and renal disease in the adolescent guinea pig after chronic placental insufficiency. Am J Obstet Gynecol 2004:847–855

Burns A, McDonald I, Thomas J, Macisaac A, Prior D (2008) Doin' the twist: new tools for an old concept of myocardial function. Heart 94(8):978–983

Calda P, Brestak M, Tomek V, Ostadal B, Sonek J (2010) Left ventricle shortening fraction: a comparison between euploid and trisomy 21 fetuses in the first trimester. Prenat Diagn 30(4):368–371

Cambonie G, Comte B, Yzydorczyk C, Ntimbane T, Germain N, Lê N, Pladys P, Gauthier C, Lahaie I, Abran D, Lavoie J, Nuyt A (2007) Antenatal antioxidant prevents adult hypertension, vascular dysfunction, and microvascular rarefaction associated with in utero exposure to a low-protein diet. Am J Physiol Regul Integr Comp Physiol 2007:R1236–R1245

Carter R, Jacobson J, Sokol R, Avison M, Jacobson S (2013) Fetal alcohol-related growth restriction from birth through young adulthood and moderating effects of maternal prepregnancy weight. Alcohol Clin Exp Res 2013:452–456

Carvalho J, O'Sullivan C, Shinebourne E, Henein M (2001) Right and left ventricular long-axis function in the fetus using angular M-mode. Ultrasound Obstet Gynecol 18(6):619–622

Catalano P, Thomas A, Huston-Presley L, Amini S (2003) Increased fetal adiposity: a very sensitive marker of abnormal in utero development. Am J Obstet Gynecol 189(6):1698–1704

Catta-Preta M, Oliveira D, Mandarim-de-Lacerda C, Aguila M (2006) Adult cardiorenal benefits from postnatal fish oil supplement in rat offspring of low-protein pregnancies. Life Sci 2006:219–229

Chaiworapongsa T, Espinoza J, Yoshimatsu J, Kalache K, Edwin S, Blackwell S, Yoon B, Tolosa J, Silva M, Behnke E, Gomez R, Romero R (2002) Subclinical myocardial injury in small-for-gestational-age neonates. J Matern Fetal Neonatal Med 11(6):385–390

Cikes M, Sutherland G, Anderson L, Bijnens B (2010) The role of echocardiographic deformation imaging in hypertrophic myopathies. Nat Rev Cardiol 7(7):384–396

Clur S, Oude Rengerink K, Ottenkamp J, Bilardo C (2011) Cardiac function in trisomy 21 fetuses. Ultrasound Obstet Gynecol 37(2):163–171

Comas M, Crispi F, Cruz-Martinez R, Martinez JM, Figueras F, Gratacós E (2010) Usefulness of myocardial tissue Doppler vs conventional echocardiography in the evaluation of cardiac dysfunction in early-onset intrauterine growth restriction. A J Obstet Gynecol 2010:45.e1–7

Comas M, Crispi F, Cruz-Martinez R, Figueras F, Gratacos E (2011) Tissue doppler echocardiographic markers of cardiac dysfunction in small-for-gestational age fetuses. Am J Obstet Gynecol 57:e1–e6

Comas M, Crispi F (2012) Assessment of fetal cardiac function using tissue Doppler techniques. Fetal Diagn Ther. 32(1-2):30–8

Committee on Practice Bulletins-Gynecology, American College of Obstetricians and Gynecologists (2001) Intrauterine growth restriction. Clinical management guidelines for obstetrician-gynecologists. Int J Gynaecol Obstet 72(1):85–96

Crispi F, Gratacós E (2012) Fetal cardiac function: technical considerations and potential research and clinical applications. Fetal Diagn Ther 2012:47–64

Crispi F, Hernandez-Andrade E, Pelsers M, Plasencia W, Benavides-Serralde JA, Eixarch E, Le Noble F, Ahmed A, Glatz JF, Nicolaides KH, Gratacos E (2008) Cardiac dysfunction and cell damage across clinical stages of severity in growth-restricted fetuses. Am J Obstet Gynecol 2008:254.e1–8

Crispi F, Bijnens B, Figueras F, Bartrons J, Eixarch E, Le Noble F, Ahmed A, Gratacós E (2010) Fetal growth restriction results in remodeled and less efficient hearts in children. Circulation 2010:2427–2436

Crispi F, Figueras F, Cruz-Lemini M, Bartrons J, Bijnens B, Gratacos E (2012a) Cardiovascular programming in children born small for gestational age and relationship with prenatal signs of severity. Am J Obstet Gynecol 2012:121.e1–9

Crispi F, Sepulveda-Swatson E, Cruz-Lemini M, Rojas-Benavente J, Garcia-Posada R, Dominguez J, Sitges M, Bijnens B, Gratacós E (2012b) Feasibility and reproducibility of a standard protocol for 2D-strain and tissue Doppler based strain analysis of the fetal heart. Fetal Diagn Ther 32(1–2):96–108

Cruz-Lemini M, Crispi F, Van Mieghem T, Pedraza D, Cruz-Martínez R, Acosta-Rojas R, Figueras F, Parra-Cordero M, Deprest J, Gratacós E (2012) Risk of perinatal death in early-onset intrauterine growth restriction according to gestational age and cardiovascular Doppler indices: a multicenter study. Fetal Diagn Ther 2012:116–122

Cruz-Lemini M, Crispi F, Valenzuela-Alcaraz B, Figueras F, Gomez O, Sitges M, Bijnens B, Gratacos E (2013) A fetal cardiovascular score to predict infant hypertension and arterial remodeling in intrauterine growth restriction. In progress

Cruz-Martinez R, Figueras F, Hernandez-Andrade E, Oros D, Gratacos E (2011) Changes in myocardial performance index and aortic isthmus and ductus venosus Doppler in term, small-for-gestational age fetuses with normal umbilical artery pulsatility index. Ultrasound Obstet Gynecol 2011:400–405

Cua C, Cooper A, Stein M, Corbitt R, Nelin L (2009) Tissue Doppler changes in three neonates with congenital diaphragmatic hernia. ASAIO J 55(4):417–419

Curhan GC, Willett WC, Rimm EB, Spiegelman D, Ascherio AL, Stampfer MJ (1996) Birth weight and adult hypertension, diabetes mellitus, and obesity in US men. Circulation 94:3246–3250

Dahl L, Kaaresen P, Tunby J, Handegård B, Kvernmo S, Rønning J (2006) Emotional, behavioral, social, and academic outcomes in adolescents born with very low birth weight. Pediatrics 118(2):e449–e459

D'hooge J, Heimdal A, Jamal F, Kukulski T, Bijnens B, Rademakers F, Hatle L, Suetens P, Sutherland G (2000) Regional strain and strain rate measurements by cardiac ultrasound: principles, implementation and limitations. Eur J Echocardiogr 1(3):154–170

Di Naro E, Cromi A, Ghezzi F, Giocolano A, Caringella A, Loverro G (2010) Myocardial dysfunction in fetuses exposed to intraamniotic infection: new insights from tissue Doppler and strain imaging. Am J Obstet Gynecol 203(5):459.e1–7

Dickinson J, Sharpe J, Warner T, Nathan E, D'Orsogna L (2010) Childhood cardiac function after severe maternal red cell isoimmunization. Obstet Gynecol 116(4):851–857

Diderholm B (2009) Perinatal energy metabolism with reference to IUGR & SGA: studies in pregnant women & newborn infants. Indian J Med Res 2009:612–617

Divanović A, Cnota J, Ittenbach R, Tan X, Border W, Crombleholme T, Michelfelder E (2011) Characterization of diastolic dysfunction in twin-twin transfusion syndrome: association between Doppler findings and ventricular hypertrophy. J Am Soc Echocardiogr 24(8):834–840

Dusick A, Poindexter B, Ehrenkranz R, Lemons J (2003) Growth failure in the preterm infant: can we catch up? Semin Perinatol 27(4):302–310

Dwyer C, Stickland N (1992) The effects of maternal undernutrition on maternal and fetal serum insulin-like growth factors, thyroid hormones and cortisol in the guinea pig. J Dev Physiol 1992:303–313

Edwards C, Benediktsson R, Lindsay R, Seckl J (1993) Dysfunction of placental glucocorticoid barrier: link between fetal environment and adult hypertension? Lancet 1993:355–357

Ehrenkranz R, Younes N, Lemons J, Fanaroff A, Donovan E, Wright L, Katsikiotis V, Tyson J, Oh W, Shankaran S, Bauer C, Korones S, Stoll B, Stevenson D, Papile L (1999) Longitudinal growth of hospitalized very low birth weight infants. Pediatrics 104(2):280–289

Eixarch E, Meler E, Iraola A, Illa M, Crispi F, Hernandez-Andrade E, Gratacos E, Figueras F (2008) Neurodevelopmental outcome in 2-year-old infants who were small-for-gestational age term fetuses with cerebral blood flow redistribution. Ultrasound Obstet Gynecol 32(7):894–899

Eixarch E, Hernandez-Andrade E, Crispi F, Illa M, Torre I, Figueras F, Gratacos E (2011) Impact on fetal mortality and cardiovascular Doppler of selective ligature of uteroplacental vessels compared with undernutrition in rabbit model of IUGR. Placenta :304–309, 17 Feb 2011

Ergaz Z, Avgil M, Ornoy A (2005) Intrauterine growth restriction – etiology and consequences: what do we know about the human situation and experimental animal models? Reprod Toxicol 2005:301–322

Eriksson J, Forsén T, Tuomilehto J, Winter P, Osmond C, Barker D (1999) Catch-up growth in childhood and death from coronary heart disease: longidutinal study. Bone Miner J 1999:427–431

Eriksson J, Forsén T, Tuomilehto J, Osmond C, Barker D (2001) Early growth amd coronary heart disease in later life: longitudinal study. Bone Miner J 2001:949–953

Fesslova V, Villa L, Nava S, Mosca F, Nicolini U (1998) Fetal and neonatal echocardiographic findings in twin-twin transfusion syndrome. Am J Obstet Gynecol 179(4):1056–1062

Figueras F, Gardosi J (2011) Intrauterine growth restriction: new concepts in antenatal surveillance, diagnosis, and management. Am J Obstet Gynecol 2011:288–300

Figueras F, Puerto B, Martinez J, Cararach V, Vanrell J (2003) Cardiac function monitoring of fetuses with growth restriction. Eur J Obstet Gynecol Reprod Biol 110(2):159–163

Figueras F, Eixarch E, Gratacos E, Gardosi J (2008) Predictiveness of antenatal umbilical artery Doppler for adverse pregnancy outcome in small-for-gestational-age babies according to customised birthweight centiles: population-based study. BJOG 2008:590–594

Figueras F, Oros D, Cruz-Martinez R, Padilla N, Hernandez-Andrade E, Botet F, Costas-Moragas C, Gratacos E (2009) Neurobehavior in term, small-for-gestational age infants with normal placental function. Pediatrics 124(5):e934–e941

Fishman S, Pelaez L, Baergen R, Carroll S (2011) Parvovirus-mediated fetal cardiomyopathy with atrioventricular nodal disease. Pediatr Cardiol 32(1):84–86

Forsén T, Eriksson J, Tuomilehto J, Teramo K, Osmond C, Barker D (1997) Mother's weight in pregnancy and coronary heart disease in a cohort of Finish men: follow up study. Bone Miner J 1997:837–840

Forsén T, Eriksson J, Tuomilehto J, Osmond C, Barker D (1999) Growth in utero and during childhood among women who develop coronary heart disease: longitudinal study. Bone Miner J 1999:1403–1407

Friedman KG1, MRGD, Harrild D, Emani S, Wilkins-Haug L, McElhinney D, Tworetzky W (2011) Postnatal left ventricular diastolic function after fetal aortic valvuloplasty. Am J Cardiol 108(4):556–560

Gandhi J, Zhang X, Maidman J (1995) Fetal cardiac hypertrophy and cardiac function in diabetic pregnancies. Am J Obstet Gynecol 173(4):1132–1136

Gardiner H, Pasquini L, Wolfenden J, Barlow A, Li W, Kulinskaya E, Henein M (2006a) Myocardial tissue Doppler and long axis function in the fetal heart. Int J Cardiol 113(1):39–47

Gardiner H, Pasquini L, Wolfenden J, Kulinskaya E, Li W, Henein M (2006b) Increased periconceptual maternal glycated haemoglobin in diabetic mothers reduces fetal long axis cardiac function. Heart 92(8):1125–1130

Garofano A, Czernichow P, Bréant B (1998) Postnatal somatic growth and insulin contents in moderate or severe intrauterine growth retardation in the rat. Biol Neonate 1998:89–98

Gennser G, Rymark P, Isberg P (1988) Low birth weight and risk of high blood pressure in adulthood. Bone Miner J 1988:1498–1500

Germanakis I, Gardiner H (2012) Assessment of fetal myocardial deformation using speckle tracking techniques. Fetal Diagn Ther 32(1–2):39–46

Germanakis I, Matsui H, Gardiner H (2012) Myocardial strain abnormalities in fetal congenital heart disease assessed by speckle tracking echocardiography. Fetal Diagn Ther 31(1–2):123–130

Gillman M, Rifas-Shiman S, Berkey C, Field A, Colditz G (2003) Maternal gestational diabetes, birth weight, and adolescent obesity. Pediatrics 111(3):e221–e226
Girsen A, Ala-Kopsala M, Mäkikallio K, Vuolteenaho O, Räsänen J (2007) Cardiovascular hemodynamics and umbilical artery N-terminal peptide of proB-type natriuretic peptide in human fetuses with growth restriction. Ultrasound Obstet Gynecol 29(3):296–303
Gluckman PD (2004) Developmental origins of disease paradigm: a mechanistic and evolutionary perspective. Pediatr Res 2004:311–317
Gluckman PD, Hanson MA, Cooper C, Thornburg KL (2008) Effect of in utero and early-life conditions on adult health. N Engl J Med 2008:61–73
Godfrey M, Messing B, Cohen S, Valsky D, Yagel S (2012a) Functional assessment of the fetal heart: a review. Ultrasound Obstet Gynecol 39(2):131–144
Godfrey M, Messing B, Valsky D, Cohen S, Yagel S (2012b) Fetal cardiac function: M-mode and 4D spatiotemporal image correlation. Fetal Diagn Ther 32(1–2):17–21
Gratacós E, Van Schoubroeck D, Carreras E, Devlieger R, Roma E, Cabero L, Deprest J (2002) Impact of laser coagulation in severe twin-twin transfusion syndrome on fetal Doppler indices and venous blood flow volume. Ultrasound Obstet Gynecol 20(2):125–130
Gray P, Cincotta R, Chan F, Soong B (2006) Perinatal outcomes with laser surgery for twin-twin transfusion syndrome. Twin Res Hum Genet 9(3):438–443
Gungor S, Glosemeyer P, Huber A, Hecher K, Baschat A (2008) Umbilical venous volume flow in twin-twin transfusion syndrome. Ultrasound Obstet Gynecol 32(6):800–806
Guyton A, Hall J (2006) Textbook of medical physiology, 11th edn. Elsvier Saunder, Philadelphia
Hack M (2006) Young adult outcomes of very-low-birth-weight children. Semin Fetal Neonat Med 11(2):127–137
Hack M, Youngstrom E, Cartar L, Schluchter M, Taylor H, Flannery D, Klein N, Borawski E (2004) Behavioral outcomes and evidence of psychopathology among very low birth weight infants at age 20 years. Pediatrics 114(4):932–940
Hales C, Barker D, Clark P, Cox L, Fall C, Osmond C, Winter P (1991) Fetal and infant growth and impaired glucose tolerance at age 64. BMJ 1991:1019–1022
Halvorsen C, Bilock S, Pilo C, Sonesson S, Norman M (2009) Childhood cardiac function after twin-to-twin transfusion syndrome–a 10-year follow up. Acta Paediatr 98(9):1468–1474
Han J, Lawlor D, Kimm S (2010) Childhood obesity. Lancet 375:1737–1748
Harder T, Rodekamp E, Schnellong K, Dudenhausen J, Plagemann A (2007) Birth weight and subsequent risk of type 2 diabetes: a meta-analysis. Am J Epidemiol 165:849–857
Harvey D, Prince J, Bunton J, Parkinson C, Campbell S (1982) Abilities of children who were small-for-gestational-age babies. Pediatrics 69(3):296–300
Hatém M, Zielinsky P, Hatém D, Nicoloso L, Manica J, Piccoli A, Zanettini J, Oliveira V, Scarpa F, Petracco R (2008) Assessment of diastolic ventricular function in fetuses of diabetic mothers using tissue Doppler. Cardiol Young 18(3):297–302
Hecher K, Campbell S, Doyle P, Harrington K, Nicolaides K (1995) Assessment of fetal compromise by Doppler ultrasound investigation of the fetal circulation: arterial, intracardiac, and venous blood flow velocity studies. Circulation 1995:129–138
Hecher K, Bilardo CM, Stigter RH, Ville Y, Hackelöer BJ, Kok HJ, Senat MV, Visser GH (2001) Monitoring of fetuses with intrauterine growth restriction: a longitudinal study. Ultrasound Obstet Gynecol 2001:564–570
Heling K, Chaoui R, Bollmann R (2000) Prenatal diagnosis of an aneurysm of the vein of Galen with three-dimensional color power angiography. Ultrasound Obstet Gynecol 15(4):333–336
Hernandez-Andrade E, Crispi F, Benavides-Serralde J, Plasencia W, Diesel H, Eixarch E, Acosta-Rojas R, Figueras F, Nicolaides K, Gratacós E (2009) Contribution of the myocardial performance index and aortic isthmus blood flow index to predicting mortality in preterm growth-restricted fetuses. Ultrasound Obstet Gynecol 34(4):430–436
Hernandez-Andrade E, Benavides-Serralde J, Cruz-Martinez R, Welsh A, Mancilla-Ramirez J (2012) Evaluation of conventional Doppler fetal cardiac function parameters: E/A ratios, outflow tracts, and myocardial performance index. Fetal Diagn Ther 32(1–2):22–29

Ho C, Solomon S (2006) A clinician's guide to tissue Doppler imaging. Circulation 113(10):e396–e398
Hovi P, Andersson S, Eriksson J, Järvenpää A, Strang-Karlsson S, Mäkitie O, Kajantie E (2007) Glucose regulation in young adults with very low birth weight. N Engl J Med 356(20):2053–2063
Huhta J (2004) Guidelines for the evaluation of heart failure in the fetus with or without hydrops. Pediatr Cardiol 25(3):274–286
Huizinga C, Engelbregt M, Rekers-Mombarg L, Vaessen S, Delemarre-van de Waal H, Fodor M (2004) Ligation of the uterine artery and early postnatal food restriction – animal models for growth retardation. Horm Res 2004:233–240
Hult M, Tornhammar P, Ueda P, Chima C, Bonamy A, Ozumba B, Norman M (2010) Hypertension, diabetes and overweight: looming legacies of the Biafram famine. PLoS One 5:e13582
Huxley R, Shiell A, Law C (2000) The role of size at birth and postnatal catch-up growth in determining systolic blood pressure: a systematic review of the literature. J Hypertens 2000:815–831
Illa M, Coloma J, Eixarch E, Meler E, Iraola A, Gardosi J, Gratacós E, Figueras F (2009) Growth deficit in term small-for-gestational fetuses with normal umbilical artery Doppler is associated with adverse outcome. J Perinat Med 37(1):48–52
Iruretagoyena J, Gonzalez-Tendero A, Garcia-Canadilla P, Amat-Roldan I, Torre I, Nadal A, Crispi F, Gratacos E (2014) Cardiac dysfunction is associated with altered sarcomere ultrastructure in intrauterine growth restriction. Am J Obstet Gynecol 210(6):550.e1–7
Jackson A, Dunn R, Marchand M, Langley-Evans S (2002) Increased systolic blood pressure in rats induced by a maternal low-protein diet is reversed by dietary supplementation with glycine. Clin Sci (Lond) 2002:633–639
Jaddoe V, Verburg B, de Ridder M, Hofman A, Mackenbach J, Moll H, Steegers E, Witteman J (2007) Maternal smoking and fetal growth characteristics in different periods of pregnancy: the generation R study. Am J Epidemiol 2007:1207–1215
Jaquet D, Gaboriau A, Czernichow P, Levy-Marchal C (2000) Insulin resistance early in adulthood in subjects born with intrauterine growth retardation. J Clin Endocrinol Meta 85(4):1401–1406
Jessup M, Abraham W, Casey D, Feldman A, Francis G, Ganiats T, Konstam M, Mancini D, Rahko P, Silver M, Stevenson L, Yancy C (2009) 2009 focused update: ACCF/AHA Guidelines for the Diagnosis and Management of Heart Failure in Adults: a report of the American College of Cardiology Foundation/American Heart Association Task Force on Practice Guidelines. Circulation 119(14):1977–2016
Jones C, Lafeber H, Roebuck M (1984) Studies on the growth of the fetal guinea pig. Changes in plasma hormone concentration during normal and abnormal growth. J Dev Physiol 1984:461–472
Kajantie E, Barker D, Osmond C, Forsén T, Eriksson J (2008) Growth before 2 years of age and serum lipids 60 years later: the Helsinki Birth Cohort study. Int J Epidemiol 37(2):280–289
Karatza A, Wolfenden J, Taylor M, Wee L, Fisk N, Gardiner H (2002) Influence of twin-twin transfusion syndrome on fetal cardiovascular structure and function: prospective case–control study of 136 monochorionic twin pregnancies. Heart 88(3):271–277
Kasper D, Braunwald E, Fauci A, Hauser S, Longo D, Jameson J, Loscalzo J (2005) Harrison's principles of internal medicine, 16th edn. McGraw-Hill, New York
Kelly T, Yang W, Chen C, Reynolds K, He J (2008) Global burden of obesity in 2005 and projections in 2030. Ont J Obes 32:1431–1437
Kensara O, Wootton SA, Phillips DI, Patel M, Jackson A, Elia M, Group HS (2005) Fetal programming of body composition: relation between birth weight and body composition measured with dual-energy X-ray absorptiometry and anthropometric methods in older Englishmen. Am J Clin Nutr 82(5):980–987
Kiserud T, Acharya G (2004) The fetal circulation. Prenat Diagn 24(13):1049–1059
Kiserud T, Ebbing C, Kessler J, Rasmussen S (2006a) Fetal cardiac output, distribution to the placenta and impact of placental compromise. Ultrasound Obstet Gynecol 28(2):126–136

Kiserud T, Kessler J, Ebbing C, Rasmussen S (2006b) Ductus venosus shunting in growth-restricted fetuses and the effect of umbilical circulatory compromise. Ultrasound Obstet Gynecol 2006:143–149

Koeners M, Braam B, Joles J (2011a) Perinatal inhibition of NF-kappaB has long-term antihypertensive effects in spontaneously hypertensive rats. J Hypertens 2011:1160–1166

Koeners M, Wesseling S, Ulu A, Sepúlveda R, Morisseau C, Braam B, Hammock B, Joles J (2011b) Soluble epoxide hydrolase in the generation and maintenance of high blood pressure in spontaneously hypertensive rats. Am J Physiol Endocrinol Metab 2011:E691–E698

Kozák-Bárány A, Jokinen E, Kero P, Tuominen J, Rönnemaa T, Välimäki I (2004) Impaired left ventricular diastolic function in newborn infants of mothers with pregestational or gestational diabetes with good glycemic control. Early Hum Dev 77(1–2):13–22

Kutschera J, Tomaselli J, Urlesberger B, Maurer U, Häusler M, Gradnitzer E, Burmucic K, Müller W (2002) Absent or reversed end-diastolic blood flow in the umbilical artery and abnormal Doppler cerebroplacental ratio–cognitive, neurological and somatic development at 3 to 6 years. Early Hum Dev 69(1–2):47–56

Lafeber H, Rolph T, Jones C (1984) Studies on the growth of the fetal guinea pig. The effects of ligation of the uterine artery on organ growth and development. J Dev Physiol 1984:441–459

Lamont R, Sobel J, Vaisbuch E, Kusanovic J, Mazaki-Tovi S, Kim S, Uldbjerg N, Romero R (2011) Parvovirus B19 infection in human pregnancy. BJOG 118(2):175–186

Langley S, Jackson A (1994) Increased systolic blood pressure in adult rats induced by fetal exposure to maternal low protein diets. Clin Sci (Lond) 1994:217–222

Langley-Evans S (1997) Hypertension induced by foetal exposure to a maternal low protein diet, in the rat, is prevented by pharmacological blockage of maternal glucocorticoid synthesis. J Hypertens 1997:537–544

Langley-Evans S, Phillips G, Benediktsson R, Gardner D, Edwards C, Jackson A, Seckl J (1996a) Protein intake in pregnancy, placental glucocorticoid metabolism and the programming of hypertension in the rat. Placenta 1996:169–172

Langley-Evans S, Welham S, Sherman R, Jackson A (1996b) Weanling rats exposed to maternal low-protein diets during discrete periods of gestation exhibit differing severity of hypertension. Clin Sci 1996:607–615

Larsen L, Clausen H, Jønsson L (2002) Stereologic examination of placentas from mothers who smoke during pregnancy. Am J Obstet Gynecol 2002:531–537

Larsen L, Petersen O, Norrild K, Sorensen K, Uldbjerg N, Sloth E (2006) Strain rate derived from color Doppler myocardial imaging for assessment of fetal cardiac function. Ultrasound Obstet Gynecol 27(2):210–213

Larsen L, Sloth E, Petersen O, Pedersen T, Sorensen K, Uldbjerg N (2009) Systolic myocardial velocity alterations in the growth-restricted fetus with cerebroplacental redistribution. Ultrasound Obstet Gynecol 34(1):62–67

Larsen L, Petersen O, Sloth E, Uldbjerg N (2011) Color Doppler myocardial imaging demonstrates reduced diastolic tissue velocity in growth retarded fetuses with flow redistribution. Eur J Obstet Gynecol Reprod Biol 155(2):140–145

Law C, Shiell A (1996) Is blood pressure inversely related to birth weight? The strength of evidence from a systematic review of the literature. J Hypertens 1996:935–941

Lee W, Allan L, Carvalho J, Chaoui R, Copel J, Devore G, Hecher K, Munoz H, Nelson T, Paladini D, Yagel S, ISUOG Fetal Echocardiography Task Force (2008) ISUOG consensus statement: what constitutes a fetal echocardiogram? Ultrasound Obstet Gynecol 32(2):239–242

Lee M, Won H, Hyun M, Lee H, Shim J, Lee P, Kim A (2011) Perinatal outcome of sacrococcygeal teratoma. Prenat Diagn 31(13):1217–1221

Leichter J (1995) Decreased birth weight and attainment of postnatal catch-up growth in offspring of rats exposed to cigarette smoke during gestation. Growth Dev Aging 1995:63–66

Leon D, Lithell H, Vâgerö D, Koupilová I, Mohsen R, Berglund L, Lithell U, McKeigue P (1998) Reduced fetal growth rate and increased risk of death from ischemic heart disease: cohort study of 15,000 Swedish men and women born 1915–29. BMJ 1998:241–245

Leon D, Johansson M, Rasmussen F (2000) Gestational age and growth rate of fetal mass are inversely associated with systolic blood pressure in young adults: an epidemiologic study of 165,136 Swedish men aged 18 years. Am J Epidemol 2000:597–604

Leppänen M, Ekholm E, Palo P, Maunu J, Munck P, Parkkola R, Matomäki J, Lapinleimu H, Haataja L, Lehtonen L, Rautava P, Group PS (2010) Abnormal antenatal Doppler velocimetry and cognitive outcome in very-low-birth-weight infants at 2 years of age. Ultrasound Obstet Gynecol 36(2):178–185

Lesage J, Blondeau B, Grino M, Bréant B, Dupouy J (2001) Maternal undernutrition during late gestation induces fetal overexposure to glucocorticoids and intrauterine growth retardation, and disturbs the hypothalamo-pituitary axis in the newborn rat. Endocrinology 2001:1692–1702

Letti Müller A, Barrios Pde M, Kliemann L, Valério E, Gasnier R, Magalhães J (2010) Tei index to assess fetal cardiac performance in fetuses at risk for fetal inflammatory response syndrome. Ultrasound Obstet Gynecol 36(1):26–31

Levitt N, Lindsay R, Holmes M, Seckl J (1996) Dexamethasone in the last week of pregnancy attenuates hippocacampal glucocorticoid receptor gene expression and elevates blood pressure in the adult offspring. Neuroendocrynology 1996:412–418

Levitt N, Lambert E, Woods D, Hales C, Andrew R, Seckl J (2000) Impaired glucose tolerance and elevated blood pressure in low birth weight, nonobese, young south African adults: early programming of cortisol axis. J Clin Endocrinol Metab 2000:4611–4618

Ley D, Laurin J, Bjerre I, Marsal K (1996) Abnormal fetal aortic velocity waveform and minor neurological dysfunction at 7 years of age. Ultrasound Obstet Gynecol 8(3):152–159

Li G, Xiao Y, Estrella J, Ducsay C, Gilbert R, Zhang L (2003) Effect of fetal hypoxia on heart susceptibility to ischemia and reperfusion injury in the adult rat. J Soc Gynecol Invest 2003:265–274

Lindsay R, Lindsay R, Edwards C, Seckl J (1996) Inhibition of 11 beta-hydroxysteroid dehydrogenase in pregnant rats and the programming of blood pressure in offspring. Hypertension 1996:1200–1204

Lingohr M, Buettner R, Rhodes C (2002) Pancreatic beta-cell growth and survival: a role in obesity-linked type 2 diabetes? Trends Mol Med 8:375–384

Lisowski L, Verheijen P, Copel J, Kleinman C, Wassink S, Visser G, Meijboom E (2010) Congenital heart disease in pregnancies complicated by maternal diabetes mellitus. An international clinical collaboration, literature review, and meta-analysis. Herz 5(1):19–26

Lussana F, Painter RC, Ocke M, Buller H, Bossuyt P, Roseboom T (2008) Prenatal exposure to the Dutch famine is associated with a preference for fatty foods and a more atherogenic lipid profile. Am J Clin Nutr 88(6):1648–1652

Mahle W, Rychik J, Tian Z, Cohen M, Howell L, Crombleholme T, Flake A, Adzick N (2000) Echocardiographic evaluation of the fetus with congenital cystic adenomatoid malformation. Ultrasound Obstet Gynecol 16(7):620–624

Mäkikallio K, Vuolteenaho O, Jouppila P, Räsänen J (2002) Ultrasonographic and biochemical markers of human fetal cardiac dysfunction in placental insufficiency. Circulation 105(17):2058–2063

Manderson J, Mullan B, Patterson C, Hadden D, Traub A, McCance D (2002) Cardiovascular and metabolic abnormalities in the offspring of diabetic pregnancy. Diabetologia 45(7):991–996

Martínez J, Comas M, Borrell A, Bennasar M, Gómez O, Puerto B, Gratacós E (2010) Abnormal first-trimester ductus venosus blood flow: a marker of cardiac defects in fetuses with normal karyotype and nuchal translucency. Ultrasound Obstet Gynecol 35(3):267–272

Martyn C, Barker D, Jespersen S, Greenwald S, Osmond C, Berry C (1995) Growth in utero, adult blood pressure, and arterial compliance. Br Heart J 1995:116–121

Matsui H, Germanakis I, Kulinskaya E, Gardiner H (2011) Temporal and spatial performance of vector velocity imaging in the human fetal heart. Ultrasound Obstet Gynecol 37(2):150–157

McCowan L, Pryor J, Harding J (2002) Perinatal predictors of neurodevelopmental outcome in small-for-gestational-age children at 18 months of age. Am J Obstet Gynecol 186(5):1069–1075

McElhinney D, Vogel M, Benson C, Marshall A, Wilkins-Haug L, Silva V, Tworetzky W (2010) Assessment of left ventricular endocardial fibroelastosis in fetuses with aortic stenosis and evolving hypoplastic left heart syndrome. Am J Cardiol 106(12):1792–1797

McLean M, Chipps D, Cheung N (2006) Mother to child transmission of diabetes mellitus: does gestational diabetes program Type 2 diabetes in the next generation? Diabet Med 23(11):1213–1215

McMillen I, Robinson J (2005) Developmental origins of the metabolic syndrome: prediction, plasticity, and programming. Physiol Rev 2005:571–633

Mercanti I, Boivin A, Wo B, Vlieghe V, Le Ray C, Audibert F, Fouron J, Leduc L, Nuyt A (2011) Blood pressures in newborns with twin-twin transfusion syndrome. J Perinatol 31(6):417–424

Meriki N, Welsh A (2012) Technical considerations for measurement of the fetal left modified myocardial performance index. Fetal Diagn Ther 31(1):76–80

Messing B, Cohen S, Valsky D, Shen O, Rosenak D, Lipschuetz M, Yagel S (2011) Fetal heart ventricular mass obtained by STIC acquisition combined with inversion mode and VOCAL. Ultrasound Obstet Gynecol 38(2):191–197

Michelfelder E, Gottliebson W, Border W, Kinsel M, Polzin W, Livingston J, Khoury P, Crombleholme T (2007) Early manifestations and spectrum of recipient twin cardiomyopathy in twin-twin transfusion syndrome: relation to Quintero stage. Ultrasound Obstet Gynecol 30(7):965–971

Moore L (1987) Altitude-aggravated illness: examples from pregnancy and prenatal life. Ann Emerg Med 1987:965–973

Morrison J (2008) Sheep models of intrauterine growth restriction: fetal adaptations and consequences. Clin Exp Pharmacol Physiol 2008:730–743

Nagueh S, Middleton K, Kopelen H, Zoghbi W, Quiñones M (1997) Doppler tissue imaging: a noninvasive technique for evaluation of left ventricular relaxation and estimation of filling pressures. J Am Coll Cardiol 30(6):1527–1533

Naujorks A, Zielinsky P, Beltrame P, Castagna R, Petracco R, Busato A, Nicoloso A, Piccoli A, Manica J (2009) Myocardial tissue Doppler assessment of diastolic function in the growth-restricted fetus. Ultrasound Obstet Gynecol 34(1):68–73

Neubauer A, Voss W, Kattner E (2008) Outcome of extremely low birth weight survivors at school age: the influence of perinatal parameters on neurodevelopment. Eur J Pediatr 167(1):87–95

Newhham J (2001) Is prenatal glucocorticoid administration another origin of adult disease? Clin Exp Pharmacol Physiol 2001:957–961

Niewiadomska-Jarosik K, Lipecka-Kidawska E, Kowalska-Koprek U, Kedziora P, Tomecka D, Krajewski PSJ (2005) Assessment of cardiac function in fetuses with intrauterine growth retardation using the Tei Index. Med Wieku Rozwoj 9(2):153–160

Nilsson P, Ostergren P, Nyberg P, Söderström M, Allebeck P (1997) Low birth weight is associated with elevated systolic blood pressure in adolescence: a prospective study of a birth cohort of 149378 Swedish boys. J Hypertens 1997:1627–1631

Nishina H, Green L, McGarrigle H, Noakes D, Poston L, Hanson M (2003) Effect of nutritional restriction in early pregnancy on isolated femoral artery function in mid-gestation fetal sheep. J Physiol 2003:637–647

Nizard J, Bonnet D, Fermont L, Ville Y (2001) Acquired right heart outflow tract anomaly without systemic hypertension in recipient twins in twin-twin transfusion syndrome. Ultrasound Obstet Gynecol 18(6):669–672

Nuyt A, Alexander B (2009) Developmental programming and hypertension. Curr Opin Nephrol Hypertens 2009:144–152

Nyirenda M, Lindsay R, Kenyon C, Burchell A, Seckl J (1998) Glucocorticoid exposure in late gestation permanently programs rat hepatic phosphoenolpyruvate carboxykinase and glucocorticoid receptor expression and causes glucose intolerance in adult offspring. J Clin Invest 1998:2174–2181

Opie L, Commerford P, Gersh B, Pfeffer M (2006) Controversies in ventricular remodelling. Lancet 367(9507):356–367

Oran B, Cam L, Başpinar O, Baysal T, Reisli I, Peru H, Karaaslan S, Koç H, Gürbilek M (2003) Cardiac troponin-I in the serum of infants of diabetic mothers. Cardiol Young 13(3):248–252

Osmond C, Barker D, Winter P, Fall C, Simmonds S (1993) Early growth and death from cardiovascular disease in women. BMJ 1993:1519–1524

Ozaki T, Nishina H, Hanson M, Poston L (2001) Dietary restriction in pregnant rats causes gender-related hypertension and vascular dysfunction in offspring. J Physiol 2001:141–152

Padilla N, Perapoch J, Carrascosa A, Acosta-Rojas R, Botet F, Gratacós E (2010) Twelve-month neurodevelopmental outcome in preterm infants with and without intrauterine growth restriction. Acta Paediatr 99(10):1498–1503

Palinski W, Napoli C (2008) Impaired fetal growth, cardiovascular disease, and the need to move on. Circulation 2008:341–343

Parkinson C, Scrivener R, Graves L, Bunton J, Harvey D (1986) Behavioural differences of school-age children who were small-for-dates babies. Dev Med Child Neurol 28(4):498–505

Parsons T, Power C, Manor O (2001) Fetal and early life growth and body mass index from birth to early adulthood in 1958 British cohort: longitudinal study. BMJ 323(7325):1331–1335

Pedra S, Smallhorn J, Ryan G, Chitayat D, Taylor G, Khan R, Abdolell M, Hornberger L (2002) Fetal cardiomyopathies: pathogenic mechanisms, hemodynamic findings, and clinical outcome. Circulation 106(5):585–591

Pedra S, Hornberger L, Leal S, Taylor G, Smallhorn J (2005) Cardiac function assessment in patients with family history of nonhypertrophic cardiomyopathy: a prenatal and postnatal study. Pediatr Cardiol 26(5):543–552

Pereira S, Ganapathy R, Syngelaki A, Maiz N, Nicolaides K (2011) Contribution of fetal tricuspid regurgitation in first-trimester screening for major cardiac defects. Obstet Gynecol 117(6):1384–1391

Phillips D (1996) Insulin resistance as a programmed response to fetal undernutrition. Diabetologia 1996:1119–1122

Phillips D, Barker D, Hales C, Hirst S, Osmond C (1994) Thinness at birth and insulin resistance in adult life. Diabetologia 37(2):150–154

Ponnappa B, Rubin E (2000) Modeling alcohol's effects on organs in animal models. Alcohol Res Health 2000:93–104

Quintero R, Morales W, Allen M, Bornick P, Johnson P, Kruger M (1999) Staging of twin-twin transfusion syndrome. J Perinatol 19(8 Pt 1):550–555

Raboisson M, Fouron J, Lamoureux J, Leduc L, Grignon A, Proulx F, Gamache S (2004) Early intertwin differences in myocardial performance during the twin-to-twin transfusion syndrome. Circulation 110(19):3043–3048

Racasan S, Braam B, Koomans H, Joles J (2005) Programming blood pressure in adult SHR by shifting perinatal balance of NO and reactive oxygen species toward NO: the inverted Barker phenomenon. Am J Physiol Renal Physiol 2005:F626–F636

Rehn A, Van Den Buuse M, Copolov D, Briscoe T, Lambert G, Rees S (2004) An animal model of chronic placental insufficiency: relevance to neurodevelopmental disorders including schizophrenia. Neuroscience 2004:381–391

Rich-Edwards J, Stampfer M, Manson J, Rosner B, Hankinson S, Colditz G, Willett W, Hennekens C (1997) Birth weight and risk of cardiovascular disease in a cohort of women followed up since 1976. BMJ 1997:369–400

Rizzo G, Arduini D, Romanini C (1992) Accelerated cardiac growth and abnormal cardiac flow in fetuses of type I diabetic mothers. Obstet Gynecol 80(3 Pt 1):369–376

Romero R, Espinoza J, Gonçalves L, Gomez R, Medina L, Silva M, Chaiworapongsa T, Yoon B, Ghezzi F, Lee W, Treadwell M, Berry S, Maymon E, Mazor M, DeVore G (2004) Fetal cardiac dysfunction in preterm premature rupture of membranes. J Matern Fetal Neonatal Med 16(3):146–157

Rossi P, Tauzin L, Marchand E, Boussuges A, Gaudart J, Frances Y (2011) Respective roles of preterm birth and fetal growth restriction in blood pressure and arterial stiffness in adolescence. J Adolesc Health 2011:520–522

Rychik J, Tian Z, Bebbington M, Xu F, McCann M, Mann S, Wilson R, Johnson M (2007) The twin-twin transfusion syndrome: spectrum of cardiovascular abnormality and development of a cardiovascular score to assess severity of disease. Am J Obstet Gynecol 197(4):392.e1–8

Rychik J, McCann M, Tian Z, Bebbington M, Johnson M (2010) Fetal cardiovascular effects of lower urinary tract obstruction with giant bladder. Ultrasound Obstet Gynecol 36(6):682–686

Savchev S, Figueras F, Cruz-Martinez R, Illa M, Botet F, Gratacos E (2012) Estimated weight centile as a predictor of perinatal outcome in small-for-gestational-age pregnancies with normal fetal and maternal Doppler indices. Ultrasound Obstet Gynecol 2012:299–303

Sedmera D (2011) Function and form in the developing cardiovascular system. Cardiovasc Res 91(2):252–259

Sengupta P, Tajik A, Chandrasekaran K, Khandheria B (2008) Twist mechanics of the left ventricle: principles and application. JACC Cardiovasc Imaging 1(3):366–376

Simpson J (2011) Speckle tracking for the assessment of fetal cardiac function. Ultrasound Obstet Gynecol 37(2):133–134

Skilton M, Ayer J, Harmer J, Webb K, Leeder S, Marks G, Celermajer D (2012) Impaired fetal growth and arterial wall thickening: a randomized trial of ω-3 supplementation. Pediatrics 2012:e698–e703

Skilton M, Mikkilä V, Würtz P, Ala-Korpela M, Sim K, Soininen P, Kangas A, Viikari J, Juonala M, Laitinen T, Lehtimäki T, Taittonen L, Kähönen M, Celermajer D, Raitakari O (2013a) Fetal growth, omega-3 (n-3) fatty acids, and progression of subclinical atherosclerosis: preventing fetal origins of disease? The cardiovascular risk in Young Finns Study. Am J Clin Nutr 2013:58–65

Skilton M, Raitakari O, Celermajer D (2013b) High intake of dietary long-chain ω-3 fatty acids is associated with lower blood pressure in children born with low birth weight: NHANES 2003–2008. Hypertension 2013:972–976

Snoeck A, Remacle C, Reusens B, Hoet J (1990) Effect of a low protein diet during pregnancy on the fetal rat endocrine pancreas. Biol Neonate 1990:107–118

Soothill P, Bobrow C, Holmes R (1999) Small for gestational age is not a diagnosis. Ultrasound Obstet Gynecol 1999:225–228

Stefanutti G, Filippone M, Tommasoni N, Midrio P, Zucchetta P, Moreolo G, Toffolutti T, Baraldi E, Gamba P (2004) Cardiopulmonary anatomy and function in long-term survivors of mild to moderate congenital diaphragmatic hernia. J Pediatr Surg 39(4):526–531

Stirnemann J, Mougeot M, Proulx F, Nasr B, Essaoui M, Fouron J, Ville Y (2010) Profiling fetal cardiac function in twin-twin transfusion syndrome. Ultrasound Obstet Gynecol 35(1):19–27

Stressig R, Fimmers R, Eising K, Gembruch U, Kohl T (2010) Preferential streaming of the ductus venosus and inferior caval vein towards the right heart is associated with left heart underdevelopment in human fetuses with left-sided diaphragmatic hernia. Heart 96(19):1564–1568

Stuart A, Amer-Wåhlin I, Gudmundsson S, Marsál K, Thuring A, Källen K (2010) Ductus venosus blood flow velocity waveform in diabetic pregnancies. Ultrasound Obstet Gynecol 36(3):344–349

Sutherland G, Hatle L, Claus P, D'hooge J, Bijnens B (2006) Doppler myocardial imaging, 1st edn. BSWK, Hasselt

Szwast A, Tian Z, McCann M, Donaghue D, Bebbington M, Johnson M, Wilson R, Rychik J (2007) Impact of altered loading conditions on ventricular performance in fetuses with congenital cystic adenomatoid malformation and twin-twin transfusion syndrome. Ultrasound Obstet Gynecol 30(1):40–46

Takahashi H, Takahashi S, Tsukamoto K, Ito Y, Nakamura T, Hayashi S, Sago H (2012) Persistent pulmonary hypertension of the newborn in twin-twin transfusion syndrome following fetoscopic laser surgery. J Matern Fetal Neonatal Med 25(5):543–545

Tarry-Adkins J, Ozanne S (2011) Mechanisms of early life programming: current knowledge amd future directions. J Clin Nutr 2011:1765S–1771S

Taveras E, Rifas-Shiman S, Belfort M, Kleinman K, Oken E, Gillman M (2009) Weight status in the first 6 months of life and obesity at 3 years of age. Pediatrics 123(4):1177–1183

Tei C, Nishimura R, Seward J, Tajik A (1997) Noninvasive Doppler-derived myocardial performance index: correlation with simultaneous measurements of cardiac catheterization measurements. J Am Soc Echocardiogr 10(2):169–178
Timmerman E, Clur S, Pajkrt E, Bilardo C (2010) First-trimester measurement of the ductus venosus pulsatility index and the prediction of congenital heart defects. Ultrasound Obstet Gynecol 36(6):668–675
Tintu A, Rouwet E, Verlohren S, Brinkmann J, Ahmad S, Crispi F, van Bilsen M, Carmeliet P, Staff AC, Tjwa M, Cetin I, Gratacos E, Hernandez-Andrade E, Hofstra L, Jacobs M, Lamers WH, Morano I, Safak E, Ahmed A, le Noble F (2009) Hypoxia induces dilated cardiomyopathy in the chick embryo: mechanism, intervention, and long-term consequences. PLoS ONE 2009, e5155
Tobita K, Garrison J, Liu L, Tinney J, Keller B (2005) Three-dimensional myofiber architecture of the embryonic left ventricle during normal development and altered mechanical loads. Anat Rec A: Discov Mol Cell Evol Biol 283(1):193–201
Torrance H, Bloemen M, Mulder E, Nikkels P, Derks J, de Vries L, Visser G (2010) Predictors of outcome at 2 years of age after early intrauterine growth restriction. Ultrasound Obstet Gynecol 36(2):171–177
Torrens C, Brawley L, Anthony F, Dance C, Dunn R, Jackson A, Poston L, Hanson M (2006) Folate supplementation during pregnancy improves offspring cardiovascular dysfunction induced by protein restriction. Hypertension 2006:982–987
Turan S, Turan O, Miller J, Harman C, Reece E, Baschat A (2011) Decreased fetal cardiac performance in the first trimester correlates with hyperglycemia in pregestational maternal diabetes. Ultrasound Obstet Gynecol 38(3):325–331
Van Mieghem T, DeKoninck P, Steenhaut P, Deprest J (2009a) Methods for prenatal assessment of fetal cardiac function. Prenat Diagn 29(13):1193–1203
Van Mieghem T, Gucciardo L, Doné E, Van Schoubroeck D, Graatsma E, Visser G, Verhaeghe J, Deprest J (2009b) Left ventricular cardiac function in fetuses with congenital diaphragmatic hernia and the effect of fetal endoscopic tracheal occlusion. Ultrasound Obstet Gynecol 34(4):424–429
Van Mieghem T, Klaritsch P, Doné E, Gucciardo L, Lewi P, Verhaeghe J, Lewi L, Deprest J (2009) Assessment of fetal cardiac function before and after therapy for twin-to-twin transfusion syndrome. Am J Obstet Gynecol 200(4):400.e1–7
Van Mieghem T, Giusca S, DeKoninck P, Gucciardo L, Doné E, Hindryckx A, D'Hooge J, Deprest J (2010) Prospective assessment of fetal cardiac function with speckle tracking in healthy fetuses and recipient fetuses of twin-to-twin transfusion syndrome. J Am Soc Echocardiogr 23(3):301–308
Vela-Huerta M, San Vicente-Santoscoy E, Guizar-Mendoza J, Amador-Licona N, Aldana-Valenzuela C, Hernnández J (2008) Leptin, insulin, and glucose serum levels in large-for-gestational-age infants of diabetic and non-diabetic mothers. J Pediatr Endocrinol Metab 21(1):17–22
Verburg B, Jaddoe V, Wladimiroff J, Hofman A, Witteman J, Steegers E (2008) Fetal hemodynamic adaptive changes related to intrauterine growth: the Generation R Study. Circulation 2008:649–659
Vickers M, Breier B, Cutfield W, Hofman P, Gluckman P (2000) Fetal origins of hyperphagia, obesity, and hypertension and postnatal amplification by hypercaloric nutrition. Am J Physiol Endocrinol Metab 2000:E83–E87
Villar J, Belizan J (1982) The timing factor in the pathophysiology of the intrauterine growth retardation syndrome. Obstet Gynecol Surv 1982:499–506
Wang Y, Gao E, Wu J, Zhou J, Yang Q, Walker M, Mbikay M, Sigal R, Nair R, Wen S (2009) Fetal macrosomia and adolescence obesity: results from a longitudinal cohort study. Int J Obestat (Lond) 33(8):923–928

Ward H, Johnson E, Salm A, Birkle D (2000) Effects of prenatal stress on defensive withdrawal behavior and corticotropin releasing factor systems in rat brain. Physiol Behav 70(3–4):359–366

Watanabe S, Hashimoto I, Saito K, Watanabe K, Hirono K, Uese K, Ichida F, Saito S, Miyawaki T, Niemann P, Sahn D (2009) Characterization of ventricular myocardial performance in the fetus by tissue Doppler imaging. Circ J 73(5):943–947

Watkins A, Lucas E, Torrens C, Cleal J, Green L, Osmond C, Eckert J, Gray W, Hanson M, Fleming T (2010) Maternal low-protein diet during mouse pre-implantation development induces vascular dysfunction and altered renin-angiotensin-system homeostasis in the offspring. Br J Nutr 2010:1762–1770

Wieczorek A, Hernandez-Robles J, Ewing L, Leshko J, Luther S, Huhta J (2008) Prediction of outcome of fetal congenital heart disease using a cardiovascular profile score. Ultrasound Obstet Gynecol 31(3):284–288

Wienerroither H, Steiner H, Tomaselli J, Lobendanz M, Thun-Hohenstein L (2001) Intrauterine blood flow and long-term intellectual, neurologic, and social development. Obstet Gynecol 97(3):449–453

Wigglesworth J (1964) Experimental growth retardation in the foetal rat. Am J Obstet Gynecol 1964:1–13

Williams C, Hayman L, Daniels S, Robinson T, Steinberger J, Paridon S, Bazzarre T (2002) Cardiovascular health in childhood: a statement for health professionals from the Committee on Atherosclerosis, Hypertension, and Obesity in the Young (AHOY) of the Council on Cardiovascular Disease in the Young, American Heart Association. Circulation 2002:143–160

Willruth A, Geipel A, Fimmers R, Gembruch U (2011) Assessment of right ventricular global and regional longitudinal peak systolic strain, strain rate and velocity in healthy fetuses and impact of gestational age using a novel speckle/feature-tracking based algorithm. Ultrasound Obstet Gynecol 37(2):143–149

Yagel S, Silverman N, Gembruch U (2009) Fetal cardiology: embryology, genetics, physiology, echocariographic evaluation, diagnosis and perinatal management of cardiac diseases, 2nd edn. Informa Healthcare USA, New York

Younoszai M, Peloso J, Haworth J (1969) Fetal growth retardation in rats exposed to cigarette smoke during pregnancy. Am J Obstet Gynecol 1969:1207–1213

Yu C, Sanderson J, Marwick T, Oh J (2007) Tissue Doppler imaging a new prognosticator for cardiovascular diseases. J Am Coll Cardiol 49(19):1903–1914

Zanardo V, Visentin S, Trevisanuto D, Bertin M, Cavallin F, Cosmi E (2013) Fetal aortic wall thickness: a marker of hypertension in IUGR children? Hypertens Res 2013

Zielinsky P, Piccoli AJ, Teixeira L, Gus E, Mânica J, Satler F, Vaz H, Nicoloso L, Luchese S, Sheid M, Marcantonio S, Hatém D (2003) Pulmonary vein pulsatility in fetuses of diabetic mothers: prenatal Doppler echocardiographic study. Arq Bras Cardiol 81(6):604–607

Zielinsky P, Marcantonio S, Nicoloso L, Luchese S, Hatem D, Scheid M, Mânica J, Gus E, Satler F, Piccoli AJ (2004) Ductus venosus flow and myocardial hypertrophy in fetuses of diabetic mothers. Arq Bras Cardiol 83(1):51–56; 45–50

Zosmer N, Bajoria R, Weiner E, Rigby M, Vaughan J, Fisk N (1994) Clinical and echographic features of in utero cardiac dysfunction in the recipient twin in twin-twin transfusion syndrome. Br Heart J 72(1):74–79

Part III
Developmental Risks of Exposures and Potential Translational Toxicology Therapeutics

Chapter 7
Ovarian Toxicity of Environmental Contaminants: 50 Shades of Grey

M.A. Dominguez, J.C. Sadeu, M.T. Guerra, H.C. Furlong, Sharnjit Baines, and Warren G. Foster

Abstract Exposure to environmental contaminants is thought to be important in the development of adverse effects on reproductive health. While the adverse effects of environmental contaminants on semen quality and testicular function have been well studied, effects on ovarian function are less well defined. Epidemiological studies have linked exposure to environmental contaminants with adverse effects on menstrual cycle characteristics, infertility, and earlier age of menopause onset; yet direct evidence of effects on ovarian function is lacking. Environmental contaminant concentrations have been quantified in human ovarian follicular fluid establishing target tissue exposure; however, such data is sporadic and limited to women undergoing assisted reproductive therapies making generalization of results to the broader population of women difficult. We note that the relationship between serum and follicular fluid concentrations can be orders of magnitude different and thus target tissue distribution requires further study. Animal studies revealed effects of environmental contaminants on ovarian follicle dynamics, oocyte maturation, steroidogenesis, and epigenetic changes. Issues of dosing such as concentration of test chemicals used, route of administration, and use of multiple dose groups remain important limitations of the current literature. While animal studies establish a basis for biological plausibility of effects and support conclusions of reproductive hazard, we conclude that exposures in the general human population are too low to present a demonstrable risk to human ovarian function.

M.A. Dominguez
Facultad de Medicina Veterinaria y Zootecnia, Universidad Autónoma de Tamaulipas, Cd. Victoria, Tamaulipas, Mexico

J.C. Sadeu • H.C. Furlong • S. Baines • W.G. Foster (✉)
Department of Obstetrics & Gynaecology, McMaster University, 1280 Main Street West, Hamilton, ON L8S 4K1, Canada
e-mail: fosterw@mcmaster.ca

M.T. Guerra
Department of Morphology, Biosciences Institute, UNESP, Botucatu, São Paulo CEP 18618-970, Brazil

C.L. Hughes, M.D. Waters (eds.), *Translational Toxicology*, Molecular and Integrative Toxicology, DOI 10.1007/978-3-319-27449-2_7

Keywords Ovary • Toxicology • Toxicants • Chemical contaminants Steroidogenesis

7.1 Introduction

Worldwide, estimates of infertility range widely from 50 to 186 million couples of which 7.3 million American women alone are infertile (Boivin et al. 2007; Chandra et al. 2005; Mascarenhas et al. 2012; Rutstein and Shah 2004; Stephen and Chandra 2006). Established risk factors for infertility include: advanced age, diet, prescription medication use, pre-existing health status, genetic mutations, and infections; however, for many women the cause of their infertility is unknown. Thus, exposures to environmental contaminants are thought to adversely affect human fertility. Indeed, an association between farm or industrial work and infertility was demonstrated using a questionnaire on reproductive and occupational health histories, lifestyle and demographics (Fuortes et al. 1997), while residue levels of specific compounds were not measured. Infertile women had higher odds of working in industry (OR, 95 % CI = 6.7, 2.3–19.6) or farms (1.8, 1.2–2.7), adjusted for age, date of outcome and smoking compared to a fertile control group. Among specific infertility types, ovulatory dysfunction was highest and shows an association with industry/occupation (10.9, 3.5–33.7). In a National survey conducted in French in vitro fertilization (IVF) clinics between 2011 and 2012 (Alvarez and Devouche 2012), one in eight women reported sexual problems (mostly dyspareunia, insensitivity and bleeding). Of the 348 responding women (mean age: 34.9 ± 5.3 years), 20 % smoked cigarettes, 1.4 % cannabis, and 23 % consumed alcohol, whereas 8.6 % reported occupational exposure to chemicals. Work stress was reported by 41 % and family stress by 14 %. Taken together, these data suggest that exposures to environmental contaminants are potentially associated with reduced fertility, however, the strength of this association and potential mechanisms are poorly defined.

While the effects of environmental contaminants on male reproductive health have received research attention, the effects on female fertility and ovarian function in particular remain to be elucidated. We propose that environmental contaminants affect reproductive health in part via impaired ovarian function (Augood et al. 1998; Foster 2003; Pocar et al. 2003; Sadeu et al. 2010). Moreover, we postulate that the ovary is potentially vulnerable to the adverse effects of environmental contaminant exposure owing to its rich blood supply, rapid cell division associated with follicle development, and ovulation. Moreover, environmental contaminant effects on the ovary have far reaching consequences on reproduction and general health owing to its central role in the production of the gonadal steroids estradiol (E_2) and progesterone (P_4), the finite and non-renewable number of follicles present in the ovary at birth, and development of female gametes (oocytes).

7.2 Epidemiology

Determining the effect of environmental contaminants on reproductive health in women and on ovarian function in particular is difficult to assess. The relationship between environmental contaminant exposure and reproductive function in women has primarily been limited to measurement of time-to-pregnancy (TTP), infertility, menstrual cycle characteristics, duration of menstrual bleeding, amenorrhea, dysmenorrhea, age at menopause, and premature ovarian failure (POF). Measurement of circulating follicle stimulating hormone (FSH), ovarian hormones (activin and inhibin), gonadal steroid hormones (E_2 and P_4), and more recently quantification of anti-Müllerian hormone (AMH) concentrations have been employed as markers of impaired reproductive health that provide indirect insight into ovarian function.

A questionnaire on reproductive health and lifestyle histories of women who presented with natural menopause revealed that only women who had been breastfed as babies showed a significantly earlier natural menopause ($p=0.014$), while alcohol consumption ($p=0.080$) and smoking ($p=0.081$) had no significant effect on timing of natural menopause (Dvornyk et al. 2006). Other factors such as the use of oral contraceptives also had no significant effect. In contrast, a survey of 443 hairdressers, aged 21–55, revealed a greater likelihood of premature ovarian failure (POF; 3.2 vs. 1.4 %, $p=0.06$) compared to 508 women employed in other occupations (Gallicchio et al. 2009). While hairdressers also smoked significantly more than the controls, the overall relative risk (RR, 95 % CI) for POF was 1.90, (0.76–4.72) and thus was non-significant. Hairdressers are exposed to several chemicals such as solvents, bleaches, hair dyes, non-lye relaxers, alcohols, ethylene glycol, methacrylate and phthalates; all of which are potentially hazardous to ovarian function. Sub-analysis of the data revealed that Caucasians over 40 years had a RR of 5.58 (1.24–25.22) illustrating that age is an important variable in assessing the impact of environmental factors on ovarian function. In another study, ovarian failure was reported (Koh et al. 1998) in 16 laborers exposed to cleaning solvent containing 2-bromopropane (n-propyl bromide). Amenorrhea was reported in 26 women after occupational exposure in South Korea, of which 16 were diagnosed with primary ovarian failure. Between these two groups, there were significant differences ($P<0.05$) in levels of LH: 32.9 mIU/ml (range = 10.1–93.0) vs. 8.5 (1.1–13.9); FSH: 87.7 mIU/ml (31.8–119.7) vs. 9.8 (3.3–28.3); and E2: 11.0 pg/ml (7.0–28.0) vs. 48.0 (12.0–205.0). While the effects of occupational exposure to environmental contaminants are well known, effects arising from exposure to environmental contaminants at exposure levels representative of the concentrations reported in contemporary biomonitoring studies involving the general population are less clear.

Persistent Organic Pollutants (POPs) are a group of chemicals including the chlorinated organic chemicals that have been used in the manufacture of a diverse group of chemical compounds such as pesticides, chlorine, bleaches, plastics, flame-retardants and metal production. Since they are stable and lipophilic, they easily accumulate in the food chain, and humans may acquire them via food or direct (occupational or accidental) exposure (Hombach-Klonisch et al. 2005). Polybrominated diphenyl ethers (PBDEs) are a family of structurally related chemicals used as flame retardants, and have been detected in human follicular fluid (Johnson et al. 2012). Several organochlorine compounds including the polychlorinated biphenyls (PCBs) and metabolites of the pesticide DDT, have also been quantified in follicular fluid (De Felip et al. 2004; Foster 1995; Mahalingaiah et al. 2012; Meeker et al. 2009; Petro et al. 2012; Schlebusch et al. 1989; Younglai et al. 2002) (Table 7.1). In a longitudinal study of 501 couples who discontinued contraception to achieve pregnancy (Bloom et al. 2007) the fecundability odds ratios (FOR) was decreased by 18–21 % in women with PCB congeners 118, 167 & 219. The strongest effect found was for PCB167 (FOR = 0.79; 95 % CI = 0.64–0.97). These correlations suggest that PCB exposure is associated with adverse effects on fertility although the direct ovarian effects of PCBs remain equivocal. Mean total PCB concentrations (3.1 ± 1.9 μg/L) were positively correlated with menstrual cycle length reported by 2314 women ($p = 0.02$) (Cooper et al. 2005). The majority of 37 follicular fluid samples from an IVF clinic contained 0–1 and 1–2 μg/kg of PCB138, 153, or 180, but the highest levels reached 15–16 μg/kg (Schlebusch et al. 1989). In women attending a fertility clinic, follicular fluid concentrations of PCB180 were lower in women who became pregnant vs. those who did not (85 ± 14 vs. 147 ± 11 pg/ml, respectively, $p < 0.05$). German women had higher serum PCB concentrations (0.35 ± 0.049 vs. 0.17 ± 0.02 μg/kg, $p < 0.01$) compared to women in Tanzania (Weiss et al. 2006). In contrast, follicular fluid levels did not show regional differences (0.26 ± 0.02 vs. 0.22 ± 0.06 μg/kg). These studies suffer from relatively small sample sizes and recruitment exclusively from women attending fertility clinics and thus generalization of results to the entire population is difficult. Although 2,3,7,8 tetrachlorodibenzo-*p*-dioxin (TCDD) is widely regarded as one of the most toxic environmental contaminants, adverse effects on ovarian function are poorly defined. A long-term study was carried out in Italy after a chemical plant explosion in 1976 that exposed a town to TCDD (Eskenazi et al. 2005). Serum residue concentrations of 43.7 ppt (range, 2.5–6320) were found in 616 women that were exposed between 1 month and 40 years of age. While there was a trend toward earlier menopause with increasing TCDD concentrations, suggesting potential effects on ovarian function, significance could not be documented.

The common pesticide 1,1-bis-(4-chlorophenyl)-2,2,2-trichloroethane (*p,p′*-DDT) has been banned in many industrialized countries, but continues to be used in developing nations. Although widely studied, the relationship between exposure to this pesticide and its metabolites on fertility and ovarian function remains equivocal. The DDT metabolite 1,1-bis-(4-chlorophenyl) 2,2-dichloroethene (*p,p′*-DDE) is frequently detected in serum and follicular fluid (De et al. 2004; Mahalingaiah et al. 2012; Petro et al. 2012; Younglai et al. 2002) although effects on fertility have been mixed (Table 7.1). A study that compared the fertility of 289 women born in the USA between 1960 and 1963 to that of their daughters' 30 years later showed a 32 % decrease in the probability of pregnancy of the daughters for every 10 μg/L in maternal serum DDT, but also an unexpected 16 % increase in the probability of pregnancy for every 10 μg/L in maternal serum DDE (Cohn et al. 2003). While in Brazil, sales

Table 7.1 Concentration of persistent organic pollutants (POPs) in the blood or follicular fluid of women participating in assisted reproductive (ART) programs

Study design	N	Age	Location	Exposure	Tissue	Concentration	Citation
Observational	28	28–38	Canada	Background	FF	hexachloroethane 232 ± 27	Younglai et al. (2002)
						1,2,4-trichlorobenzene 212 ± 20 pg/ml	
						mirex 3595 ± 590 pg/ml	
						cotinine 606 ± 147 ng/ml	
						PCB-49 62.4 ± 6.8 pg/ml	
						PCB-153 73.3 ± 7.0 pg/ml	
						PCB-180 62.2 ± 5.2 pg/ml	
Observational	31	Not reported	Germany & Tanzania	Background	S	PCBs 0.35 ± 0.05 vs. 0.17 ± 0.02 μg/kg	Weiss et al. (2006)
	21				FF	0.26 ± 0.02 vs. 0.22 ± 0.06 μg/kg	
Prospective	79	26–34	United States	Background	S	PCBs 5.27 ng/g serum	Bloom et al. (2007)
Case-control	765	35.9 ± 4.21[a]	United States	Background	S[b]	HCB 18 ng/g lipid	Mahalingaiah et al. (2012)
						p,p′-DDE 226 ng/g lipid	
						DDT 251 ng/g lipid	

[a]mean ± SEM
[b]Follicular fluid concentrations were quantified but not reported

of pesticides in 1985 were correlated with reproductive alterations and cancer rates a decade later (Koifman et al. 2002), no association between serum DDE levels (mean value: 30.0 ± 19.6 μg/L) and menstrual cycle dysfunction (cycle length and missed periods) was found in two American studies (Cooper et al. 2005; Mahalingaiah et al. 2012). A study comparing German and Tanzanian women attending fertility clinics (Weiss et al. 2006) found that Tanzanian women had higher DDE levels (12.77 ± 9.7 vs. 0.78 ± 0.75 μg/kg, $P<0.01$). In contrast, in a Canadian study (Younglai et al. 2002), pesticide concentrations in the follicular fluid were not related to pregnancy rates whereas in a Belgian study of infertility patients (Petro et al. 2012), an association between follicular fluid DDE levels (392 ± 348 pg/ml) and reduced fertilization rates and embryo quality was described. We postulate that equivocal findings may be related to study design, sample size, analytical methods, regional differences, and potential interactions amongst the myriad chemicals to which people are exposed.

Organophosphates are a widely used family of pesticides to increase crop productivity by eradicating pests. Despite substantial available information involving the impact of organophosphates on the environment and different animal species (Patel et al. 2007), evidence of toxicity in humans remains scarce. Exposures to high concentrations of organophosphates in a short period, especially during the manufacture, formulation, and application of chemicals can lead to toxic effects (Dyer et al. 2001). In Iowa and North Carolina, the association between pesticide use and menstrual function among 3103 women living on farms was examined (Farr et al. 2004). These women were premenopausal, between the ages of 21–40 years of age, not pregnant or breastfeeding, and not taking hormonal contraceptives. Results revealed that women who were exposed to pesticides experienced a non-significant increase in menstrual cycle length (odds ratio = 1.2, 95 % CI: 0.66, 2.2) and increased odds of missed periods (odds ratio = 1.5, 95 % CI: 1.2, 1.9) compared with women who have never used pesticides (Farr et al. 2004). Women who were exposed to organophosphates experienced longer menstrual cycles (odds ratio = 1.3, 95 % CI: 0.55, 2.9) and increased odds of missed periods (odds ratio = 1.4, 95 % CI: 0.93, 2.2) compared with women who had never used pesticides. Women with suspected exposure to hormonally active or ovotoxic pesticides (parathion and trichlorfon) had fewer cases of irregular menstrual cycles (odds ratio = 0.53, 95 % CI: 0.37, 0.78) but 60–100 % increased odds of long cycles (odds ratio = 1.6, 95 % CI: 1.0, 2.5), missed periods (odds ratio = 1.7, 95 % CI: 1.3, 2,2), and intermenstrual bleeding (odds ratio = 1.3, 95 % CI: 1.0, 1.6) than women who had never used pesticides (Farr et al. 2004). The findings suggest a possible but weak correlation between the use of certain hormonally active pesticides and decreased menstrual function among women living on farms. Further detailed data on menstrual cycle and organophosphate exposure is required to strengthen the association between organophosphate exposure and menstrual cycle characteristics.

Of the metals, exposure to cadmium (Cd), mercury (Hg), lead (Pb), selenium (Se) and zinc (Zn) are the most commonly encountered and linked with adverse reproductive effects (Dickerson et al. 2011; Jackson et al. 2011; Pollack et al. 2011; Yang et al. 2002). Cd, Hg and Pb have been measured in follicular fluid (Al-Saleh et al. 2008; Bloom et al. 2012b) providing direct evidence of target tissue exposure (Table 7.2). The effects of metals including Cd on ovarian function have been

Table 7.2 Concentration of metals in the blood or follicular fluid of women

Study design	N	Age	Location	Exposure	Tissue	Concentration	Citation
Observational	394	18–44	China	Occupational	Air at workplace	0.001 to 0.2 mg/m^3	Yang et al. (2002)
Case-control	619	19–50	Saudi Arabia	Background	Follicular fluid	Cd = 0.29 ± 0.30 μg/L	Al-Saleh et al. (2008)
						Pb = 0.61 ± 1.17 μg/dL	
						Hg = 2.12 ± 2.47 μg/L	
Case-control	252	18–44	United States	Background	Whole blood	Cd = 0.30 μg/L	Jackson et al. (2011)
						Pb = 0.87 μg/dL	
						Hg = 1.10 μg/L	
Case-control	30	26–42	United Kingdom	Background	Hair samples (Hg)	Hg = 0.89 ± 0.3 μg/g	Dickerson et al. (2011)
					Serum (Zn & Se)	Zn = 897.3 ± 386.4 μg/g	
						Se = 100.2 ± 19.7 μg/g	
Case-control	46	28–43	United States	Background	Blood & Follicular fluid	Pb = 0.82 ± 0.31 μg/dL	Bloom et al. (2012a, b)
						0.25 ± 0.31 μg/dL	

evaluated by several research groups (Davis et al. 2001; Laughlin et al. 1987; Zhang and Jia 2007; Zhang et al. 2008). An inverse relationship between serum Cd concentrations and circulating concentrations of FSH were reported in the absence of any effect on circulating E_2 concentrations (Pollack et al. 2011). The relationship between menstrual cycle characteristics and circulating metal concentrations was investigated previously; however, only Cd was associated with higher serum E_2 concentrations (Jackson et al. 2011) which contrast with the results described by others who could not find an effect (Pollack et al. 2011). For each μg/l increase in Cd there was a 24.3 % increase in circulating E_2 concentrations during the follicular phase (95 % CI=1.1–52.9). Although the source of exposure was not confirmed, cigarette smoking is thought to be the primary source. Cigarette smoking is associated with elevated levels of blood Cd that are 2.5–4 times higher than in non-smokers (Batariova et al. 2006; Wong and Lye 2008; Zhang and Jia 2007). However, these results contrast with studies that have reported decreased circulating concentrations of E_2 in female smokers (Windham et al. 2005).

While a positive association between blood Pb concentrations and circulating P4 levels have been reported (Pollack et al. 2011) an inverse association between follicular fluid Pb concentrations and fertilization rates (relative risk=0.68, P=0.026) was found in women undergoing fertility treatments in another study (Bloom et al. 2012b). The mean blood and follicular fluid Pb concentrations were 0.82±0.31 and 0.25±0.31 μg/L, respectively. In another study (Jackson et al. 2011), Pb levels of 0.87 μg/dl were reported in human whole blood samples, but were not associated with toxic effects. An increased odds ratio for dysmenorrhea (1.66; 1.07–2.59) compared to non-exposed controls was found in a study of normal cycling women with occupational exposure to Hg vapor in a Chinese lamp factory (Yang et al. 2002). In another study (Dickerson et al. 2011), Hg, Zn and Se concentrations were measured in patients attending a fertility clinic. Only Hg showed a negative correlation with the number of oocytes collected and follicle number after ovarian stimulation, after adjusting for age and BMI. There were no effects on fertilization, cleavage rates or embryo quality. Despite evidence of reproductive hazard, the literature relating metal exposure with reproductive effects lacks consistency across studies and thus the evidence is judged as weak.

Unlike POPs and metals, newer generation commercial chemicals are engineered to be more labile and hydrophilic in order to obviate concerns with bioaccumulation and toxicity to mammalian systems. Of these bisphenol A (BPA) has received extensive research attention and has been linked with reproductive effects in women, yet considerable controversy continues to surround the health risk associated with this chemical. For example, urinary concentrations of BPA have recently been associated with a shorter luteal phase but not with adverse effects on the follicular phase length, TTP or early pregnancy loss (Jukic et al. 2015). In a subsequent study of women (n=44) attending a fertility clinic (Fujimoto et al. 2011), serum unconjugated (free) BPA concentrations were measured by HPLC. In this study, the median unconjugated BPA was 2.53 ng/ml and ranged from non-detectable to 67.4 ng/ml. Serum BPA concentrations were inversely associated with E_2 ($\beta=-0.16$; 95 % CI=−0.32, 0.01) and with E_2 concentration normalized to the number of mature-sized follicles at the human chorionic gonadotropin (hCG) trigger ($\beta=-0.14$; 95 % CI = −0.24, −0.03). However, the concentration of unconjugated BPA in the serum was not associated with oocyte maturation rate, but intracytoplasmic sperm injection (ICSI) or conventional IVF patients had a 55 % decrease in the probability of fertilization for each doubling in BPA levels (adjusted relative risk = 0.45, 95 % CI = 0.21–0.66). Although preliminary, these data suggest that free BPA in the circulation is potentially associated with a reduced serum E_2 response to ovarian stimulation in IVF and reduced fertility. Limitations of the above studies are the focus on study participants who are attending fertility programs and thus are not necessarily representative of the population overall.

7.3 Exposure Data

Insight into whole body and target tissue exposure is essential for dose setting in experimental studies as well as for transparent and defensible evidence-based regulatory decisions and advice to guide industry action. However, exposure data is frequently lacking for many environmental contaminants and quantification of target tissue exposure is complicated by difficulty in accessing the ovary and ovarian follicular fluid. Therefore, understanding the relationship between circulating concentrations of contaminants and the concentrations in the follicular fluid is potentially valuable. Herein we summarize the available exposure data and explore the relationship between serum and follicular fluid concentrations.

Advances in analytical chemistry have enabled the measurement of increasingly smaller concentrations of environmental contaminants, many with reproductive and developmental toxicity, in the human circulation and reproductive fluids (Calafat et al. 2008; Silva et al. 2004). Environmental contaminants have been quantified in serum (Bloom et al. 2007, 2011; Fujimoto et al. 2011; Mahalingaiah et al. 2012; Younglai et al. 2002) and ovarian follicular fluid (Baukloh et al. 1987; Foster et al. 1995; Ikezuki et al. 2002; Jarrell et al. 1993; Jirsova et al. 2010; Neal et al. 2008; Trapp et al. 1984, 1990; Younglai et al. 2002) of women participating in assisted reproductive therapy (ART) programs. Although exposure to environmental contaminants has been linked to adverse effects on fertility (Bloom et al. 2011; Fujimoto et al. 2011; Ikezuki et al. 2002; Mahalingaiah et al. 2012; Takeuchi et al. 2004; Younglai et al. 2002), the available literature is limited to studies involving patients attending ART programs and thus is not representative of the general population. Thus improved methods of estimating target tissue exposure are needed.

Organofluorine compounds include the surfactants perfluorooctane sulfonic acid (PFOS) and perfluorooctanoic acid (PFOA); used as stain repellents in the manufacture of textiles, carpets and furniture (La Rocca et al. 2012), have been measured in follicular fluid (Petro et al. 2014). A study on the effects of long-term PFOS exposure in drinking water (Knox et al. 2011) found serum levels of 16.9 ± 9.9 19.1 ± 12.9 and 24.8 ± 16.3 ng/ml for 3 age groups (18–42, 43–51 and 52–65 years, respectively). PFOS exposure was negatively associated with estrogen levels (E_2 decreased from

~88 to 72 pg/ml, $p < 0.0001$ and ~32 to 22 pg/ml, $p < 0.007$, for the 1st compared to the 5th quintile of PFOS levels, respectively) in the two oldest age groups described above. Similarly there was an increased odds of experiencing early menopause for the 5th vs. the 1st quintile of PFOS concentrations (OR, 95 % CI = 1.4, 1.1–1.8 and 2.1, 1.6–2.8 for both age groups, respectively). A positive effect of PFOA levels in follicular fluid on fertilization rate and proportion of viable embryos collected from IVF patients has been described (Petro et al. 2014); however, the mechanism for this effect is unclear. A study of 53 couples attending a fertility clinic in Italy (La Rocca et al. 2012) revealed that patients with unexplained infertility had detectable PFOS (>0.5 ng/g whole blood). Of those exhibiting infertility factors (PCOS, endometriosis or male-factor infertility), 30 women had >20-fold lower limit (3–144 ng/g). Whole blood PFOA concentrations were below the detection limit in 90 % of samples. It is worth noting that serum and follicular fluid concentrations of the perfluorinated compounds are close to identical (Petro et al. 2014) suggesting that serum concentrations are useful surrogate markers of target tissue exposure.

The concentrations of total BPA (free + conjugated) in the serum from healthy non-pregnant and pregnant women, and follicular fluid from women attending an IVF clinic have been reported (Ikezuki et al. 2002). Concentrations ranged from 1.4 ± 0.9 to 2.4 ± 0.8 ng/ml in most cases suggesting exposure to BPA in women. While these data suggest exposure of estrogen sensitive target tissues at critical stages of steroid regulated development, these reports are limited to measures of total BPA and thus the concentration of the bioactive form of BPA, free BPA, in ovarian follicular fluid remains unknown. Polycystic ovarian syndrome (PCOS) and serum concentrations of BPA were linked through two independent studies, one conducted in Japan and the other in Greece (Kandaraki et al. 2011; Tsutsumi 2005). Serum BPA concentrations were 0.64 ± 0.1 for normal cycling women compared to 1.04 ± 0.1 ng/ml for women with PCOS (Tsutsumi 2005). In the second study (Kandaraki et al. 2011), women referred to a PCOS clinic were stratified by BMI and serum BPA concentrations were again higher in women with PCOS and a BMI ≥25 (0.96 ± 0.46 *vs.* 0.72 ± 0.39 ng/ml, $p < 0.05$) or <25 (1.13 ± 0.63 *vs.* 0.70 ± 0.36 ng/ml, $p < 0.001$) compared to healthy cycling women, respectively. However, the antibody based ELISA method employed to quantify circulating BPA concentrations in these studies employed an antibody with significant cross reactivity with BPA metabolites and thus the assay cannot discriminate between free (bioactive) BPA and its metabolites. Therefore, it is unclear, on the basis of this evidence, what role if any BPA may play in the development or progression of this disease. Free BPA could not be detected in the follicular fluid of women (Krotz et al. 2012) raising further concern that the reported link between BPA and PCOS may be spurious; however, we note that the data in this study is limited to only five study subjects and thus cannot be taken as conclusive. Regardless of the findings with free BPA, the authors were able to quantify phthalate metabolites in ovarian follicular fluid of these same five women suggesting that technical limitations are unlikely to account for the failure to detect free BPA in these samples. The phthalate metabolites quantified in the follicular fluid included: mono-2-ethylhexyl phthalate (m-EHP, 9.34 ± 3.33 ng/mL), monoethyl phthalate (m-EP, 3.19 ± 2.97 ng/mL), mono-n-butyl phthalate (m-BuP, 1.62±0.59 ng/mL) and monomethyl phthalate (m-MP, 1.19±0.25 ng/mL) (Krotz et al. 2012) but these concentrations are considered to be 200–1000 times lower than those harmful to reproductive function in laboratory animals (Kay et al. 2013). Therefore, although human exposure is recognized as wide-spread, the concentrations present in the ovary are thought too low to have an adverse effect. While regulatory action seems inappropriate, health conscious individuals who remain concerned can minimize their personal exposures to these chemicals through consumption of fresh foods and reducing personal exposure to plastics and canned goods.

Finally, parabens are another group of contemporary chemicals used as antimicrobial agents in the production of cosmetics, food and pharmaceuticals (Ahn et al. 2012; Taxvig et al. 2008). Urinary parabens concentrations were measured in patients attending a fertility clinic (Smith et al. 2013) yielding median concentrations of 158, 35.5 and 1.53 μg/L for methyl, propyl and butyl parabens (mPB, pPB and bPB), respectively. Although serum or follicle fluid levels were not measured, the authors suggested a correlation between urinary paraben concentrations and ovarian reserve. This provocative suggestion requires additional study in both clinical samples and animal studies to confirm and explore potential mechanisms of action. Although widely appreciated, we feel compelled to remind the reader that evidence of exposure cannot be equated with evidence of an adverse health effect or increased risk of an adverse health outcome at a later date. Risk is the product of exposure, evidence of a health hazard usually derived from experimental animal studies and an uncertainty factor that accounts for differences between experimental animals and humans.

7.4 Experimental Animal Evidence

Well-controlled animal studies allow for testing with known concentrations of pure test compounds, using relevant routes of exposure, to genetically homogenous animals at well-defined developmental stages, and under carefully regulated environmental conditions that cannot be replicated in human populations. Consequently, animal studies are important to understanding the potential health hazards associated with exposure to environmental contaminants and contribute essential data to evidence based regulatory decisions.

Different reproductive and developmental toxicology test protocols have been developed by government bodies such as the Organization for Economic Cooperation and Development (OECD), the United States-Environmental Protection Agency (US-EPA), and the National Toxicology Program (NTP). While these protocols exploit good laboratory practices rendering their results highly reproducible and thus attractive for regulatory purposes, they have been criticized for the superficial nature of the outcome measures employed and lack of mechanistic insight. Animal testing for reproductive and developmental effects using these protocols provides indirect evidence for toxicant effects on ovarian function as shown by changes in estrous cycle length, sexual maturation, fertility index, weight of reproductive organs and histology, and circulating reproductive hormone concentrations. By comparison, academic scientists operate in less restrictive environments and thus are able to develop innovative protocols designed to assess reproductive function and provide mechanistic insight that is lacking in traditional regulatory toxicity testing paradigms. For example, a series of elegant studies have shown that BPA exposure can affect oogenesis and aneuploidy in mouse and non-human primate oocytes (Hunt et al. 2003, 2012), insight that could not be achieved through traditional testing paradigms. Moreover, access to primary (Haney et al. 1984) and immortalized ovarian granulosa cell lines (Kwintkiewicz et al. 2010) has also allowed for direct assessment of test chemical effects. Recent developments have seen the introduction of isolated ovarian follicle cultures employed to assess test chemical effects on stage dependent effects of follicle growth, steroidogenesis, follicle survival, and ovulation (Cortvrindt et al. 1998; Cortvrindt and Smitz 2002; Lenie et al. 2004; Lenie and Smitz 2009; Neal et al. 2007, 2010; Sadeu and Foster 2011a, b; Van Wemmel et al. 2005), a model that has been extended to three dimensional cultures of follicles (Songsasen et al. 2011). While these studies provide much needed mechanistic insight, they suffer from the need for sophisticated testing, and high level of expertise in the field. Regardless of their limitations, results arising from both regulatory and academic scientific studies contribute essential data to the risk assessment process and thus are central to evidence based regulatory decisions. Chief among the effects detected are depletion of the ovarian follicle reserve, interference with the cumulus oocyte complex (COC) communication, and steroidogenesis.

7.4.1 Effects on Ovarian Reserve

With each menstrual cycle, a cohort of primordial follicles enters one of the waves of growing follicles (Baerwald et al. 2003a, b). Typically one follicle from the growing pool of follicles is selected to ovulate while the remainder become atretic. Of the estimated 500,000 follicles present in the human ovary at the start of reproductive life, only about 400 reach the pre-ovulatory stage and ovulate whereas follicle atresia is the fate for the vast majority of female germ cells (Byskov 1978). The number of follicles formed *in utero,* the rate of recruitment into the growing pool, and rate of follicle destruction are all potentially modifiable by environmental contaminant exposure with important implications to fertility, sterility, premature ovarian failure, and menopause. Indeed, contaminant exposures that disrupt the number of follicles formed during development, affect the rate of follicle recruitment, follicle development, or follicle loss can have serious implications for circulating steroid concentrations, fertility, and ultimately the reproductive life span for a woman. Since estrogens are important mediators of cell proliferation, neurogenesis, growth and maintenance of bone, and cardiovascular health, the consequences to general health arising from ovarian toxicity are potentially profound and reach across the life span with important implications to health care needs and costs.

Contaminant exposure in adult animals has been shown to decrease ovarian follicle counts but the most sensitive follicle stage to the toxic effects of environmental contaminants is variable. For example, 4 vinylcyclohexene diepoxide (VCD) has been shown to selective target primordial and primary follicles (Appt et al. 2006; Sahambi et al. 2008; Sobinoff et al. 2010), the follicles at the earliest stages of development. Similar effects have also been shown with polycyclic aromatic hydrocarbons such as 9,10- dimethyl benzanthracene, 3-methylcholanthrene, and benzo[a] pyrene and (Borman et al. 2000), all of which involve activation of the apoptosis pathway. Of note, repeated exposure to low concentrations may be more toxic that exposure to a single high concentration of an environmental contaminant (Borman et al. 2000), an effect that could have important implications for humans who are typically exposed to low concentrations of contaminants over long periods of time. Similar effects on follicle counts have been demonstrated with several different POPs. For example, PCB treatment-related increase in ovarian follicle loss (Baldridge et al. 2003) has been documented with fewer preantral follicles (15 ± 0.6 vs. 56 ± 0.7/section, $p < 0.05$), and fewer small and large antral follicles (18 ± 0.9 vs. 41 ± 2 and 8 ± 0.9 vs. 30 ± 0.6/section, respectively, $p < 0.05$), as well as more atretic follicles (59 ± 0.9 vs. 26 ± 1.2, $p < 0.05$). Similarly, in a study on the long-term reproductive effects of dioxin exposure (Franczak et al. 2006), prepubertal rats were treated orally with vehicle or 10 μg TCDD/kg at 29 days of age (acute treatment) while pregnant rats were treated (0, 50 or 200 ng/kg) on GD 14, 21 and postnatal D7 and 14. Subsequently, the female pups were given the same weekly doses until 8 months of age (chronic treatment). The acute treatment delayed ($p < 0.05$) vaginal opening (37.8 ± 1.5 vs. 33.2 ± 1.0 d), reduced reproductive lifespan (287.2 ± 9.7 vs. 319.2 ± 4.1d), and prolonged diestrus (0.65 ± 0.01 vs. 0.58 ± 0.02 of cycle). Treatment also reduced E_2 output ($p < 0.05$). Chronic treatment had no effect on follicle and corpus luteum numbers or structure but at 8 months only 25–30 % of treated females showed normal cycles (vs. 100 % in controls, $p < 0.05$). Although the two largest doses of TCDD used accelerated reproductive ageing (fewer had normal cycles at 9 and 11 months, $p < 0.05$), there was no effect of TCDD treatment on follicle counts (Shi et al. 2007). In another study TCDD (0 or 5 μg/kg, i.p.) treatment decreased ER gene expression in the ovary and other tissues in 6-to-8-week-old mice (Tian et al. 1998). Available studies have primarily evaluated the effect of test chemicals under basal conditions. In a provocative study (Jung et al. 2010), TCDD treatment decreased ovulation rate (2.7 ± 1.9 vs. 20.3 ± 3.6, $p < 0.05$) and increased Aryl hydrocarbon Receptor (AhR) expression in granulosa cells in 25-day-old rats treated with 1 dose of TCDD (0 or 32 μg/kg, by gavage) plus hCG and LH (1 and 3 d later, respectively) to induce follicle development and ovulation. Treatment also decreased the number of follicles available for ovulation. Taken together these data suggest that TCDD may have direct effects upon the ovary although the mechanisms remain to be determined.

In addition to direct effects of TCDD in adults, evidence has been brought forward in the literature that suggests effects may be transmitted across generations. In a transgenerational study (Nilsson et al. 2012), F3 females also exhibited fewer primordial follicles than controls (~14 to 9, $p < 0.005$), with no effect on antral follicle counts, but developed cystic structures similar to PCOS following treatment of pregnant rats with TCDD (0 or 100 ng/kg BW/d i.p.) on gestation day (GD) 8–14. However, the underlying mechanisms remain unclear.

Clinical markers of adverse effects of environmental contaminants on ovarian function are sparse; however, AMH has received increasing interest as a marker of ovarian reserve and has been used in toxicology to access Methoxychlor (MXC) effects on the ovary. In a developmental study (Masutomi et al. 2003), rats were treated with MXC (0, 24, 240, or 1200 ppm in the diet) from GD 3 to PND 21. Only the highest dose (1200 ppm) decreased body and ovarian weight, accelerated vaginal opening, and caused irregular estrous cycles. Furthermore, longer estrous cycles (6.28 ± 1.91 vs. 3.76 ± 0.25 days, $p < 0.001$), fewer corpora albicantia from previous cycles, fewer corpora hemorrhagica, and more degenerate oocytes were found in MXC treated rats (Quignot et al. 2012a, b). MXC (0, 1, 10, 50, 100, or 500 mg/kg/d) exposure from 3 to 10 days of age resulted in antral follicle loss but increased the growing follicle pool, and increased AMH expression at 20 days of age (Uzumcu et al. 2006). Doses of 50, 100, and 500 led to 1.6 ± 0.2, 1.85 ± 0.6 and 2.2 ± 0.5 times the AMH expression of the control group ($p < 0.05$), as measured using immunohistochemistry. Antral follicle counts were reduced (~6/section after 100 or 500 mg, vs. ~17 in controls, $p < 0.05$), whereas preantral follicles increased (~13 and 15 after 100 and 500 mg, vs. ~8 in controls, $p < 0.05$). The effects of VCD on circulating AMH concentrations (Sahambi et al. 2008) further support the notion that AMH is a potentially useful minimally invasive marker of toxicant effect.

The adverse effects of legacy chemicals such as the POPs on ovarian follicle counts are not unique and can be extended to contemporary commercial chemicals such as BPA. In a study on prenatal exposure (Markey et al. 2003), pregnant mice were exposed to BPA on GD 9 to term (via s.c. pump delivering 0, 25 or 250 μg/kg/day). Treatment did not affect age at vaginal opening, but more mice showed persistent estrus/metestrus (≥4 days) after exposure to 25 μg (55.7 %) and 250 μg (55.6 %) than controls (37.1 %, $p < 0.05$). A proportion of exposed females showed blood-filled ovarian bursae at 3 months (11 and 16 vs. 0 %), and those exposed to 250 μg had a greater ovarian area occupied by antral follicles (15.5 ± 2.7 vs. 6.4 ± 2.8 %, $p < 0.05$). In a transgenerational study (Nilsson et al. 2012), pregnant rats were treated with BPA (0 or 50 mg/kg BW/d i.p.) on GD 8–14, F3 females exhibited fewer primordial follicles than controls (~14 to 9, $p < 0.005$), with no effect on antral follicle counts, but developed cystic structures similar to PCOS.

In another study on pre- and postnatal exposure (Signorile et al. 2012), pregnant mice were treated with BPA (0, 100 or 1000 μg/kg) on GD 1-PND 7. Exposed females showed fewer primordial follicles (median numbers: 7.75, 3.8 and 2.65, respectively, $p < 0.001$), reduced numbers of developing follicles (11.9, 5.5 and 4.5, $p < 0.001$), and increased the number of atretic follicles (1.55, 3.3 and 3.4, $p < 0.001$). In another study (Xi et al. 2011), female mice and their pups were treated with BPA (0, 12, 25 or 50 mg/kg/d, by gavage) from GD 1 to PND 20, and from PND 20 to 49, respectively. Exposure led to a dose-dependent increase in E_2 (~150, 300, 550 and 600 pMol/L), which altered feedback signals within the hypothalamic-pituitary-gonadal axis.

Neonatal exposure to BPA (PND 1, 3, 5 and 7) was examined (Rodriguez et al. 2010) in rats treated with oil (negative control), DES (positive control, 0.2 or 20 μg/kg) or BPA (0.05 or 20 mg/kg). BPA treatments (20 mg/kg) led to the highest proportion of recruited follicles on PND 8: ~30 % of follicles for control, ~45 % for DES0.2 ($p<0.001$), ~40 % for DES20 ($p<0.01$), ~55 % for BPA20 ($p<0.001$), and ~30 % for BPA0.05 (NS). The BPA (20 mg/kg) dose also led to more EsR1-positive follicles than negative controls (~65 vs. 15 %, $p<0.05$), which may explain the higher recruitment rate. In contrast, EsR2 positive staining was not affected by treatment. Taken together the data consistently show BPA exposure induced primordial follicle loss and suggest accelerated follicle recruitment into the growing pool.

In contrast to the examples of BPA, more primordial follicles and more primary follicles and altered gene expression for steroidogenic enzymes such as *StAR* and *Cyp11a1* were found (Ahn et al. 2012) in rats treated with methyl, propyl or butyl parabens (mPB, pPB and bPB, respectively, at 0, 62.5, 250 or 1000 mg/kg/d each, s.c.) or E_2 (40 μg/kg/d, positive control) for 7 days (PND 1-7). In a gestational exposure study (Taxvig et al. 2008), rats were exposed to s.c. oil, ethyl paraben (ePB, 400 mg/kg/d) or bBP (200 or 400 mg/kg/d) on GD 7-21. The females and the fetuses showed no histologic alterations in the ovaries, but treatment increased EsR2 gene expression. Additionally, female rats maternally exposed to bPB from GD 6 to PND 20 (100 mg/kg/day) presented early vaginal opening with no alterations in ovarian weight and histology (Kang et al. 2002), possibly due to the estrogenic action of parabens in the female HPG axis during development. In contrast, prepubertal mPB and isopropyl paraben exposure (250 and 1000 mg/kg from PND 21-40) leads to a delay in vaginal opening, a decrease in the number of corpora lutea and an increase in the number of cystic follicles in the ovary (Vo et al. 2010).

In summary, multiple contaminants from different chemical classes have been shown to affect ovarian follicle counts and thus establish the ovary as a potentially important target organ for adverse effects of environmental contaminant exposure. Moreover, we postulate that the accelerated loss of primordial follicles has the potential to shorten reproductive life span as well as advance the age of onset for health problems such as osteoporosis and cardiovascular disease arising from loss of estrogen. The mechanisms of ovarian follicle loss are unclear and may include decreased number of oocytes and follicles formed during development (Yin et al. 2015), increased recruitment of follicles into the growing pool (Rodriguez et al. 2010), and increased rate of follicle loss through either apoptosis (Jurisicova et al. 2007; Matikainen et al. 2001; Takai et al. 2003) or autophagy (Gannon et al. 2012, 2013). Regardless of the mechanism, the underlying initiating events remain to be elucidated and have important implications for fertility preservation as well as life-long health benefits. Tissue culture studies have shown that the adverse effects of polycyclic aromatic hydrocarbons such as benzyo[a]pyrene can be attenuated by co-treatment with AhR antagonists such as resveratrol and 3,4 dimethoxyflavone (Casper et al. 1999; Neal et al. 2010). These data raise the encouraging possibility that the ovary can be protected from the adverse effects of some contaminants on ovarian follicle loss. Fertility preservation is an exciting area of reproductive biology that is receiving renewed interest which we anticipate will uncover novel therapeutic interventions to protect ovarian function.

7.4.2 Steroidogenesis

Steroid production in the ovary is regulated by gonadotrophins and involves androgen production by theca cells and subsequent conversion of androgens to estrogens by granulosa cells. Results of animal studies demonstrate that $CdCl_2$ exposure decreased serum E_2 and P_4 concentrations by altering expression and/or activity of granulosa cell steroidogenic enzymes P450scc and StAR in adult female rats injected s.c. with $CdCl_2$ (0, 2.5, 5 and 7.5 mg/kg) (Zhang and Jia 2007; Zhang et al. 2008). $CdCl_2$ has a MW of 183, but only 112 of it is Cd. 1M would be 112 g/L; 1 μM would be 112 μg/L; 20 μM (lowest significant effect) would be 2.24 mg/L, a level much higher than the 0.29 ± 0.30 μg/l reported in follicular fluid in Saudi Arabian women for both controls and women under fertility treatment (Al-Saleh et al. 2008). Lead acetate was given to adult rhesus monkeys over several years at different doses (3.6, 5.9 and 8.1 mg/kg/d) in drinking water (Laughlin et al. 1987). Lead was detected in blood at concentrations that ranged between 51 and 88 μg/dl. Although the monkeys were still fertile, treated females showed menstrual cycles that were longer (50.9 vs. 32.9 days) and more variable, with shorter menses (2.3 vs. 3.4 d, $p<0.05$). Chronic lead exposure suppressed circulating concentrations of LH, FSH and E_2 in cynomolgus monkeys (Foster 1992) and suppressed luteal function at moderate blood levels (Foster et al. 1996).

Toxic effects of phthalates and phenols in experimental animal studies have been examined by several authors (Guerra et al. 2010; Kimura et al. 2006; Laskey and Berman 1993; Masutomi et al. 2003; Nagao et al. 2000; Willoughby et al. 2005) and have recently been reviewed elsewhere (Kay et al. 2013). Pregnant rats were exposed to di-η-butyl-phthalate (DBP) from GD 12 to PND 21 (0 or 100 mg/kg/d by gavage); the dose used did not affect ovarian development, puberty, hormone levels or fertility (Guerra et al. 2010). In another study (Masutomi et al. 2003), rats were treated with diisononyl phthalate (DINP, 0, 400, 4000, or 20,000 ppm in the diet) from GD 3 to PND 21; the highest dose decreased body and ovarian weight, whereas the lower doses had no adverse effects. Using a different approach, adult rats were treated with bis(2-diethylhexyl) phthalate (DEHP) (0 or 1500 mg/kg/day by gavage) for 10 days and their ovaries collected at different cycle stages for culture and measurement of steroid output (Laskey and Berman 1993). In diestrus, treated rats produced more E_2 and testosterone than controls, whereas in estrus, they produced more E_2. Another study (Mlynarcikova et al. 2009) also evaluated the effects of phenols and phthalates, by treating cumulus-oocyte- complexes (COCs) with 4-Chloro-3-methyl phenol (CMP), di(2-ethylhexyl) phthalate (DEHP) and benzyl butyl phthalate (BBP), in a range of 10^2–10^{-4} μM. Only the highest levels of CMP and BBP decreased COC expansion rate. The highest level of CMP decreased oocyte maturation rate (50.0 % reached metaphase II (MII) vs. 81.5 % in controls; $p<0.001$) and only the second highest level of DEHP increased P_4 output (~50 % more than control $p<0.05$). In another study from this group (Mlynarcikova et al. 2007), porcine granulosa cells were treated with 10^{-8}M to 10^{-4}M 4-octylphenol (OP), 4-nonylphenol (NP), and 4-*tert*-octylphenol (tOP), diisononyl phthalate

(DiNP), diisodecyl phthalate (DiDP) or dioctyl phthalate (DOP), with or without FSH. OP (10^{-5}M) reduced P_4 output ($p<0.05$), and NP (10^{-8} M) decreased E_2 output ($p<0.05$). FSH stimulated E_2 concentration in the media was reduced by OP, NP ($p<0.05$), as well as DiDP and DiNP ($p<0.01$) treatment. Taken together these data suggest that ovarian steroidogenesis can be affected by some of the phenols and phthalates used as plasticizers. To evaluate the effect of phthalate treatment on oocyte maturation equine COCs were incubated with di-(2-ethylhexyl) phthalate (DEHP) at 0, 0.12, 12 or 1200 μM (Ambruosi et al. 2011). The lowest concentration of DEHP (0.12 μM) inhibited oocyte maturation ($P<0.05$) with increased apoptosis and reduced ROS levels ($p<0.001$). Higher concentrations did not affect oocyte maturation, but apoptosis and ROS were higher ($p<0.0001$). Treated oocytes also exhibited higher ATP content ($p<0.05$). Taken together these data suggest that the COC is sensitive to the adverse effects of environmental contaminants; however, the underlying mechanisms are uncertain. We postulate that contaminant effects on gap junction communication are an area that may yield interesting results (Ganesan and Keating 2014; Gittens et al. 2003, 2005; Kidder and Mhawi 2002; Li et al. 2007).

While contaminant effects on ovarian steroidogenesis have been established, the mechanisms remain ill-defined and thus different culture methods have been employed in an effort to clarify this issue. Cocultures of granulosa and theca cells were isolated from pig antral follicles (Ptak et al. 2006) and treated with 4-Chlorobiphenyl (PCB3) or two of its metabolites (4-OH or 3,4-diOH PCB3, at 6 ng/ml). PCB3 and metabolite treatment decreased P_4 output by 55–65 % and 70–80 % of control ($p<0.05$), respectively. All 3 compounds increased E_2 output (130–330 % of controls, $p<0.05$), via increasing P_4 conversion to androgen and subsequently androgen aromatization. Chronic TCDD exposure upregulated the expression of 19 genes of known function (e.g. Cyp1a1, indicating activation of AHR) and downregulated the expression of 31 genes (e.g. 17α-OHase); FSH receptor, aromatase and inhibin were not affected (Valdez et al. 2009). Thus, TCDD alters ovarian steroidogenesis by affecting genes related to androgen synthesis. Similarly, rat granulosa cells from punctured follicles treated with 2,2-bis-(*p*-hydroxyphenyl)-1,1,1-trichloroethane (HPTE, the main metabolite of methoxychlor) produced a dose-dependent inhibition of P_4 output, blocked FSH-induced P450scc, 3β-HSD and P450arom mRNAs, but not StAR (Zachow and Uzumcu 2006). Thus, HPTE alters steroidogenesis by inhibiting several of the corresponding enzymes. The cells were incubated for 48 h with androgens (as aromatase substrates) and FSH (30 ng/ml) in the presence or absence of HPTE (0.1, 1, or 10 μM). The effects of methoxychlor (MXC) on ovarian steroidogenesis have been demonstrated by treating granulosa cells from pig antral follicles for 48 h with DDE (10^0–10^3 μM) or MXC (10^{-2}–10^1 μM), E_2 (0.1 μM, positive control) or negative control (10 % new born calf serum), with a final FSH exposure (Chedrese and Feyles 2001). In addition to stimulating cell proliferation, DDE affected FSH-induced cAMP production. MXC inhibited P_4 but not cAMP. The authors concluded that the mechanism appeared to be non-estrogenic. In rats, exposed prenatally (GD 19–22) and postnatally (PND 0–7) to 0, 20 or 100 μg/kg/d MXC exhibited 83–100, 78—100 and 0–11 % regular cycles and E_2 levels of ~27, 20, 18 pg/ml, respectively ($p<0.05$), at 13–15 months of age (Gore et al. 2011).

In tissue culture experiments (Grasselli et al. 2010), granulosa cells aspirated from pig antral follicles were treated with BPA (0, 0.1, 1, or 10 μM) and cell proliferation, E_2, P_4, and VEGF output were measured. Cell proliferation was not affected by BPA whereas the lowest dose of BPA increased estradiol, while the two higher ones inhibited it ($p<0.001$), and all BPA levels inhibited progesterone ($p<0.01$). VEGF output was increased after treatment with 1 and 10 μM BPA ($p<0.05$). The authors concluded that BPA alters follicle steroidogenesis and VEGF production. In one study (Mlynarcikova et al. 2005), granulosa cells from pig antral follicles were treated with BPA or BPA-dimethacrylate (DMA) at 10^{-4}–10^{-8} M and E_2 and P_4 output was determined. P_4 concentrations were increased by 10^{-6} BPA treatment ($p<0.05$) whilst 10^{-4} BPA inhibited P_4 output ($p<0.001$). At all concentrations tested, BPA inhibited FSH-induced E_2 production ($p<0.05$). In another study (Mlynarcikova et al. 2009), porcine COCs were treated with BPA (10^2–10^{-4} μM) during FSH-induced oocyte maturation. Subsequently, cumulus expansion, oocyte maturation and P_4 output were determined. Only the highest BPA concentration tested reduced the COC expansion rate compared to controls (16.9 % vs. 39.3 %, $p<0.05$), oocyte maturation rate (50.0 % reached MII vs. 81.5 %, $p<0.001$), and P_4 output (47.4 % less than controls, $p<0.01$). Although equivocal, the data lend some support to the observation that BPA is associated with an impaired response to ovulation induction (Fujimoto et al. 2011).

Human cells have also been used to study potential ovariotoxic effects. KGN cells, an ovarian granulosa-like tumour cell line that expresses a functional FSH receptor, as well as human granulosa cells (GLC) from IVF patients were treated with BPA (0, 40, 60, 80 or 100 μM), with or without FSH (Kwintkiewicz et al. 2010). In the KGN cells, BPA exhibited a dose-dependent reduction on FSH-induced IGF1, aromatase and E_2. In the GLC, only aromatase was examined, with similar effects to those seen in KGN cells (but the cells were more sensitive to lower doses). BPA also induced PPARγ, an aromatase inhibitor, which was seen as a possible mechanism for lower aromatase expression. Human fetal (21–22 weeks) ovarian explants were incubated with BPA (0 or 30 μM) for 1, 2 and 3 weeks (Brieno-Enriquez et al. 2012). Oocytes and fibroblasts were isolated for RNA extraction. BPA led to higher expression ($p<0.05$) of genes related to double-strand break generation, signaling and repair (*Rpa, Spo11, H2ax, Blm, Stra8*, *Nalp5*), onset of meiosis (*Stra8*), estrogen receptors and primordial follicle formation (*Nalp5*). However, the importance of these findings to risk assessment is unclear in the absence of accurate measures of free BPA concentrations in the follicular fluid. Assuming that the total BPA concentrations previously reported in follicular fluid were composed entirely of free BPA then the concentrations as added to the culture media used in this study would be approximately 4500 times greater than the concentrations measured in follicular fluid.

Isolated ovarian follicle culture techniques have been developed to assess the effect of test chemicals on stage-dependent follicle development and ovulation (Sun et al. 2004). In one study (Lenie and Smitz 2009), mouse follicles were isolated and cultured for 12 days to determine follicular development. Ovulation was stimulated (with recombinant hCG and recombinant epidermal growth factor), and evaluated on day 13. During development, follicles were exposed to the active metabolite of di (2-ethylhexyl) phthalate (DEHP), mono (2-ethylhexyl) phthalate (MEHP, 10–200 μM). MEHP led to a fivefold increase in P_4 output by day 12, but no change after ovulation stimulus. Testosterone/E_2 ratios showed that, despite increased testosterone output, less E_2 was produced with increasing MEHP. In another study (Treinen et al. 1990), MEHP was added to cultures of rat granulosa cells collected from freshly ruptured follicles (the rats had been treated with DES to induce granulosa cell proliferation). MEHP decreased up to 40 % cAMP production. Treatments containing 10, 25, 50 and 100 μM MEHP produced 30, 35, 40 and 70 % less P_4 than controls ($p<0.05$), and cAMP production was similarly inhibited. Human granulosa cells from IVF patients were treated with MEHP (0 or 0.6–500 μM) and aromatase mRNA, cell viability and steroid output were measured. MEHP decreased aromatase and E_2 output (Reinsberg et al. 2009) whereas P_4 output and cell viability decreased only at the highest dose. The large variety of phthalate compounds appears to have varying effects, and again, most studies involve laboratory animals, where significant effects are seen only at higher concentrations. The lowest dose with an effect was 0.12 μM of DEHP (Ambruosi et al. 2011), a dose equivalent to 47 μg/L (or ng/mL); a concentration that is substantially higher than the highest concentration of a phthalate metabolite (9.3 ng/ml) previously measured in follicular fluid (Krotz et al. 2012). In preliminary data from our lab, using an isolated mouse ovarian follicle culture system, BPA treatment significantly decreased the number of follicles as a percentage of the controls progressing to the preantral (18.04±5.5 %), antral (28±13.7 %) and preovulatory (54.5±20.4 %) stage, and follicle survival (50.6±18.8 %), but only at the highest concentration (5.0 μM) tested (Fig. 7.1). Treatments had no effect on gonadal steroid output in these cultures (Fig. 7.2). Thus, effective concentrations as added in cultures ranged between 1.14 and 2.28 mg of BPA/ml; concentrations far in excess of human exposure.

7.4.3 Genetic and Epigenetic Effects

The mutagenic effects of environmental toxicants have been well documented and effects on DNA repair mechanisms have also been described (Ganesan et al. 2013, 2014; Ganesan and Keating 2015). In addition to transcriptional regulation of genes through toxicant changes in transcription factor expression and microRNA (miRNA), contemporary studies have revealed that contaminant exposures are also associated with changes in the epigenome involving histone modifications including DNA methylation. Marks on the DNA that regulate gene expression and may be carried across generations.

Epigenetic effects of dietary factors such as phytoestrogens and environmental contaminants have been documented on the rodent uterus and lung whereas effects on the ovary remain largely unexplored (Zama and Uzumcu 2010). In one study identified in our literature search (Zama and Uzumcu 2009), MXC induced treatment-related effects on estrous cyclicity, ovarian steroidogenesis, and produced

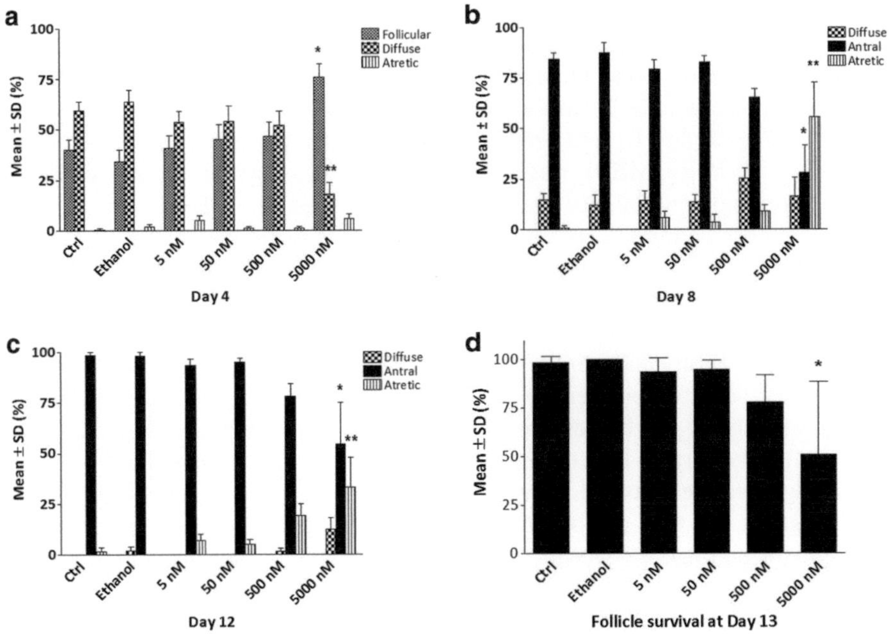

Fig. 7.1 Effect of BPA-exposure on *in vitro* follicle growth and survival. Follicular development at (**a**) 4, (**b**) 8, and (**c**) 12 days of culture and (**d**) follicular survival at 12 days of culture. All the follicles were follicular at day 0 (start) of culture. Each experiment was repeated at least 3 times in duplicate plate/condition, n=826 follicles (10–16 follicles/plate). *Bars* with different letters (within the same follicular stages) differ significantly ($p<0.05$)

changes in DNA methylation with potential effects across generations. Effects of MXC on DNA methylation have been documented in pregnant rats exposed to MXC (0, 20 μg or 100 mg/kg/d i.p.) from GD 19 to PND 7. Female pups were also treated from day 0 to PND 7 and euthanized after a cycle on PND 50-60. DNA methylation showed that 10 ovarian genes were hypermethylated after the highest MXC dose, including genes related to follicle maturation and ovulation, such as *PAPP-A*. Ovarian EsR1 was not affected, but EsR2 expression was reduced with 100 mg MXC (19.9±1.06 vs. 36.7±5.47, $p<0.05$). Environmental contaminant-induced changes in ovarian epigenetic marks is an important emerging area in reproductive biology that holds great promise for increased mechanistic insight.

7.5 Future Directions

Understanding the hazard to reproductive health and ovarian function arising from exposure to environmental contaminants is difficult to appreciate in the absence of direct measures of exposure and more sensitive markers of an adverse effect. While

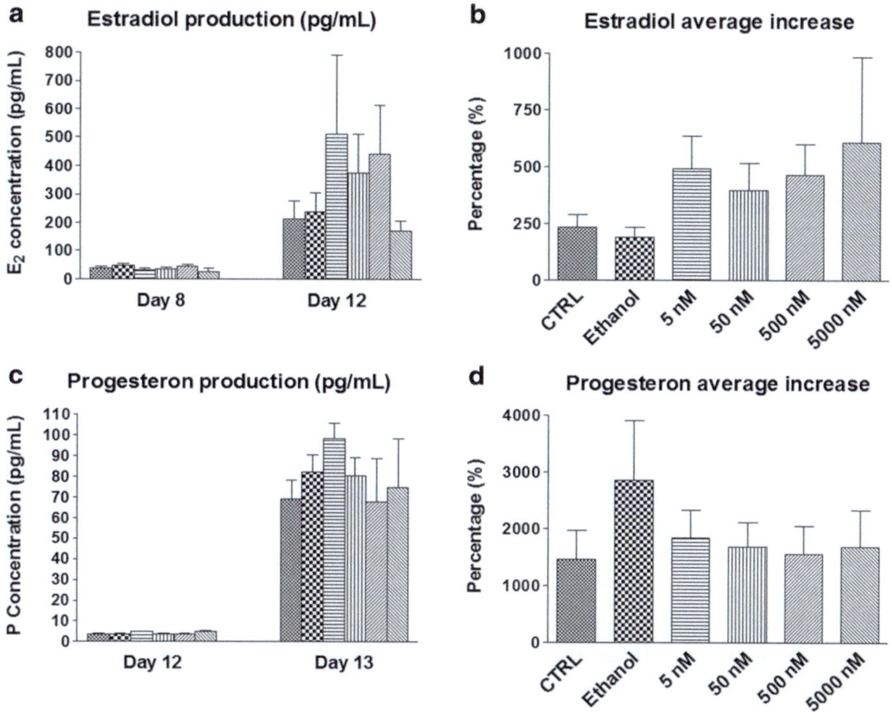

Fig. 7.2 Effect of BPA-exposure on steroid outputs during *in vitro* follicle growth. (**a**) Estradiol (E_2) and (**c**) progesterone (P_4) outputs after antral (Day 8) and preovulatory (Day 12) follicle development. Relative average increase of (**b**) E_2 between Days 8 and 12 [(Day 12/Day 8) × 100] and of (**d**) P_4 between Days 12 and 13 [(Day 13/Day 12) × 100]. Values are means ± S.E.M. E_2: n = 4–8 replicates; P_4: n = 4–5 replicates

circulating concentrations of AMH provides insight into the primordial follicle count/ovarian reserve, there are a number of limitations with this marker that have yet to be adequately resolved. Specifically, changes in circulating concentrations of AMH as women age as well as relationship with BMI and health status such as PCOS and diabetes are yet to be defined and normal ranges are only generally established.

While the concentrations of some chemicals in the serum are representative of concentrations that can be measured in ovarian follicular fluid (e.g. perfluorinated compounds), this is not always the case as shown for *p,p′*-DDE which can be three times higher in the serum than the follicular fluid (Younglai et al. 2002). It is encouraging that good correlation has been shown between serum and ovarian follicular fluid concentrations of DDT and its metabolites (Mahalingaiah et al. 2012). Therefore, we suggest that the pharmacokinetics and pharmacodynamics of environmental contaminants are essential data for establishing relevant target doses for animal and tissue culture studies as well as for risk assessment purposes.

We note that animal studies are clearly indispensable in the assessment of potential hazards to human health and reproductive function. We propose that their contribution to the risk assessment process could be enhanced with greater attention to issues of comparative endocrinology. For example, emerging evidence suggests that women may have more than one wave of follicles developing for ovulation in a given cycle (Baerwald et al. 2003a, b; Baerwald and Pierson 2004). However, many questions remain and the relevance of these observations to human health, fertility and toxicology has yet to be defined. We further postulate that while the academic investigators enjoy a much less restrictive environment that favors development of innovative testing strategies to enhance understanding of the mechanisms underlying toxic phenomenon, greater insight into the dose used, route of exposure, and use of multiple dose groups would enhance uptake of study results by regulatory bodies. Finally, there have been many calls for studies of mixtures and we add to the cacophony of voices calling for such testing.

Finally, although the literature suggests that effects of environmental contaminants at concentrations typically measured in contemporary biomonitoring studies are unlikely, precaution is never a poor option. Overall, regulatory decisions appear to be effective in protecting the health of general population against adverse effects on ovarian function; however, we note that chemicals continue to enter the environment. Manufacturing practices have improved substantially over the last decades with reductions in chemical emissions or releases into the environment. However, leaching of chemicals from finished products continues to be a problem and thus is an issue for engineering to insure safety. While calls to regulatory authorities for chemical bans continue, we caution against replacement of suspected chemicals with alternatives that are potentially more hazardous to human health such as Bisphenol S and Bisphenol F, replacements for BPA (Eladak et al. 2015; Rosenmai et al. 2014). Toxicity testing has progressed substantially over the last several decades as techniques have improved and understanding of reproductive physiology and endocrinology have advanced. We note that it will never be possible to prove absolute safety; however, enhancements to existing regulatory testing paradigms and continued interest of academic investigators promises to insure that potential health effects of environmental contaminants are identified and brought forward in the literature for use in the risk assessment process. In addition to partnering with chemical engineers to identify safer alternatives, academic investigators are encouraged to explore mechanisms of action as well as potential therapeutic interventions that can protect the ovary and aid in fertility preservation. Renewed interest in this area could have spill-over benefits for women undergoing cancer therapies that are well known to damage ovarian health.

7.6 Conclusions

Human exposure to potential ovarian toxicants has not been well defined and of greater concern is the lack of data on internal exposure dose and measurement of parent and metabolite concentrations in the ovary. While there is a robust animal literature illustrating the potential hazard to ovarian function and reproductive health from exposure to environmental contaminants, the concentrations at which effects were documented were frequently orders of magnitude beyond human exposure. While a robust literature was uncovered, the environmental contaminants addressed represent only a very small proportion of the many thousands of chemicals in commercial use that enter the environment, the food chain, and ultimately result in human exposure. Thus the available literature provides only an incomplete picture of the potential impact of environmental contaminants on reproductive health and ovarian function. We note that some environmental contaminant exposures have been linked with decreased circulating concentrations of E_2, shorter or longer menstrual cycle length, attenuated response to ovulation induction strategies, impaired fertility, and earlier menopause. However, the available literature is equivocal and highly variable with respect to study design, populations studied, sample size, exposure assessment, analytical methods employed, and outcomes assessed. Therefore, we propose that the state-of-the science is presently inadequate to conclude that exposure to environmental contaminants are or are not associated with adverse effects on ovarian function or human fertility.

References

Ahn HJ, An BS, Jung EM, Yang H, Choi KC, Jeung EB (2012) Parabens inhibit the early phase of folliculogenesis and steroidogenesis in the ovaries of neonatal rats. Mol Reprod Dev 79:626–636

Al-Saleh I, Coskun S, Mashhour A, Shinwari N, El-Doush I, Billedo G, Jaroudi K, Al-Shahrani A, Al-Kabra M, El Din Mohamed G (2008) Exposure to heavy metals (lead, cadmium and mercury) and its effect on the outcome of in-vitro fertilization treatment. Int J Hyg Environ Health 211:560–579

Alvarez S, Devouche E (2012) First French national survey on lifestyle and toxic factors in infertile couples. Gynecol Obstet Fertil 40:765–771

Ambruosi B, Uranio MF, Sardanelli AM, Pocar P, Martino NA, Paternoster MS, Amati F, Dell'Aquila ME (2011) In vitro acute exposure to DEHP affects oocyte meiotic maturation, energy and oxidative stress parameters in a large animal model. PLoS One 6:e27452

Appt SE, Kaplan JR, Clarkson TB, Cline JM, Christian PJ, Hoyer PB (2006) Destruction of primordial ovarian follicles in adult cynomolgus macaques after exposure to 4-vinylcyclohexene diepoxide: a nonhuman primate model of the menopausal transition. Fertil Steril 86(Suppl 4):1210–1216

Augood C, Duckitt K, Templeton AA (1998) Smoking and female infertility: a systematic review and meta-analysis. Hum Reprod 13:1532–1539

Baerwald AR, Pierson RA (2004) Endometrial development in association with ovarian follicular waves during the menstrual cycle. Ultrasound Obstet Gynecol 24:453–460

Baerwald AR, Adams GP, Pierson RA (2003a) Characterization of ovarian follicular wave dynamics in women. Biol Reprod 69:1023–1031

Baerwald AR, Adams GP, Pierson RA (2003b) A new model for ovarian follicular development during the human menstrual cycle. Fertil Steril 80:116–122

Baldridge MG, Stahl RL, Gerstenberger SL, Tripoli V, Hutz RJ (2003) Modulation of ovarian follicle maturation in Long-Evans rats exposed to polychlorinated biphenyls (PCBs) in-utero and lactationally. Reprod Toxicol 17:567–573

Batariova A, Spevackova V, Benes B, Cejchanova M, Smid J, Cerna M (2006) Blood and urine levels of Pb, Cd and Hg in the general population of the Czech Republic and proposed reference values. Int J Hyg Environ Health 209:359–366

Baukloh V, Bohnet HG, Trapp M, Heeschen W, Feichtinger W, Kemeter P (1987) Biocides in human follicular fluid. Ann N Y Acad Sci 442:240–250

Bloom MS, Buck Louis GM, Schisterman EF, Liu A, Kostyniak PJ (2007) Maternal serum polychlorinated biphenyl concentrations across critical windows of human development. Environ Health Perspect 115:1320–1324

Bloom MS, Kim D, vom Saal FS, Taylor JA, Cheng G, Lamb JD, Fujimoto VY (2011) Bisphenol A exposure reduces the estradiol response to gonadotropin stimulation during in vitro fertilization. Fertil Steril 96:672–677

Bloom MS, Fujimoto VY, Steuerwald AJ, Cheng G, Browne RW, Parsons PJ (2012a) Background exposure to toxic metals in women adversely influences pregnancy during in vitro fertilization (IVF). Reprod Toxicol 34:471–481

Bloom MS, Kim K, Kruger PC, Parsons PJ, Arnason JG, Steuerwald AJ, Fujimoto VY (2012b) Associations between toxic metals in follicular fluid and in vitro fertilization (IVF) outcomes. J Assist Reprod Genet 29:1369–1379

Boivin J, Bunting L, Collins JA, Nygren KG (2007) International estimates of infertility prevalence and treatment-seeking: potential need and demand for infertility medical care. Hum Reprod 22:1506–1512

Borman SM, Christian PJ, Sipes IG, Hoyer PB (2000) Ovotoxicity in female Fischer rats and B6 mice induced by low-dose exposure to three polycyclic aromatic hydrocarbons: comparison through calculation of an ovotoxic index. Toxicol Appl Pharmacol 167:191–198

Brieno-Enriquez MA, Reig-Viader R, Cabero L, Toran N, Martinez F, Roig I, Garcia Caldes M (2012) Gene expression is altered after bisphenol A exposure in human fetal oocytes in vitro. Mol Hum Reprod 18:171–183

Byskov AGS (1978) Follicle atresia. In: Jones R (ed) The vertebrate ovary. Plenum Press, New York, pp 533–562

Calafat AM, Ye X, Wong LY, Reidy JA, Needham LL (2008) Exposure of the U.S. population to bisphenol A and 4-tertiary-octylphenol: 2003–2004. Environ Health Perspect 116:39–44

Casper RF, Quesne M, Rogers IM, Shirota T, Jolivet A, Milgrom E, Savouret JF (1999) Resveratrol has antagonist activity on the aryl hydrocarbon receptor: implications for prevention of dioxin toxicity. Mol Pharmacol 56:784–790

Chandra A, Martinez GM, Mosher WD, Abma JC, Jones J (2005) Fertility, family planning, and reproductive health of U.S. women: data from the 2002 National Survey of Family Growth. Vital Health Stat 23:1–160

Chedrese PJ, Feyles F (2001) The diverse mechanism of action of dichlorodiphenyldichloroethylene (DDE) and methoxychlor in ovarian cells in vitro. Reprod Toxicol 15:693–698

Cohn BA, Cirillo PM, Wolff MS, Schwingl PJ, Cohen RD, Sholtz RI, Ferrara A, Christianson RE, van den Berg BJ, Siiteri PK (2003) DDT and DDE exposure in mothers and time to pregnancy in daughters. Lancet 361:2205–2206

Cooper GS, Klebanoff MA, Promislow J, Brock JW, Longnecker MP (2005) Polychlorinated biphenyls and menstrual cycle characteristics. Epidemiology 16:191–200

Cortvrindt RG, Smitz JE (2002) Follicle culture in reproductive toxicology: a tool for in-vitro testing of ovarian function? Hum Reprod Update 8:243–254

Cortvrindt RG, Hu Y, Liu J, Smitz JE (1998) Timed analysis of the nuclear maturation of oocytes in early preantral mouse follicle culture supplemented with recombinant gonadotropin. Fertil Steril 70:1114–1125

Davis BJ, Price HC, O'Connor RW, Fernando R, Rowland AS, Morgan DL (2001) Mercury vapor and female reproductive toxicity. Toxicol Sci 59:291–296

De FE, Porpora MG, Di DA, Ingelido AM, Cardelli M, Cosmi EV, Donnez J (2004) Dioxin-like compounds and endometriosis: a study on Italian and Belgian women of reproductive age. Toxicol Lett 150:203–209

De Felip E, di Domenico A, Miniero R, Silvestroni L (2004) Polychlorobiphenyls and other organochlorine compounds in human follicular fluid. Chemosphere 54:1445–1449

Dickerson EH, Sathyapalan T, Knight R, Maguiness SM, Killick SR, Robinson J, Atkin SL (2011) Endocrine disruptor & nutritional effects of heavy metals in ovarian hyperstimulation. J Assist Reprod Genet 28:1223–1228

Dvornyk V, Long JR, Liu PY, Zhao LJ, Shen H, Recker RR, Deng HW (2006) Predictive factors for age at menopause in Caucasian females. Maturitas 54:19–26

Dyer SM, Cattani M, Pisaniello DL, Williams FM, Edwards JW (2001) Peripheral cholinesterase inhibition by occupational chlorpyrifos exposure in Australian termiticide applicators. Toxicology 169:177–185

Eladak S, Grisin T, Moison D, Guerquin MJ, N'Tumba-Byn T, Pozzi-Gaudin S, Benachi A, Livera G, Rouiller-Fabre V, Habert R (2015) A new chapter in the bisphenol A story: bisphenol S and bisphenol F are not safe alternatives to this compound. Fertil Steril 103:11–21

Eskenazi B, Warner M, Marks AR, Samuels S, Gerthoux PM, Vercellini P, Olive DL, Needham L, Patterson D Jr, Mocarelli P (2005) Serum dioxin concentrations and age at menopause. Environ Health Perspect 113:858–862

Farr SL, Cooper GS, Cai J, Savitz DA, Sandler DP (2004) Pesticide use and menstrual cycle characteristics among premenopausal women in the Agricultural Health Study. Am J Epidemiol 160:1194–1204

Foster WG (1992) Reproductive toxicity of chronic Lead exposure in the female cynomolgus monkey. Reprod Toxicol 6:123–131

Foster WG (1995) The reproductive toxicity of Great Lakes contaminants. Environ Health Perspect 103(Suppl 9):63–69

Foster WG (2003) Do environmental contaminants adversely affect human reproductive physiology? J Obstet Gynaecol Can 25:33–44

Foster WG, McMahon A, Younglai EV, Jarrell JF, Lecavalier P (1995) Alterations in circulating ovarian steroids in hexachlorobenzene-exposed monkeys. Reprod Toxicol 9:541–548

Foster WG, McMahon A, Rice DC (1996) Subclinical changes in luteal function in cynomolgus monkeys with moderate blood lead levels. J Appl Toxicol 16:159–163

Franczak A, Nynca A, Valdez KE, Mizinga KM, Petroff BK (2006) Effects of acute and chronic exposure to the aryl hydrocarbon receptor agonist 2,3,7,8-tetrachlorodibenzo-p-dioxin on the transition to reproductive senescence in female Sprague-Dawley rats. Biol Reprod 74:125–130

Fujimoto VY, Kim D, vom Saal FS, Lamb JD, Taylor JA, Bloom MS (2011) Serum unconjugated bisphenol A concentrations in women may adversely influence oocyte quality during in vitro fertilization. Fertil Steril 95:1816–1819

Fuortes L, Clark M, Kirchner H, Smith E (1997) Association between female infertility and agricultural work history. Am J Ind Med 31:445–451

Gallicchio L, Miller S, Greene T, Zacur H, Flaws JA (2009) Premature ovarian failure among hairdressers. Hum Reprod 24:2636–2641

Ganesan S, Keating AF (2014) Impact of 7,12-dimethylbenz[a]anthracene exposure on connexin gap junction proteins in cultured rat ovaries. Toxicol Appl Pharmacol 274:209–214

Ganesan S, Keating AF (2015) Phosphoramide mustard exposure induces DNA adduct formation and the DNA damage repair response in rat ovarian granulosa cells. Toxicol Appl Pharmacol 282:252–258

Ganesan S, Bhattacharya P, Keating AF (2013) 7,12-Dimethylbenz[a]anthracene exposure induces the DNA repair response in neonatal rat ovaries. Toxicol Appl Pharmacol 272:690–696

Ganesan S, Nteeba J, Keating AF (2014) Enhanced susceptibility of ovaries from obese mice to 7,12-dimethylbenz[a]anthracene-induced DNA damage. Toxicol Appl Pharmacol 281:203–210

Gannon AM, Stampfli MR, Foster WG (2012) Cigarette smoke exposure results in significant follicle loss via an alternative ovarian cell death pathway in mice. Toxicol Sci 125:274–284

Gannon AM, Stampfli MR, Foster WG (2013) Cigarette smoke exposure elicits increased autophagy and dysregulation of mitochondrial dynamics in murine granulosa cells. Biol Reprod 88:1–11

Gittens JE, Mhawi AA, Lidington D, Ouellette Y, Kidder GM (2003) Functional analysis of gap junctions in ovarian granulosa cells: distinct role for connexin 43 in early stages of folliculogenesis. Am J Physiol Cell Physiol 284:C880–C887
Gittens JE, Barr KJ, Vanderhyden BC, Kidder GM (2005) Interplay between paracrine signaling and gap junctional communication in ovarian follicles. J Cell Sci 118:113–122
Gore AC, Walker DM, Zama AM, Armenti AE, Uzumcu M (2011) Early life exposure to endocrine-disrupting chemicals causes lifelong molecular reprogramming of the hypothalamus and premature reproductive aging. Mol Endocrinol 25:2157–2168
Grasselli F, Baratta L, Baioni L, Bussolati S, Ramoni R, Grolli S, Basini G (2010) Bisphenol A disrupts granulosa cell function. Domest Anim Endocrinol 39:34–39
Guerra MT, Scarano WR, de Toledo FC, Franci JA, Kempinas Wde G (2010) Reproductive development and function of female rats exposed to di-eta-butyl-phthalate (DBP) in utero and during lactation. Reprod Toxicol 29:99–105
Haney AF, Hughes SF, Hughes CL Jr (1984) Screening of potential reproductive toxicants by use of porcine granulosa cell cultures. Toxicology 30:227–241
Hombach-Klonisch S, Pocar P, Kietz S, Klonisch T (2005) Molecular actions of polyhalogenated arylhydrocarbons (PAHs) in female reproduction. Curr Med Chem 12:599–616
Hunt PA, Koehler KE, Susiarjo M, Hodges CA, Ilagan A, Voigt RC, Thomas S, Thomas BF, Hassold TJ (2003) Bisphenol a exposure causes meiotic aneuploidy in the female mouse. Curr Biol 13:546–553
Hunt PA, Lawson C, Gieske M, Murdoch B, Smith H, Marre A, Hassold T, Vandevoort CA (2012) Bisphenol A alters early oogenesis and follicle formation in the fetal ovary of the rhesus monkey. Proc Natl Acad Sci U S A 109:17525–17530
Ikezuki Y, Tsutsumi O, Takai Y, Kamei Y, Taketani Y (2002) Determination of bisphenol A concentrations in human biological fluids reveals significant early prenatal exposure. Hum Reprod 17:2839–2841
Jackson LW, Howards PP, Wactawski-Wende J, Schisterman EF (2011) The association between cadmium, lead and mercury blood levels and reproductive hormones among healthy, premenopausal women. Hum Reprod 26:2887–2895
Jarrell JF, Villeneuve D, Franklin C, Bartlett S, Wrixon W, Kohut J, Zouves CG (1993) Contamination of human ovarian follicular fluid and serum by chlorinated organic compounds in three Canadian cities. CMAJ 148:1321–1327
Jirsova S, Masata J, Jech L, Zvarova J (2010) Effect of polychlorinated biphenyls (PCBs) and 1,1,1-trichloro-2,2,-bis (4-chlorophenyl)-ethane (DDT) in follicular fluid on the results of in vitro fertilization-embryo transfer (IVF-ET) programs. Fertil Steril 93:1831–1836
Johnson PI, Altshul L, Cramer DW, Missmer SA, Hauser R, Meeker JD (2012) Serum and follicular fluid concentrations of polybrominated diphenyl ethers and in-vitro fertilization outcome. Environ Int 45:9–14
Jukic AM, Calafat AM, McConnaughey DR, Longnecker MP, Hoppin JA, Weinberg CR, Wilcox AJ, Baird DD (2015) Urinary concentrations of phthalate metabolites and bisphenol a and associations with follicular-phase length, luteal-phase length, fecundability, and early pregnancy loss. Environ Health Perspect. doi:10.1289/ehp.1408164
Jung NK, Park JY, Park JH, Kim SY, Park JK, Chang WK, Lee HC, Kim SW, Chun SY (2010) Attenuation of cell cycle progression by 2,3,7,8-tetrachlorodibenzo-p-dioxin eliciting ovulatory blockade in gonadotropin-primed immature rats. Endocr J 57:863–871
Jurisicova A, Taniuchi A, Li H, Shang Y, Antenos M, Detmar J, Xu J, Matikainen T, Benito HA, Nunez G, Casper RF (2007) Maternal exposure to polycyclic aromatic hydrocarbons diminishes murine ovarian reserve via induction of Harakiri. J Clin Invest 117:3971–3978
Kandaraki E, Chatzigeorgiou A, Livadas S, Palioura E, Economou F, Koutsilieris M, Palimeri S, Panidis D, Diamanti-Kandarakis E (2011) Endocrine disruptors and polycystic ovary syndrome (PCOS): elevated serum levels of bisphenol A in women with PCOS. J Clin Endocrinol Metab 96:E480–E484

Kang KS, Che JH, Ryu DY, Kim TW, Li GX, Lee YS (2002) Decreased sperm number and motile activity on the F1 offspring maternally exposed to butyl p-hydroxybenzoic acid (butyl paraben). J Vet Med Sci 64:227–235
Kay VR, Chambers C, Foster WG (2013) Reproductive and developmental effects of phthalate diesters in females. Crit Rev Toxicol 43:200–219
Kidder GM, Mhawi AA (2002) Gap junctions and ovarian folliculogenesis. Reproduction 123:613–620
Kimura N, Kimura T, Suzuki M, Totsukawa K (2006) Effect of gestational exposure to nonylphenol on the development and fertility of mouse offspring. J Reprod Dev 52:789–795
Knox SS, Jackson T, Javins B, Frisbee SJ, Shankar A, Ducatman AM (2011) Implications of early menopause in women exposed to perfluorocarbons. J Clin Endocrinol Metab 96:1747–1753
Koh JM, Kim CH, Hong SK, Lee KU, Kim YT, Kim OJ, Kim GS (1998) Primary ovarian failure caused by a solvent containing 2-bromopropane. Eur J Endocrinol 138:554–556
Koifman S, Koifman RJ, Meyer A (2002) Human reproductive system disturbances and pesticide exposure in Brazil. Cad Saude Publica 18:435–445
Krotz SP, Carson SA, Tomey C, Buster JE (2012) Phthalates and bisphenol do not accumulate in human follicular fluid. J Assist Reprod Genet 29:773–777
Kwintkiewicz J, Nishi Y, Yanase T, Giudice LC (2010) Peroxisome proliferator-activated receptor-gamma mediates bisphenol A inhibition of FSH-stimulated IGF-1, aromatase, and estradiol in human granulosa cells. Environ Health Perspect 118:400–406
La Rocca C, Alessi E, Bergamasco B, Caserta D, Ciardo F, Fanello E, Focardi S, Guerranti C, Stecca L, Moscarini M, Perra G, Tait S, Zaghi C, Mantovani A (2012) Exposure and effective dose biomarkers for perfluorooctane sulfonic acid (PFOS) and perfluorooctanoic acid (PFOA) in infertile subjects: preliminary results of the PREVIENI project. Int J Hyg Environ Health 215:206–211
Laskey JW, Berman E (1993) Steroidogenic assessment using ovary culture in cycling rats: effects of Bis(2-Diethylhexyl) phthalate on ovarian steroid production. Reprod Toxicol 7:25–33
Laughlin N, Bowman R, Franks P, Dierschke D (1987) Altered menstrual cycles in rhesus monkeys induced by lead. Fundam Appl Toxicol 9:722–729
Lenie S, Smitz J (2009) Steroidogenesis-disrupting compounds can be effectively studied for major fertility-related endpoints using in vitro cultured mouse follicles. Toxicol Lett 185:143–152
Lenie S, Cortvrindt R, Adriaenssens T, Smitz J (2004) A reproducible two-step culture system for isolated primary mouse ovarian follicles as single functional units. Biol Reprod 71:1730–1738
Li TY, Colley D, Barr KJ, Yee SP, Kidder GM (2007) Rescue of oogenesis in Cx37-null mutant mice by oocyte-specific replacement with Cx43. J Cell Sci 120:4117–4125
Mahalingaiah S, Missmer SA, Maity A, Williams PL, Meeker JD, Berry K, Ehrlich S, Perry MJ, Cramer DW, Hauser R (2012) Association of hexachlorobenzene (HCB), dichlorodiphenyltrichloroethane (DDT), and dichlorodiphenyldichloroethylene (DDE) with in vitro fertilization (IVF) outcomes. Environ Health Perspect 120:316–320
Markey CM, Coombs MA, Sonnenschein C, Soto AM (2003) Mammalian development in a changing environment: exposure to endocrine disruptors reveals the developmental plasticity of steroid-hormone target organs. Evol Dev 5:67–75
Mascarenhas MN, Flaxman SR, Boerma T, Vanderpoel S, Stevens GA (2012) National, regional, and global trends in infertility prevalence since 1990: a systematic analysis of 277 health surveys. PLoS Med 9:e1001356
Masutomi N, Shibutani M, Takagi H, Uneyama C, Takahashi N, Hirose M (2003) Impact of dietary exposure to methoxychlor, genistein, or diisononyl phthalate during the perinatal period on the development of the rat endocrine/reproductive systems in later life. Toxicology 192:149–170
Matikainen T, Perez GI, Jurisicova A, Pru JK, Schlezinger JJ, Ryu HY, Laine J, Sakai T, Korsmeyer SJ, Casper RF, Sherr DH, Tilly JL (2001) Aromatic hydrocarbon receptor-driven Bax gene

expression is required for premature ovarian failure caused by biohazardous environmental chemicals. Nat Genet 28:355–360

Meeker JD, Missmer SA, Altshul L, Vitonis AF, Ryan L, Cramer DW, Hauser R (2009) Serum and follicular fluid organochlorine concentrations among women undergoing assisted reproduction technologies. Environ Health 8:32

Mlynarcikova A, Kolena J, Fickova M, Scsukova S (2005) Alterations in steroid hormone production by porcine ovarian granulosa cells caused by bisphenol A and bisphenol A dimethacrylate. Mol Cell Endocrinol 244:57–62

Mlynarcikova A, Fickova M, Scsukova S (2007) The effects of selected phenol and phthalate derivatives on steroid hormone production by cultured porcine granulosa cells. Altern Lab Anim 35:71–77

Mlynarcikova A, Nagyova E, Fickova M, Scsukova S (2009) Effects of selected endocrine disruptors on meiotic maturation, cumulus expansion, synthesis of hyaluronan and progesterone by porcine oocyte-cumulus complexes. Toxicol In Vitro 23:371–377

Nagao T, Saito Y, Usumi K, Nakagomi M, Yoshimura S, Ono H (2000) Disruption of the reproductive system and reproductive performance by administration of nonylphenol to newborn rats. Hum Exp Toxicol 19:284–296

Neal MS, Zhu J, Holloway AC, Foster WG (2007) Follicle growth is inhibited by benzo-[a]-pyrene, at concentrations representative of human exposure, in an isolated rat follicle culture assay. Hum Reprod 22:961–967

Neal MS, Zhu J, Foster WG (2008) Quantification of benzo[a]pyrene and other PAHs in the serum and follicular fluid of smokers versus non-smokers. Reprod Toxicol 25:100–106

Neal MS, Mulligan Tuttle AM, Casper RF, Lagunov A, Foster WG (2010) Aryl hydrocarbon receptor antagonists attenuate the deleterious effects of benzo[a]pyrene on isolated rat follicle development. Reprod Biomed Online 21:100–108

Nilsson E, Larsen G, Manikkam M, Guerrero-Bosagna C, Savenkova MI, Skinner MK (2012) Environmentally induced epigenetic transgenerational inheritance of ovarian disease. PLoS One 7:e36129

Patel S, Bajpayee M, Pandey AK, Parmar D, Dhawan A (2007) In vitro induction of cytotoxicity and DNA strand breaks in CHO cells exposed to cypermethrin, pendimethalin and dichlorvos. Toxicol In Vitro 21:1409–1418

Petro EM, Leroy JL, Covaci A, Fransen E, De ND, Dirtu AC, De PI, Bols PE (2012) Endocrine-disrupting chemicals in human follicular fluid impair in vitro oocyte developmental competence. Hum Reprod 27:1025–1033

Petro EM, D'Hollander W, Covaci A, Bervoets L, Fransen E, De Neubourg D, De Pauw I, Leroy JL, Jorssen EP, Bols PE (2014) Perfluoroalkyl acid contamination of follicular fluid and its consequence for in vitro oocyte developmental competence. Sci Total Environ 496:282–288

Pocar P, Brevini TA, Fischer B, Gandolfi F (2003) The impact of endocrine disruptors on oocyte competence. Reproduction 125:313–325

Pollack AZ, Schisterman EF, Goldman LR, Mumford SL, Albert PS, Jones RL, Wactawski-Wende J (2011) Cadmium, lead, and mercury in relation to reproductive hormones and anovulation in premenopausal women. Environ Health Perspect 119:1156–1161

Ptak A, Ludewig G, Robertson L, Lehmler HJ, Gregoraszczuk EL (2006) In vitro exposure of porcine prepubertal follicles to 4-chlorobiphenyl (PCB3) and its hydroxylated metabolites: effects on sex hormone levels and aromatase activity. Toxicol Lett 164:113–122

Quignot N, Arnaud M, Robidel F, Lecomte A, Tournier M, Cren-Olive C, Barouki R, Lemazurier E (2012a) Characterization of endocrine-disrupting chemicals based on hormonal balance disruption in male and female adult rats. Reprod Toxicol 33:339–352

Quignot N, Tournier M, Pouech C, Cren-Olive C, Barouki R, Lemazurier E (2012b) Quantification of steroids and endocrine disrupting chemicals in rat ovaries by LC-MS/MS for reproductive toxicology assessment. Anal Bioanal Chem 403:1629–1640

Reinsberg J, Wegener-Toper P, van der Ven K, van der Ven H, Klingmueller D (2009) Effect of mono-(2-ethylhexyl) phthalate on steroid production of human granulosa cells. Toxicol Appl Pharmacol 239:116–123

Rodriguez HA, Santambrosio N, Santamaria CG, Munoz-de-Toro M, Luque EH (2010) Neonatal exposure to bisphenol A reduces the pool of primordial follicles in the rat ovary. Reprod Toxicol 30:550–557

Rosenmai AK, Dybdahl M, Pedersen M, Alice van Vugt-Lussenburg BM, Wedebye EB, Taxvig C, Vinggaard AM (2014) Are structural analogues to bisphenol a safe alternatives? Toxicol Sci 139:35–47

Rutstein SO, Shah IH (2004) Infecundity, infertility, and childlessness in developing countries. ORC Macro, Calverton

Sadeu JC, Foster WG (2011a) Cigarette smoke condensate exposure delays follicular development and function in a stage-dependent manner. Fertil Steril 95:2410–2417

Sadeu JC, Foster WG (2011b) Effect of in vitro exposure to Benzo[a]pyrene, a component of cigarette smoke, on folliculogenesis, steroidogenesis, and oocyte maturation. Reprod Toxicol 31:402–408

Sadeu JC, Hughes CL, Agarwal S, Foster WG (2010) Alcohol, drugs, caffeine, tobacco, and environmental contaminant exposure: reproductive health consequences and clinical implications. Crit Rev Toxicol 40:633–652

Sahambi SK, Visser JA, Themmen AP, Mayer LP, Devine PJ (2008) Correlation of serum anti-Mullerian hormone with accelerated follicle loss following 4-vinylcyclohexene diepoxide-induced follicle loss in mice. Reprod Toxicol 26:116–122

Schlebusch H, Wagner U, van der Ven H, Al-Hasani S, Diedrich K, Krebs D (1989) Polychlorinated biphenyls: the occurrence of the main congeners in follicular and sperm fluids. J Clin Chem Clin Biochem 27:663–667

Shi Z, Valdez KE, Ting AY, Franczak A, Gum SL, Petroff BK (2007) Ovarian endocrine disruption underlies premature reproductive senescence following environmentally relevant chronic exposure to the aryl hydrocarbon receptor agonist 2,3,7,8-tetrachlorodibenzo-p-dioxin. Biol Reprod 76:198–202

Signorile PG, Spugnini EP, Citro G, Viceconte R, Vincenzi B, Baldi F, Baldi A (2012) Endocrine disruptors in utero cause ovarian damages linked to endometriosis. Front Biosci (Elite Ed) 4:1724–1730

Silva MJ, Slakman AR, Reidy JA, Preau JL Jr, Herbert AR, Samandar E, Needham LL, Calafat AM (2004) Analysis of human urine for fifteen phthalate metabolites using automated solid-phase extraction. J Chromatogr B Analyt Technol Biomed Life Sci 805:161–167

Smith KW, Souter I, Dimitriadis I, Ehrlich S, Williams PL, Calafat AM, Hauser R (2013) Urinary paraben concentrations and ovarian aging among women from a fertility center. Environ Health Perspect 121:1299–1305

Sobinoff AP, Pye V, Nixon B, Roman SD, McLaughlin EA (2010) Adding insult to injury: effects of xenobiotic-induced preantral ovotoxicity on ovarian development and oocyte fusibility. Toxicol Sci 118:653–666

Songsasen N, Woodruff TK, Wildt DE (2011) In vitro growth and steroidogenesis of dog follicles are influenced by the physical and hormonal microenvironment. Reproduction 142:113–122

Stephen EH, Chandra A (2006) Declining estimates of infertility in the United States: 1982–2002. Fertil Steril 86:516–523

Sun F, Betzendahl I, Shen Y, Cortvrindt R, Smitz J, Eichenlaub-Ritter U (2004) Preantral follicle culture as a novel in vitro assay in reproductive toxicology testing in mammalian oocytes. Mutagenesis 19:13–25

Takai Y, Canning J, Perez GI, Pru JK, Schlezinger JJ, Sherr DH, Kolesnick RN, Yuan J, Flavell RA, Korsmeyer SJ, Tilly JL (2003) Bax, caspase-2, and caspase-3 are required for ovarian follicle loss caused by 4-vinylcyclohexene diepoxide exposure of female mice in vivo. Endocrinology 144:69–74

Takeuchi T, Tsutsumi O, Ikezuki Y, Takai Y, Taketani Y (2004) Positive relationship between androgen and the endocrine disruptor, bisphenol A, in normal women and women with ovarian dysfunction. Endocr J 51:165–169

Taxvig C, Vinggaard AM, Hass U, Axelstad M, Boberg J, Hansen PR, Frederiksen H, Nellemann C (2008) Do parabens have the ability to interfere with steroidogenesis? Toxicol Sci 106:206–213

Tian Y, Ke S, Thomas T, Meeker RJ, Gallo MA (1998) Regulation of estrogen receptor mRNA by 2,3,7,8-tetrachlorodibenzo-p-dioxin as measured by competitive RT-PCR. J Biochem Mol Toxicol 12:71–77

Trapp M, Baukloh V, Bohnet H-G, Heeschen W (1984) Pollutants in human follicular fluid. Fertil Steril 42(1):146–148

Treinen KA, Dodson WC, Heindel JJ (1990) Inhibition of FSH-stimulated cAMP accumulation and progesterone production by mono(2-ethylhexyl) phthalate in rat granulosa cell cultures. Toxicol Appl Pharmacol 106:334–340

Tsutsumi O (2005) Assessment of human contamination of estrogenic endocrine-disrupting chemicals and their risk for human reproduction. J Steroid Biochem Mol Biol 93:325–330

Uzumcu M, Kuhn PE, Marano JE, Armenti AE, Passantino L (2006) Early postnatal methoxychlor exposure inhibits folliculogenesis and stimulates anti-Mullerian hormone production in the rat ovary. J Endocrinol 191:549–558

Valdez KE, Shi Z, Ting AY, Petroff BK (2009) Effect of chronic exposure to the aryl hydrocarbon receptor agonist 2,3,7,8-tetrachlorodibenzo-p-dioxin in female rats on ovarian gene expression. Reprod Toxicol 28:32–37

Van Wemmel K, Gobbers E, Eichenlaub-Ritter U, Smitz J, Cortvrindt R (2005) Ovarian follicle bioassay reveals adverse effects of diazepam exposure upon follicle development and oocyte quality. Reprod Toxicol 20:183–193

Vo TT, Yoo YM, Choi KC, Jeung EB (2010) Potential estrogenic effect(s) of parabens at the prepubertal stage of a postnatal female rat model. Reprod Toxicol 29:306–316

Weiss JM, Bauer O, Bluthgen A, Ludwig AK, Vollersen E, Kaisi M, Al-Hasani S, Diedrich K, Ludwig M (2006) Distribution of persistent organochlorine contaminants in infertile patients from Tanzania and Germany. J Assist Reprod Genet 23:393–399

Willoughby KN, Sarkar AJ, Boyadjieva NI, Sarkar DK (2005) Neonatally administered tert-octylphenol affects onset of puberty and reproductive development in female rats. Endocrine 26:161–168

Windham GC, Mitchell P, Anderson M, Lasley BL (2005) Cigarette smoking and effects on hormone function in premenopausal women. Environ Health Perspect 113:1285–1290

Wong SL, Lye EJ (2008) Lead, mercury and cadmium levels in Canadians. Health Rep 19:31–36

Xi W, Lee CK, Yeung WS, Giesy JP, Wong MH, Zhang X, Hecker M, Wong CK (2011) Effect of perinatal and postnatal bisphenol A exposure to the regulatory circuits at the hypothalamus-pituitary-gonadal axis of CD-1 mice. Reprod Toxicol 31:409–417

Yang JM, Chen QY, Jiang XZ (2002) Effects of metallic mercury on the perimenstrual symptoms and menstrual outcomes of exposed workers. Am J Ind Med 42:403–409

Yin S, Song C, Wu H, Chen X, Zhang Y (2015) Adverse effects of high concentrations of fluoride on characteristics of the ovary and mature oocyte of mouse. PLoS One 10:e0129594

Younglai EV, Foster WG, Hughes EG, Trim K, Jarrell JF (2002) Levels of environmental contaminants in human follicular fluid, serum, and seminal plasma of couples undergoing in vitro fertilization. Arch Environ Contam Toxicol 43:121–126

Zachow R, Uzumcu M (2006) The methoxychlor metabolite, 2,2-bis-(p-hydroxyphenyl)-1,1,1-trichloroethane, inhibits steroidogenesis in rat ovarian granulosa cells in vitro. Reprod Toxicol 22:659–665

Zama AM, Uzumcu M (2009) Fetal and neonatal exposure to the endocrine disruptor methoxychlor causes epigenetic alterations in adult ovarian genes. Endocrinology 150:4681–4691

Zama AM, Uzumcu M (2010) Epigenetic effects of endocrine-disrupting chemicals on female reproduction: an ovarian perspective. Front Neuroendocrinol 31:420–439

Zhang W, Jia H (2007) Effect and mechanism of cadmium on the progesterone synthesis of ovaries. Toxicology 239:204–212

Zhang W, Pang F, Huang Y, Yan P, Lin W (2008) Cadmium exerts toxic effects on ovarian steroid hormone release in rats. Toxicol Lett 182:18–23

Chapter 8
The "Toxic" Effects of a Perinatal Obesogenic Environment: Maternal Obesity and Impacts on Future Generations

Leon Chalil and Deborah M. Sloboda

Abstract In recent decades, epidemic levels of metabolic disease states have become global issues. Among the most well-known of these are obesity, diabetes, and cardiovascular diseases, whose social and economic impacts have prompted a global fervor of investigation into their causes. Though initially considered to be determined largely by genetic and lifestyle factors, such as diet and exercise, this paradigm would ultimately be insufficient to explain the continued propagation of non-communicable diseases. In the late 1980s, a new and exciting field of research described the roots of disease to be founded early in developmental life. These early studies showed that perturbations to the organism during critical developmental windows had long-term impacts on disease risk even after controlling for lifestyle and genetics. By extension, alterations in maternal physiology are implicated by this discovery, as it is the primary determinant of the fetal environment. Indeed, extensive data derived from animal models and clinical studies have been essential to defining the nature and extent of the influence that the mother's own metabolic status has on the developing fetus. An altered substrate profile in cases of maternal obesity is said to "program" the offspring, resulting in a maladapted physiology and increased disease risk. A compelling case is then made for a metabolically abnormal intrauterine environment acting as a "toxin" for the fetus, initiating and driving later-in-life disease. The focus of this chapter will be to review the known complications of maternal high fat diet and obesity on offspring health in later life, and to describe recent mechanistic insights into its origins.

L. Chalil
Department of Biochemistry and Biomedical Sciences, Obstetrics and Gynecology, and Pediatrics, McMaster University, HSC4H21, 1280 Main St W, Hamilton, ON L8N 3Z5, Canada

D.M. Sloboda (✉)
Department of Biochemistry and Biomedical Sciences, Obstetrics and Gynecology, and Pediatrics, McMaster University, HSC4H21, 1280 Main St W, Hamilton, ON L8N 3Z5, Canada

Farncombe Family Digestive Health Research Institute, Hamilton, ON, Canada
e-mail: sloboda@mcmaster.ca

C.L. Hughes, M.D. Waters (eds.), *Translational Toxicology*, Molecular and Integrative Toxicology, DOI 10.1007/978-3-319-27449-2_8

Keywords Placenta • Intrauterine • DOHaD • Obesity • Metabolic syndrome • Diabetes • Programming • Toxicant

8.1 Introduction

Rising obesity rates have been a ballooning problem in the West, and now globally, for several decades, with the prevalence of the disease exceeding 35 % in adults as of 2012 (Ogden et al. 2014). The incidence of obesity and its chronic disease morbidities has plateaued here for over a decade, heavily straining health systems (Jensen et al. 2014; Ogden et al. 2014). Sedentary lifestyles and poor dietary practices have been historically cited as the primary culprits (Dabney 1964) (Beaton 1967), and have been the focus of an entire subculture dedicated to changing the behavioral and psychological approach the epidemic of obesity. Many healthy-living programs have been primarily aimed at weight loss as the means to improving the health of individuals with obesity, and thereby increasing the health of society as a whole (Brown et al. 2009; Flodgren et al. 2010; Lau et al. 2007). However, the modest success of these initiatives leaves much to be desired (Kirk et al. 2012). Like many other general indicators of health, that an individual is obese is not the problem on its own; like many other general indicators, taken in context, obesity can be a useful predictor for other co-morbidities that arise alongside – if not as a result of an individual being obese. The increased risk of developing many other chronic diseases including hypertension and Type 2 diabetes mellitus has made obesity a leading candidate for health research.

Although the undesirable nature of obesity is well recognized for its impacts on the health of the individual, the story does not end here. An emerging body of evidence has begun to shed light on obesity and overweight in a very specific subset of the population; pregnant women. It has been long understood that maternal obesity carries an increased risk of complications in pregnancy, a phenomenon which is well-reviewed in the literature (Stupin and Arabin 2014) (Faucher and Barger 2015). However, in the last 20 years, we have seen a paradigm expansion – if not a shift – towards a greater understanding of the maternal-fetal relationship and how this relationship is modulated by the maternal diet and her metabolic physiology. Indeed, one might consider whether in a state of poor maternal nutrition (high fat, sugar and artificial sweeteners) or maternal metabolic dysfunction (like in the case of gestational diabetes or diabetes) the fetus may be exposed to "toxins" of a natural origin, including glucose, free fatty acids and fructose. The focus of this chapter will be to provide a summary of the identified impacts of maternal obesity and poor nutrient intake to date, and to explore future directions that can help reconcile the major challenges facing offspring born to pregnancies complicated by obesity.

8.2 Obesity Where for Art Thou?

As a disease, obesity may be characterized by any number of features, the most predominant of which is increased body mass index (BMI; kg/m^2). The World Health Organization (WHO) BMI classifications are divided into overweight (BMI 25–29.99); Obese class I (BMI 30–34.99), Obese class II (BMIT 35–39.99) and Obese class III (BMI$\geq$40) (Organization 2015) and although BMI gives no indication of body composition, BMI and waist circumference measurements are useful indicators of obesity and its associated complications (Jensen et al. 2014).

Dietary, lifestyle, and genetic factors constitute an enormous part of the molecular basis of obesity (Pate et al. 2013). The most common culprit is the consumption of an "obesogenic" diet. Prolonged intake of fatty foods in large quantities contributes to fatty acid uptake into circulation and the tissues of the body (Brunner et al. 1979). As circulating fatty acid levels steadily rise over time, internal clearance systems fail to clear them rapidly enough, resulting in widespread ectopic deposition in metabolically important tissues, including the liver, skeletal muscle, pancreas, and adipose; known to be associated with the metabolic syndrome (Rasouli et al. 2007). This accumulation is undesirable for several reasons. Fatty acid by-products have a direct influence on glucose homeostasis. Specifically, at increased concentrations, metabolites of fatty acid oxidation such as diacylglycerides and ceramides can activate inhibitory "stress" serine kinases, as well as decreasing the activity of Akt (de la Monte et al. 2010; Szendroedi et al. 2014). These effects work synergistically to inhibit insulin signalling (Summers et al. 1998; Yu et al. 2002), leading to impaired insulin sensitivity and dysregulation of blood glucose level maintenance. Impairment of this system by these substrates is a potent contributory factor to insulin resistance in obesity, elevating blood glucose levels and increasing the risk of Type 2 Diabetes Mellitus (Li and Zhang 2000).

Compounding the issue, obesity is characterized by a state of chronic, low-grade inflammation (Wellen and Hotamisligil 2005). As a major cytokine producer, adipose tissue plays a contributory role to inflammation throughout the body by increasing production of proinflammatory cytokines as the tissue mass expands (notably TNFα) (Hotamisligil et al. 1993, 1995). Furthermore, saturated free fatty acids demonstrate partial specificity of pattern recognition receptors including Toll-like receptors (TLRs) (Fessler et al. 2009; Shi et al. 2006) where the principal activation of an interleukin-1-pathway, dependent upon the adapter MyD88 through a series of kinases and scaffolding proteins, leads to activation of NF-κB ultimately initiating pro-inflammatory cytokine secretion (Sabroe et al. 2008). The nearly ubiquitous presence of these TLRs in tissues means that inflammation is almost a certainty following lipid accumulation (Ahmad et al. 2012). Obesity related inflammation then, is associated with stress kinase activation, and is an aggravating factor for insulin resistance (Lee et al. 2014). Leptin regulation and responsiveness is perhaps the most important example of this. Cytokine-mediated inflammatory profiles are managed in part through the signalling of Suppressor of Cytokine Signalling (SOCS) proteins (J. Wang and Campbell 2002). Some SOCS proteins can have the undesirable effect of inhibiting the tyrosine kinase activity of the leptin receptor

(Bjorbaek et al. 1999). In a healthy state, leptin acts at the arcuate nucleus of the hypothalamus to generate a feeling of satiety and terminate eating (Satoh et al. 1997). However, when leptin signalling becomes dysfunctional, circulating leptin levels increase futilely and leptin resistance develops (Bjorbaek et al. 1998).

Though fairly rare (~5–10 % of obese cases) inborn errors of metabolism as well as predisposing risk factors for obesity can arise from single nucleotide polymorphisms (SNPs) (Choquet and Meyre 2011). Defects in leptin signalling pathways, encompassing deficiencies in leptin production, deficiencies in the leptin receptor itself, or any number of mutations which block the ability of the leptin receptor to transduce signals (Lubis et al. 2008) are associated with obesity. An obesogenic model of leptin knockout was first demonstrated quite dramatically in mice by removing their ability to produce leptin entirely. These ob/ob mice eat ravenously becoming extremely obese (Pelleymounter et al. 1995). A wide host of defective components in the leptin signalling pathway can manifest with similar effects (Bjorbaek et al. 1999). Other genetically based receptor defects include mutations of the GPR120 receptor, which regulates adipogenesis and appetite. GPR120 activity is linked to insulin desensitization and adipocyte hypertrophy, suggesting further genetic bases for obese conditions (Hirasawa et al. 2005; Ichimura et al. 2012).

The genetic bases for obesity may also be transmitted through more complex inheritance patterns, related to multiple allelic dysfunctions. Because of their often widespread inputs, individual alleles may be partially implicated in a variety of disease states. Accordingly, specific "polygenic" aggravators of obesity risk have been described in the literature, the best characterized of which include the "fatso" gene (FTO), and the melanocortin 4 receptor gene (MC4R).

The *FTO* gene product is a DNA demethylase (Gerken et al. 2007) with a strong preference for single-stranded DNA (Han et al. 2010). Several SNPs in this gene have been directly linked to increased BMI, increased odds of obesity, and increased fat mass from a young age (Dina et al. 2007; Frayling et al. 2007). Indeed, initial estimates suggested that certain SNPs of *Fto* might be responsible for as much as 22 % of the attributable risk for obesity (Dina et al. 2007). Evidence from animal models has further demonstrated that increased *Fto* induces hyperphagia and obesity (Church et al. 2010), while the loss of *Fto* is protective against the development of obesity (Fischer et al. 2009). The exact mechanisms by which *Fto* regulates energy homeostasis are not well understood, though its high concentrations in the hypothalamus and ability to modulate the expression of STAT3 (a key factor in leptin signalling) suggest that this as a possible point at which *Fto* modifies the dietary habits of the organism (Tung et al. 2010). This is further corroborated by evidence of *Fto* co-localization with the leptin receptor long isoform in the arcuate nucleus (Wang et al. 2011). The demethylase activity of *Fto*, combined with its proximity to leptin signalling activity, have suggested a role for it as a transcriptional co-activator which may be involved in the regulation of fat tissue (Wu et al. 2010).

The *MC4R* gene product is a G-protein coupled receptor which is expressed in the brain, and is part of the melanocortin receptor family of proteins (Huszar et al. 1997). *In vitro* inactivation of *MC4R* is known to induce the development of obesity, hyperphagia, and disruptions to glucose/insulin regulation (Huszar et al. 1997).

More recently, genetic analyses have revealed a sizeable involvement for *MC4R* in heritable cases of obesity, naming specific mutations as primary culprits (Vaisse et al. 1998; Yeo et al. 1998). Further data from genome-wide studies have linked mutations in *MC4R* to increased waist circumference, insulin resistance, and fat mass (Chambers et al. 2008; Loos et al. 2008). These findings have established *MC4R* inactivating mutations as leading candidates for many cases of early onset, genetic obesity (Loos et al. 2008). Interestingly, though the function of *MC4R* in regulating appetite and obesity is not well understood, it is known that individuals heterozygous for a null mutation in this gene develop obesity, which is suggestive of an expression-level sensitivity to *MC4R* in this system (Yeo et al. 2000).

Despite great efforts in managing the growing population of individuals that develop obesity, we have been unsuccessful. No doubt a multi-system approach is needed, where multiple causative and indirect associations can be targeted at once. But even with this focused approach, "symptomatic" treatment of the epidemic through dietary and lifestyle has left much to be desired at least in part due to compliance issues (Kirk et al. 2012). New ways to predict and manage obesity demand a new way of understanding where the problem is coming from and how it can be handled.

8.3 A New Perspective on Obesity Development: The Early Life Environment

Over 20 years ago, the first glimmer of a new approach to understanding disease risk emerged in an unsuspecting area of science: perinatal development. In the late eighties, Dr. David Barker published an observational study of seniors in England whose birth weights had been recorded in the early twentieth century (Barker et al. 1989). The data showed a novel association between birth weight and disease risk in later life. Though birth weight is an imperfect surrogate measure of intrauterine development patterns, these studies provided the first look into the possibility of the continued influence of the *in utero* environment into later life. In 1992 this would be fleshed out as the hypothesis that disease programmed at birth is a side effect of inaccurate signaling between the fetus and the mother about the extrauterine environment, leading to long term maladaptations (Hales and Barker 1992). This exciting data paved the path for the more modern hypothesis of the Developmental Origins of Health and Disease (Barker 2004), and has provided powerful tools for evaluating the basic problems concerning the prediction and management of obesity.

8.4 Maternal Metabolic Adaptations, Gestational Weight Gain and Obesity

During fetal development, the maternal environment acts as the portal through which all fetal requirements are channeled. All nutrients that the fetus requires – and toxic substances the fetus may be exposed to – are derived from the mother's circulation. The growing fetus is in a very vulnerable state, and is extremely susceptible to perturbations, however subtle, that may induce developmental adaptations that have long-term effects. For this reason, the mother undergoes a set of physiological changes to accommodate the growing needs of the fetus in a buffered environment where the balances of nutrient distribution are drastically altered. Nearly all organ systems demonstrate significantly altered profiles during pregnancy; musculoskeletal, neuroendocrine, and cardiovascular (Chapman et al. 1998); herein we will focus on metabolic alterations (Ryan et al. 1985).

Maternal metabolic adaptations are largely the result of endocrine-mediated changes in carbohydrate, protein and fat metabolism. Early in pregnancy, the mother is in an anabolic state, where placental lactogen, progesterone and estrogens alter hepatic glucose production, lipogenesis and gluconeogenesis (Desoye et al. 1987). These early adaptations serve to increase adipose deposition, which later in pregnancy can be used as a fuel supply primarily for the mother, since glucose is preferentially diverted to the placenta and to the growing fetus. As such in the last third of pregnancy the mother is catabolic; defined by a state of insulin and leptin resistance, where elevated circulating glucose, triglycerides and fatty acids (Punnonen 1977), can accommodate the increasing demands placed on the maternal system from both the fetus and the metabolically active placenta. Although changes are largely mediated by placental lactogen, estradiol and progesterone (Beck and Daughaday 1967), new emerging data suggest that maternal – bacterial relationships could also potentially impose changes on maternal metabolic function during pregnancy. Maternal gut bacterial load increases over the course of pregnancy (Collado et al. 2008) and there is a significant change in gut microbiome composition from the first to the third trimester of pregnancy, accompanied by increased bacterial diversity between mothers (Koren et al. 2012). Although the composition and diversity of the gut microbiome during the first trimester is similar to the that of nonpregnant healthy women, over the course of pregnancy, there is an increase in the abundance of *Proteobacteria* and *Actinobacteria* (Koren et al. 2012). Since an increase in *Proteobacteria* has been observed in inflammatory bowel disease, it has been suggested that a similar dysbiosis of the gut occurs during the third trimester of pregnancy. The concept that maternal gut bacterial populations change over the course of pregnancy to influence maternal and fetal metabolic development is novel, and hasn't yet been proven, but presents an interesting deviation from the current concept that metabolic function is entirely regulated by placental hormones (Gohir et al. 2015).

We can briefly pause here to consider the points upon which we have touched thus far. Obesity is defined as the state of being overweight, typically denoted by a BMI of 30 or greater. In many cases, obesity presents with a phenotypic profile consistent with metabolic syndrome, characterized by insulin and leptin resistance, elevated blood pressure, high blood triglycerides, and low blood HDL cholesterol levels. The comorbidities associated with obesity include Type 2 diabetes mellitus, hypertension, Non-Alcoholic Fatty Liver Disease, and cardiovascular diseases. Adjacently, maternal pregnancy induced metabolic adaptations are oriented to ensure satisfactory growth and development of the fetus, whilst meeting the increased metabolic demands place upon the mother during pregnancy. By adopting a state of relative insulin resistance and increasing the role of fat in supplying the metabolic demands, these adaptations improve glucose delivery to the fetus. Indeed, many of the metabolic adaptations that accompany late pregnancy mimic the characteristics that define the pathophysiology of obesity including hypertriglyceridemia (Punnonen 1977), hypercholesterolemia (Bartels et al. 2012), insulin and leptin resistance (Catalano et al. 1991), and increased adipose depots (Kopp-Hoolihan et al. 1999). Even shifts in the gut microbes that occur with obesity are not inconsistent with the few data on maternal pregnancy induced microbial shifts (Koren et al. 2012). Pregnancy, however, is a normal physiological state of metabolic adaptation largely based upon the secretion of placental hormones, and upon removal of the placenta – the maternal metabolic function returns to normal non-pregnant values. It is not surprising then, that if this fine balance between maternal metabolic adaptation and fetal nutrient demand is disrupted, that long term deleterious impacts on physiological function occur.

8.5 Maternal Obesity: Is This a Toxic in Utero Environment?

The fetus is less isolated than once thought. After decades of research, we have begun to understand the intrauterine environment as an active responder to the external environment, and the placenta a more forgiving filter. In this spirit, an examination of the fetal interaction with the outside world grows increasingly warranted, and what better place to look than the maternal metabolic environment. The characterization of the maternal obesity phenotype is an important step towards a deeper understanding of the interplay at work here and much of what we know today comes from animal studies: from rodents to sheep and non-human primates.

The use of animal models circumvents many of the ethical issues that would be associated with performing many equivalent experiments on humans. The ones of interest range from the manipulation of diets to the harvesting and analysis of otherwise inaccessible tissues. Further, different animal models offer unique advantages that may favor their use in answering certain research questions. For example, the use of rodent (mostly rat and mouse) models offers the unique advantage of a shorter gestation time and an ease of manipulation compared to higher order animals. However, studies in rodents are limited since their biology is far from human. Conversely, non-human primates are a close evolutionary ancestor to humans, but non-human primate models have long gestation times, ethical constraints and often prohibitive expenses. Independent of the model that is used, the use of animals has been indispensible to the elucidation of many mechanisms underlying the effects of maternal obesity, both in the mother and in the offspring. Where mechanisms currently escape us, animal models further our understanding by offering a convenient inroad to forward thinking and hypothesis generation.

8.6 Impacts of Maternal Obesity

8.6.1 Maternal Outcomes

In rodents fed a high fat diet (HFD) during and before pregnancy, consistent effects are observed in maternal physiology and metabolic profiles. Several reviews have repeatedly shown metabolic dysregulation, an aggravated systemic pro-inflammatory profile, and an increased risk of obstetric complications (Athukorala et al. 2010; Dube et al. 2012; Stupin and Arabin 2014).

In both rat and mouse models, HFD-induced obesity both before and during pregnancy demonstrated the generation of a phenotype which was consistent with the metabolic syndrome. In mice, dams fed a HFD during pregnancy demonstrate increased adiposity, hyperleptinemia (Jones et al. 2009), insulin resistance, and serum free fatty acids (Mao et al. 2010). This is similar to observations made in a rat model increased, where a HFD resulted in maternal hyperleptinemia (Mark et al. 2011), as well as increased plasma glucose, insulin, and triglyceride levels (Srinivasan et al. 2006) (Hayes et al. 2012). In mice, increased levels of inflammatory markers such as interleukin 6 (IL6), interleukin 10 (IL10), and interferon gamma (IFNγ) (Kepczynska et al. 2013) are observed. However, there does not appear to be increased macrophage infiltration into adipose tissue by gestational day 18 (E18) (term E21-22) (Ingvorsen et al. 2014).

The data in non-human primate models is somewhat limited in nature, but are nevertheless consistent with rodent data. In Japanese macaques, there is evidence to suggest that maternal HFD – independent of obesity – can generate an increase in circulating pro-inflammatory cytokine levels (notably IL1ß and MCP-1), a condition which was aggravated in animals that developed insulin resistance. In macaques, a HFD decreased uterine blood flow independent of obesity or other morbidities (Frias et al. 2011). Perhaps the greatest importance of this finding lies in that it provides a potential mechanistic link between intrauterine growth restriction (IUGR), a strong indicator for later life complications, and maternal HFD intake through changes in placental function.

Evidence describing an interaction between a HFD and pregnancy in humans is fairly well reported, largely from epidemiological and mechanistic studies, is are consistent with the reports that have been made in animal models. Epidemiological data suggests that obesity increases the likelihood of experiencing obstetric complications in pregnancy (P. M. Catalano and Ehrenberg 2006). Notably, there is an increased risk of pre-eclampsia, the development of Gestational Diabetes Mellitus, delivering large for gestational age babies (LGA), a greater incidence of miscarriage (Metwally et al. 2008), and caesarean section (Athukorala et al. 2010; Lutsiv et al. 2015). In humans, obesity is linked to the downregulation of the leptin receptor (Farley et al. 2010), hyperleptinemia (Farley et al. 2010; Ramsay et al. 2002), hyperinsulemia, increased plasma triglycerides, and decreased vascular responses (Ramsay et al. 2002). A state of chronic, systemic low-grade inflammation has been documented, attributable to the release of greater levels of inflammatory adipokines;

of special note is TNFα, which is released by adipose tissue and is a known modulator of insulin resistance (Hotamisligil et al. 1993). However, this observation is not entirely consistent across studies as evidence exists that adiposity is not related to levels of pro-inflammatory cytokines in pregnancy (Friis et al. 2013). This understanding of the changes to maternal physiology is important to set the context of further discussions regarding fetal development and the environment which shape this development.

8.6.2 Fetal/Offspring Outcomes

The introduction of a high fat and/or obesogenic dietary insult is linked in varying degrees of certainty to adverse outcomes both in first generation offspring, as well as continued deviations from healthy states in subsequent generations (Dabelea and Crume 2011; Vickers 2014). However, our current understanding of the mechanisms underpinning these adverse outcomes and transgenerational transmission are unclear. Early life exposure to an obesogenic environment perturbs a wide range of physiological parameters in offspring, including the promotion of phenotypes consistent with metabolic syndrome, low-grade inflammation, hepatic and endocrine dysregulation, and cardiovascular impairment. We will explore here the current paradigms surrounding these features, and the data that have improved our understanding of them over time.

In mice, a high fat, obesogenic maternal diet is shown to direct the development of metabolic syndrome in offspring (Samuelsson et al. 2008). This finding is supported by the characterization of offspring phenotypes which showcase insulin resistance (Dunn and Bale 2009; Graus-Nunes et al. 2015; Gregorio et al. 2010; Murabayashi et al. 2013; Samuelsson et al. 2008; Volpato et al. 2012; Vuguin et al. 2013), increased blood pressure (Elahi et al. 2009; Masuyama and Hiramatsu 2012; Samuelsson et al. 2008), increased serum triglycerides (Masuyama and Hiramatsu 2012), and increased adiposity and body mass (Samuelsson et al. 2008; Volpato et al. 2012); (Dunn and Bale 2009; Masuyama and Hiramatsu 2012). The concurrent presence of hyperglycemia and increased serum triglycerides is proposed to aggravate the dysregulation of glucose management by promoting "gluco-lipotoxicity" in the beta cells of the pancreas (Cerf 2010). Other findings include increased serum cholesterol levels (Elahi et al. 2009), and increased serum leptin (Graus-Nunes et al. 2015; Howie et al. 2009) despite the apparent absence of central leptin resistance (Graus-Nunes et al. 2015). An analysis of adipose tissue in these offspring suggests that there is increased macrophage infiltration (as measured by CD68 levels), increased TNFα mRNA levels, and a decrease in GLUT4 expression, suggesting that the insulin resistance is a contributing factor to the observed low-grade inflammation that is present (Murabayashi et al. 2013). Of interest is the fact that these offspring were never fed a high fat diet postnatally, though it is very well established that doing so produces an aggravated form of metabolic syndrome (Elahi et al. 2009).

In many cases, the effects of a maternal HFD are sexually dimorphic; where for example in mice, only male offspring showed increase fat mass and serum insulin (Ashino et al. 2012), while other data show that the magnitude of hypertension in female offspring was more pronounced (Elahi et al. 2009). A notable difference was increased serum TNFα and IL1ß in adult offspring born to dams fed a HFD, suggesting the presence of an inflamed state (Ashino et al. 2012). Using transgenic mice heterozygous for the expression of GLUT4 to induce a model of Type 2 diabetes mellitus, it has been shown that G4+/− offspring born to mothers fed a high fat, obesogenic diet, are hypertensive (Vuguin et al. 2013), and males further demonstrated insulin resistance, glucose intolerance, and increased adiposity (Hartil et al. 2009).

Adverse hepatic outcomes have also been shown in offspring born to mothers fed a HFD and/or obesogenic diet including Non-Alcoholic Fatty Liver Disease (NAFLD) and Non-Alcoholic Steatohepatitis (NASH). NAFLD and NASH are characterized by an accumulation of fat in hepatocytes and have been referred to as "the hepatic manifestation of metabolic syndrome" (Musso et al. 2009). This lipid deposition is the result of an inability to correctly regulate the balance between the deposition and synthesis of fats, and their removal by oxidative processes (Matherly and Puri 2012). Changes in factors known to be involved in this balance are often mechanistically linked to disease development in the offspring. In offspring of mothers fed a HFD, increased liver triglycerides suggest the development of NAFLD (Ashino et al. 2012; Bruce et al. 2009) and decreased phosphorylation of c-Jun terminal kinases (JNK) and I kappa B kinase (IκK) – important regulators of insulin signaling – suggest a decrease in hepatic insulin signaling (Ashino et al. 2012). Decreased hepatic functionality of electron chain complexes has been implicated in the process (Bruce et al. 2009), as well as an upregulation of hepatic genes involved in glycolysis, gluconeogenesis, oxidative stress, and inflammation (Vuguin et al. 2013).

Rat models investigating offspring outcomes in response to maternal HFD largely paint the same picture. Rat models show maternal obesity results in offspring with hyperglycemia (Franco et al. 2012), insulin resistance (Burgueno et al. 2013) (Nivoit et al. 2009), altered pancreatic signalling (Howie et al. 2013) and increased adiposity (Howie et al. 2009; White et al. 2009). In one study, the mortality of offspring from high fat fed obese mothers was three times that of control fed mothers (Hayes et al. 2012).

In male offspring born to high fat fed obese mothers, glucose intolerance, weight gain, insulin resistance and hyperglycemia are observed (Srinivasan et al. 2006). There is conflicting evidence regarding maternal obesity effects on offspring birth weight with some observing increased birth weights (in males) (Srinivasan et al. 2006), while others report growth restriction (Connor et al. 2012; Howie et al. 2009; Mark et al. 2011).

There has also been extensive research on rats regarding leptin as a mediator between early life exposure to an obesogenic environment and long-term obesity risk. Almost ubiquitously, a maternal HFD produces hyperleptinemia in offspring (Mark et al. 2011) (Franco et al. 2012) (White et al. 2009) (Burgueno et al. 2013; Howie et al. 2009b). Leptin resistance is thought to occur centrally where a decrease in STAT3 and SOCS signalling in the arcuate nucleus of the hypothalamus is indicative of impaired leptin signalling, and therefore a state of leptin resistance (Franco et al. 2012). Indeed, in diet-induced obese mice, the ability of leptin to activate signalling in arcuate neurons and promote arcuate neurite outgrowth is significantly reduced (Bouret et al. 2008). Similarly, in offspring born to diabetic mothers (i.e., state of hyperglycemia *in utero*), decreased hypothalamic leptin signaling in arcuate neurons was also associated with decreased neural projections from the arcuate nucleus to the paraventricular nucleus (Steculorum and Bouret 2011). The impacts of maternal obesity on hypothalamic signalling have been shown to be exacerbated in the presence of a post-weaning HFD where a combination of an *in utero* + post-weaning HFD results in greater levels of hyperphagia, adiposity, hyperlipidemia, and glucose intolerance in offspring (Chen et al. 2009; Vickers et al. 2000). This was linked to increased hypothalamic NPY signaling and leptin resistance in adult offspring (Chen et al. 2009). More recently maternal obesity has been shown to dampen in vivo hypothalamic NPY response to acute hyperglycemia and decrease glucose uptake and lactate release in hypothalamic cell culture models (Chen et al. 2014). Taken together, the stressors of a maternal HFD and/or obesity on offspring in rodent models demonstrate an increased propensity towards the development of a phenotype that is consistent with The Metabolic Syndrome.

The data on physiological outcomes relevant to the fetus in non-human primates (NHP) is limited, though what is known offers valuable insights which are arguably more relevant in informing human studies than rodent models. The major phenotypic finding in the NHPs involves hepatic injury. Several studies demonstrate that a maternal HFD produces offspring with increased liver triglycerides (Aagaard-Tillery et al. 2008; Grant et al. 2011; Grayson et al. 2010) (McCurdy et al. 2009) and predisposition to NAFLD (Aagaard-Tillery et al. 2008; McCurdy et al. 2009; Thorn et al. 2014). Interestingly, offspring did not develop The Metabolic Syndrome unless *in utero* exposure was combined with a postweaning HFD (Fan et al. 2013; Thorn et al. 2014). Further, despite no change in birth weight, offspring showed increased adiposity (Grant et al. 2011). In the livers of these offspring, there is also evidence of irregularities in key genes that regulate circadian rhythmicity (Suter et al. 2011).

It has been shown that in NHP, a maternal HFD alone is sufficient to significantly reduce uterine blood flow. This condition is aggravated in the presence of a maternal hyperinsulinemic state, which also restricts the blood flow to the placenta. This drives, in part, a higher incidence of stillbirths among these offspring (Frias et al. 2011). Other cardiovascular insults include endothelial artery damage, measured by the presence of inflammatory cytokines, which may play a role in the development of later life cardiovascular complications (Fan et al. 2013), though more investigation is warranted. Maternal HFD produces (postnatally) impaired thyroid function in offspring, where decreased free thyroxine levels, as well as expression of thyroid releasing hormone, thyroid stimulating hormone, and the precursor thyroglobulin in the hypothalamus and thyroid suggest slowed metabolic processes and lower energy expenditure (Suter et al. 2012). An increase in thyroid hormone receptor ß may be compensatory in these animals, consistent with a deficiency in thyroid hormone signalling. This is potentially an aggravating factor in the discussion of altered substrate utilization, resulting in increased fatty circulation and deposition (Suter et al. 2012).

Owing to the earlier discussed difficulties in conducting experiments on humans, the data regarding fetal outcomes in humans is, on the whole, less than satisfactory. The evidence is largely epidemiological, often drawn retrospectively from observational studies. Multi-factor analyses of the factors predisposing human fetuses to later life disease have largely established a role for obesity, and glucose/insulin dysregulation. It is clear that obesity has an impact in later life outcomes, demonstrated by an increase in offspring obesity (Rooney and Ozanne 2011; Yang and Huffman 2013) and weight gain (Jansson et al. 2013). Maternal obesity is associated with childhood obesity (Schack-Nielsen et al. 2010; Starling et al. 2015) although in some studies the exact role of genetic and lifestyle associations in not clear (Lau et al. 2014). It is clear however, that maternal obesity and excessive weight gain during pregnancy is significantly associated with fetal macrosomia (overgrowth, LGA) (Gaudet et al. 2014; Mamun et al. 2014) and that LGA babies are at a greater risk of metabolic complications (Boney et al. 2005). It is interesting to note that a subset of obese women also deliver small for gestational age (SGA) babies (Leung et al. 2008; Rajasingam et al. 2009). Further, the risk of offspring obesity is positively correlated to socioeconomic status – a useful consideration for identifying at-risk populations (Whitaker 2004). Indeed, maternal hyperglycemia is a well-established insult predicting adverse fetal and later life outcomes (Tenenbaum-Gavish and Hod 2013) such as offspring obesity and diabetes (Dabelea et al. 2000).

In humans, maternal obesity is associated with premature death in the adult offspring (Reynolds et al. 2013), indeed birth weight shows a "u-shaped" association with mortality, where both low and high birth weight is correlated with premature death (Baker et al. 2008). Maternal obesity is highly correlated with fatty liver in offspring (Brumbaugh et al. 2013), increased incidence of asthma (Forno et al. 2014), and cardiovascular disease (Fraser et al. 2010). Recently, maternal obesity has been associated with changes in immune cell profiling in umbilical cord blood (Wilson et al. 2015), and although more investigation into immune development of offspring is needed, these data could suggest one mechanism linking risk of immune diseases including asthma and atopy (Harpsoe et al. 2013; Lowe et al. 2011) to the intrauterine environment.

A serious issue with human investigations is that outcomes can become difficult to track over time, requiring staggering sample sizes and careful measurements to accurately and validly capture them (Yang and Huffman 2013). Though human data is the most relevant, it is also difficult to come by in good quality thus the most pressing matter in understanding the maternal-fetal interplay is improving the quality and quantity of human data in this field.

8.6.3 Placental Outcomes

As the interface between the mother and fetus, the placenta plays a pivotal role in fetal growth and development. All of the nutrients and blood that reach the fetus pass through by way of the placenta (Harding 2001). Acting as a first-line filter, the placenta plays a major role in the composition of the fetal environment, and is therefore of great interest in discussions concerning the impacts of diet on fetal outcomes. Many of the notable findings in the placenta arising from high fat and obesogenic diets concern changes in nutrient transporters, as well as a tendency towards an inflammatory state. Though conclusions thus far are largely inferential, there is an incontrovertible link between the placental changes and offspring outcomes.

In mice, placental responses to a HFD diet are marked by altered nutrient transporter activity levels, as well as sexually dimorphic changes to gene expression levels. In mothers fed a HFD, the transport of glucose and amino acids are increased, as well as an accompanying increase in the levels of their transporters. At E18.5 (near term), glucose transport across the placenta is increased fivefold, and amino acid transport increased tenfold (Mark et al. 2011; Sferruzzi-Perri et al. 2013). This is accompanied by a fivefold increase in the expression of the insulin-independent glucose transporter 1 (GLUT1), and a ninefold increase in the expression of sodium-coupled neutral amino acid transporter 2 (SNAT2) (Jones et al. 2009; Sferruzzi-Perri et al. 2013). In cases of a maternal obesogenic diet, which incorporates both high fat and high sugar, expression of fatty acid transport protein (FATP) was also increased in the placenta (Sferruzzi-Perri et al. 2013). Furthermore, overall gene expression has been demonstrated to be sex-dependent, showing greater differences in the placentas of female fetuses (Mao et al. 2010). Maternal diabetes and nutrient intake modulate placental gene expression differentially and interact where dietary changes in placental growth were modulated by maternal diabetes (Kappen et al. 2012). These data support the hypothesis that the maternal metabolic state sets the stage for placental growth and development and that these processes are modulated by dietary intake of fat.

Rat models of HFD/obesity in pregnancy produce a phenotype of altered placental growth, endocrine signaling, and inflammation. Early trophoblast invasion is increased twofold, accompanied by an increase in metalloprotease 9 – a key protein in placental remodeling, in a model of maternal obesity (Hayes et al. 2014). This is relevant because altered placental establishment can lead to complications including (but not limited to) pre-eclampsia (Goldman-Wohl and Yagel 2002). While there is evidence to suggest that in a rat model, a HFD does not change placental VEGFα and PPARγ (markers of vascular growth) (Mark et al. 2011), the vasculature may nevertheless be altered, producing a hypoxic environment in the labyrinth zone (Hayes et al. 2012) and restricting junctional zone growth (Mark et al. 2011). Evidence of increased cyclooxygenase 2 (COX2) expression in the placenta of rat dams fed a HFD is suggestive of an inflammatory environment, possible compensating for impaired blood flow (Saben et al. 2014a).

Though the data in non-human primates is somewhat limited, a HFD model of pregnancy in Japanese macaques has provided valuable information regarding placental alterations. Thirteen inflammatory cytokines were elevated in the placentas of animals, which demonstrated sensitivity to a HFD (by developing obesity and hyperinsulemia). However, even where particular sensitivity was not observed, proinflammatory cytokines IL1ß and MCP-1 were upregulated. These animals also display increased placental triglycerides, and altered uteroplacental perfusion.

Taken together, the data from non-human primates is consistent with those data in rodents, in painting a picture of an inflamed and hemodynamically challenged placenta (Frias et al. 2011; Thorn et al. 2014).

Human placenta data is consistent with those observed in animals, in terms of hypoxic states, and alterations in endocrine signaling and nutrient transporter expression patterns. However, it is notable that in human studies, changes to nutrient transport protein levels are opposite in directionality to what was seen in animals. Obese women exhibiting hyperleptinemia and leptin resistance were also observed to have decreased SNAT, SNAT4, and leptin receptor expression in the syncytiotrophoblast of their placenta (Farley et al. 2010). Placental GLUT4 expression is decreased (Colomiere et al. 2009), while GLUT3 levels on the maternal side of the placenta (trophoblast) increased (Janzen et al. 2013). There is evidence of impaired insulin signaling, both at the level of the receptor and downstream molecules; insulin receptor (IRß) and PI3Kp85α expression (downstream mediator) were decreased in placenta from obese women – a state that was aggravated with existing maternal diabetes (Colomiere et al. 2009). Further to this, there is evidence of increased SNAT2 activity in the placental sodium-dependent amino acid transport "system A" of obese women. Placental nutrient sensor, mTOR has been linked to obesity where increased mTOR and IGF/insulin signaling suggests a role for these molecules in determining placental nutrient transport expression, which is relevant to the predisposition of offspring to adverse states in adulthood (Jansson et al. 2013).

It appears that placental vascular development is also impaired in placenta of obese women; where increased levels of hypoxia induced factor 1 alpha (HIF-1α), accompanied by increased NFκB and JNK signaling, and the promotion of proinflammatory cytokine expression (Saben et al. 2014b). The transcription factor FOXO4 is down-regulated, which may be important on account of its activation in oxidative environments, demonstrating anti-hypoxic and anti-inflammatory activity (Saben et al. 2014b). This potentially suggests a compensatory role for FOXO4 in combatting the state of inflammation in the placenta. Implicit to this idea, placental inflammation appears to play a pivotal role in human pregnancies associated with HFD and obesity.

Taken together, these data support the role the placenta plays in mediating fetal outcomes, and many indeed modulate the exposure of the fetus to these "toxic" metabolic environments and metabolites. An improved understanding of the mechanistic signaling pathways involved in these changes will be pivotal to an improved description of the dietary impact on the placenta, and correspondingly the fetus.

8.7 Concluding Remarks

There is no doubt that the intrauterine environment modulates fetal growth and development and that key fetal adaptations as a result of exposure to a multitude of agents will change offspring phenotype. This chapter has expanded the use of the term "toxin" to include changes in the metabolic profile of the mother; this may include high circulating levels of glucose, free fatty acids and/or pro-inflammatory cytokines. Indeed, phenotypic outcomes in offspring born to mothers exposed to environmental toxicants including pthlalates (Lee et al. 2015), BPA (Alonso-Magdalena et al. 2010), and cigarette smoke (Behl et al. 2013) are consistent with what this chapter presents in cases of maternal obesity/overweight. This may be suggestive of a common driver of phenotype, and that the developmentally plastic fetus responds to varying cues with similar outcomes. Whether these are adaptive in nature however is difficult to imagine currently, as exposure to these "toxicants" is a relatively recent event in our evolutionary past. However, what is clear is that the fetus is amenable to a number of environmental factors including those that one normally does not term "toxic". As this field continues to evolve and grow, future investigations should be aimed at discerning the fine details of the signaling pathways involved, placing a larger emphasis on human studies.

References

Aagaard-Tillery KM, Grove K, Bishop J, Ke X, Fu Q, McKnight R, Lane RH (2008) Developmental origins of disease and determinants of chromatin structure: maternal diet modifies the primate fetal epigenome. J Mol Endocrinol 41(2):91–102. doi:10.1677/JME-08-0025

Ahmad R, Al-Mass A, Atizado V, Al-Hubail A, Al-Ghimlas F, Al-Arouj M, … Behbehani K (2012) Elevated expression of the toll like receptors 2 and 4 in obese individuals: its significance for obesity-induced inflammation. J Inflamm (Lond) 9(1):48. doi:10.1186/1476-9255-9-48

Alonso-Magdalena P, Vieira E, Soriano S, Menes L, Burks D, Quesada I, Nadal A (2010) Bisphenol a exposure during pregnancy disrupts glucose homeostasis in mothers and adult male offspring. Environ Health Perspect 118(9):1243–1250. doi:10.1289/ehp.1001993

Ashino NG, Saito KN, Souza FD, Nakutz FS, Roman EA, Velloso LA, … Torsoni MA (2012) Maternal high-fat feeding through pregnancy and lactation predisposes mouse offspring to molecular insulin resistance and fatty liver. J Nutr Biochem 23(4):341–348. doi:10.1016/j.jnutbio.2010.12.011

Athukorala C, Rumbold AR, Willson KJ, Crowther CA (2010) The risk of adverse pregnancy outcomes in women who are overweight or obese. BMC Pregnancy Childbirth 10:56. doi:10.1186/1471-2393-10-56

Baker JL, Olsen LW, Sorensen TI (2008) Weight at birth and all-cause mortality in adulthood. Epidemiology 19(2):197–203. doi:10.1097/EDE.0b013e31816339c6

Barker DJ (2004) Developmental origins of adult health and disease. J Epidemiol Community Health 58(2):114–115

Barker DJ, Winter PD, Osmond C, Margetts B, Simmonds SJ (1989) Weight in infancy and death from ischaemic heart disease. Lancet 2(8663):577–580

Bartels Ä, Egan N, Broadhurst DI, Khashan AS, Joyce C, Stapleton M, … O'Donoghue K (2012) Maternal serum cholesterol levels are elevated from the 1st trimester of pregnancy: a cross-sectional study. J Obstet Gynaecol 32(8):747–752. doi:10.3109/01443615.2012.714017

Beaton JR (1967) Energy balance and obesity. Can J Public Health 58(11):479–482

Beck P, Daughaday WH (1967) Human placental lactogen: studies of its acute metabolic effects and disposition in normal man*. J Clin Invest 46(1):103–110. doi:10.1172/JCI105503

Behl M, Rao D, Aagaard K, Davidson TL, Levin ED, Slotkin TA, … Holloway AC (2013) Evaluation of the association between maternal smoking, childhood obesity, and metabolic disorders: a national toxicology program workshop review. Environ Health Perspect 121(2):170–180. doi:10.1289/ehp.1205404

Bjorbaek C, Elmquist JK, Frantz JD, Shoelson SE, Flier JS (1998) Identification of SOCS-3 as a potential mediator of central leptin resistance. Mol Cell 1(4):619–625

Bjorbaek C, El-Haschimi K, Frantz JD, Flier JS (1999) The role of SOCS-3 in leptin signaling and leptin resistance. J Biol Chem 274(42):30059–30065

Boney CM, Verma A, Tucker R, Vohr BR (2005) Metabolic syndrome in childhood: association with birth weight, maternal obesity, and gestational diabetes mellitus. Pediatrics 115(3):e290–e296. doi:10.1542/peds.2004-1808

Bouret SG, Gorski JN, Patterson CM, Chen S, Levin BE, Simerly RB (2008) Hypothalamic neural projections are permanently disrupted in diet-induced obese rats. Cell Metab 7(2):179–185. doi:10.1016/j.cmet.2007.12.001

Brown T, Avenell A, Edmunds LD, Moore H, Whittaker V, Avery L, Summerbell C (2009) Systematic review of long-term lifestyle interventions to prevent weight gain and morbidity in adults. Obes Rev 10(6):627–638. doi:10.1111/j.1467-789X.2009.00641.x

Bruce KD, Cagampang FR, Argenton M, Zhang J, Ethirajan PL, Burdge GC, … Byrne CD (2009) Maternal high-fat feeding primes steatohepatitis in adult mice offspring, involving mitochondrial dysfunction and altered lipogenesis gene expression. Hepatology 50(6):1796–1808. doi:10.1002/hep.23205

Brumbaugh DE, Tearse P, Cree-Green M, Fenton LZ, Brown M, Scherzinger A, … Barbour LA (2013) Intrahepatic fat is increased in the neonatal offspring of obese women with gestational diabetes. J Pediatr 162(5):930–936. doi:10.1016/j.jpeds.2012.11.017, e931

Brunner D, Weissbort J, Fischer M, Bearman JE, Loebl K, Schwartz S, Levin S (1979) Serum lipid response to a high-caloric, high-fat diet in agricultural workers during 12 months. Am J Clin Nutr 32(6):1342–1349

Burgueno AL, Cabrerizo R, Gonzales Mansilla N, Sookoian S, Pirola CJ (2013) Maternal high-fat intake during pregnancy programs metabolic-syndrome-related phenotypes through liver mitochondrial DNA copy number and transcriptional activity of liver PPARGC1A. J Nutr Biochem 24(1):6–13. doi:10.1016/j.jnutbio.2011.12.008

Catalano PM, Ehrenberg HM (2006) The short- and long-term implications of maternal obesity on the mother and her offspring. BJOG 113(10):1126–1133. doi:10.1111/j.1471-0528.2006.00989.x

Catalano PM, Tyzbir ED, Roman NM, Amini SB, Sims EAH (1991) Longitudinal changes in insulin release and insulin resistance in nonobese pregnant women. Am J Obstet Gynecol 165(6, Part 1):1667–1672. doi:http://dx.doi.org/10.1016/0002-9378(91)90012-G

Cerf ME (2010) High fat programming of beta-cell failure. Adv Exp Med Biol 654:77–89. doi:10.1007/978-90-481-3271-3_5

Chambers JC, Elliott P, Zabaneh D, Zhang W, Li Y, Froguel P, … Kooner JS (2008) Common genetic variation near MC4R is associated with waist circumference and insulin resistance. Nat Genet 40(6):716–718. doi:10.1038/ng.156

Chapman AB, Abraham WT, Zamudio S, Coffin C, Merouani A, Young D, … Schrier RW (1998) Temporal relationships between hormonal and hemodynamic changes in early human pregnancy. Kidney Int 54(6):2056–2063. doi:10.1046/j.1523-1755.1998.00217.x

Chen H, Simar D, Morris MJ (2009) Hypothalamic neuroendocrine circuitry is programmed by maternal obesity: interaction with postnatal nutritional environment. PLoS One 4(7):e6259. doi:10.1371/journal.pone.0006259

Chen H, Simar D, Morris MJ (2014) Maternal obesity impairs brain glucose metabolism and neural response to hyperglycemia in male rat offspring. J Neurochem 129(2):297–303. doi:10.1111/jnc.12623

Choquet H, Meyre D (2011) Genetics of obesity: what have we learned? Curr Genomics 12(3):169–179. doi:10.2174/138920211795677895

Church C, Moir L, McMurray F, Girard C, Banks GT, Teboul L, … Cox RD (2010) Overexpression of Fto leads to increased food intake and results in obesity. Nat Genet 42(12):1086–1092. doi:10.1038/ng.713

Collado MC, Isolauri E, Laitinen K, Salminen S (2008) Distinct composition of gut microbiota during pregnancy in overweight and normal-weight women. Am J Clin Nutr 88(4):894–899

Colomiere M, Permezel M, Riley C, Desoye G, Lappas M (2009) Defective insulin signaling in placenta from pregnancies complicated by gestational diabetes mellitus. Eur J Endocrinol 160(4):567–578. doi:10.1530/EJE-09-0031

Connor KL, Vickers MH, Beltrand J, Meaney MJ, Sloboda DM (2012) Nature, nurture or nutrition? Impact of maternal nutrition on maternal care, offspring development and reproductive function. J Physiol 590(9):2167–2180. doi:10.1113/jphysiol.2011.223305

Dabelea D, Crume T (2011) Maternal environment and the transgenerational cycle of obesity and diabetes. Diabetes 60(7):1849–1855. doi:10.2337/db11-0400

Dabelea D, Hanson RL, Lindsay RS, Pettitt DJ, Imperatore G, Gabir MM, ... Knowler WC (2000) Intrauterine exposure to diabetes conveys risks for type 2 diabetes and obesity: a study of discordant sibships. Diabetes 49(12):2208–2211

Dabney JM (1964) Energy balance and obesity. Ann Intern Med 60:689–699

de la Monte SM, Tong M, Nguyen V, Setshedi M, Longato L, Wands JR (2010) Ceramide-mediated insulin resistance and impairment of cognitive-motor functions. J Alzheimers Dis 21(3):967–984. doi:10.3233/JAD-2010-091726

Desoye G, Schweditsch MO, Pfeiffer KP, Zechner R, Kostner GM (1987) Correlation of hormones with lipid and lipoprotein levels during normal pregnancy and postpartum. J Clin Endocrinol Metab 64(4):704–712. doi:10.1210/jcem-64-4-704

Dina C, Meyre D, Gallina S, Durand E, Korner A, Jacobson P, ... Froguel P (2007) Variation in FTO contributes to childhood obesity and severe adult obesity. Nat Genet 39(6):724–726. doi:10.1038/ng2048

Dube E, Gravel A, Martin C, Desparois G, Moussa I, Ethier-Chiasson M, ... Lafond J (2012) Modulation of fatty acid transport and metabolism by maternal obesity in the human full-term placenta. Biol Reprod 87(1):14, 11–11. doi:10.1095/biolreprod.111.098095

Dunn GA, Bale TL (2009) Maternal high-fat diet promotes body length increases and insulin insensitivity in second-generation mice. Endocrinology 150(11):4999–5009. doi:10.1210/en.2009-0500

Elahi MM, Cagampang FR, Mukhtar D, Anthony FW, Ohri SK, Hanson MA (2009) Long-term maternal high-fat feeding from weaning through pregnancy and lactation predisposes offspring to hypertension, raised plasma lipids and fatty liver in mice. Br J Nutr 102(4):514–519. doi:10.1017/S000711450820749X

Fan L, Lindsley SR, Comstock SM, Takahashi DL, Evans AE, He GW, ... Grove KL (2013) Maternal high-fat diet impacts endothelial function in nonhuman primate offspring. Int J Obes (Lond) 37(2):254–262. doi:10.1038/ijo.2012.42

Farley DM, Choi J, Dudley DJ, Li C, Jenkins SL, Myatt L, Nathanielsz PW (2010) Placental amino acid transport and placental leptin resistance in pregnancies complicated by maternal obesity. Placenta 31(8):718–724. doi:10.1016/j.placenta.2010.06.006

Faucher MA, Barger MK (2015) Gestational weight gain in obese women by class of obesity and select maternal/newborn outcomes: a systematic review. Women Birth. doi:10.1016/j.wombi.2015.03.006

Fessler MB, Rudel LL, Brown JM (2009) Toll-like receptor signaling links dietary fatty acids to the metabolic syndrome. Curr Opin Lipidol 20(5):379–385. doi:10.1097/MOL.0b013e32832fa5c4

Fischer J, Koch L, Emmerling C, Vierkotten J, Peters T, Bruning JC, Ruther U (2009) Inactivation of the Fto gene protects from obesity. Nature 458(7240):894–898. doi:10.1038/nature07848

Flodgren G, Deane K, Dickinson HO, Kirk S, Alberti H, Beyer FR, ... Eccles MP (2010) Interventions to change the behaviour of health professionals and the organisation of care to promote weight reduction in overweight and obese people. Cochrane Database Syst Rev 3:CD000984. doi:10.1002/14651858.CD000984.pub2

Forno E, Young OM, Kumar R, Simhan H, Celedon JC (2014) Maternal obesity in pregnancy, gestational weight gain, and risk of childhood asthma. Pediatrics 134(2):e535–e546. doi:10.1542/peds.2014-0439

Franco JG, Fernandes TP, Rocha CP, Calvino C, Pazos-Moura CC, Lisboa PC, ... Trevenzoli IH (2012) Maternal high-fat diet induces obesity and adrenal and thyroid dysfunction in male rat offspring at weaning. J Physiol 590(Pt 21):5503–5518. doi:10.1113/jphysiol.2012.240655

Fraser A, Tilling K, Macdonald-Wallis C, Sattar N, Brion MJ, Benfield L, … Lawlor DA (2010) Association of maternal weight gain in pregnancy with offspring obesity and metabolic and vascular traits in childhood. Circulation 121(23):2557–2564. doi:10.1161/CIRCULATIONAHA.109.906081

Frayling TM, Timpson NJ, Weedon MN, Zeggini E, Freathy RM, Lindgren CM, … McCarthy MI (2007) A common variant in the FTO gene is associated with body mass index and predisposes to childhood and adult obesity. Science 316(5826):889–894. doi:10.1126/science.1141634

Frias AE, Morgan TK, Evans AE, Rasanen J, Oh KY, Thornburg KL, Grove KL (2011) Maternal high-fat diet disturbs uteroplacental hemodynamics and increases the frequency of stillbirth in a nonhuman primate model of excess nutrition. Endocrinology 152(6):2456–2464. doi:10.1210/en.2010-1332

Friis CM, Paasche Roland MC, Godang K, Ueland T, Tanbo T, Bollerslev J, Henriksen T (2013) Adiposity-related inflammation: effects of pregnancy. Obesity (Silver Spring) 21(1):E124–E130. doi:10.1002/oby.20120

Gaudet L, Ferraro ZM, Wen SW, Walker M (2014) Maternal obesity and occurrence of fetal macrosomia: a systematic review and meta-analysis. Biomed Res Int 2014:640291. doi:10.1155/2014/640291

Gerken T, Girard CA, Tung YC, Webby CJ, Saudek V, Hewitson KS, … Schofield CJ (2007) The obesity-associated FTO gene encodes a 2-oxoglutarate-dependent nucleic acid demethylase. Science 318(5855):1469–1472. doi:10.1126/science.1151710

Gohir W, Ratcliffe EM, Sloboda DM (2015) Of the bugs that shape us: maternal obesity, the gut microbiome, and long-term disease risk. Pediatr Res 77(1–2):196–204. doi:10.1038/pr.2014.169

Goldman-Wohl D, Yagel S (2002) Regulation of trophoblast invasion: from normal implantation to pre-eclampsia. Mol Cell Endocrinol 187(1–2):233–238

Grant WF, Gillingham MB, Batra AK, Fewkes NM, Comstock SM, Takahashi D, … Marks DL (2011) Maternal high fat diet is associated with decreased plasma n-3 fatty acids and fetal hepatic apoptosis in nonhuman primates. PLoS One 6(2):e17261. doi:10.1371/journal.pone.0017261

Graus-Nunes F, Dalla Corte Frantz E, Lannes WR, da Silva Menezes MC, Mandarim-de-Lacerda CA, Souza-Mello V (2015) Pregestational maternal obesity impairs endocrine pancreas in male F1 and F2 progeny. Nutrition 31(2):380–387. doi:10.1016/j.nut.2014.08.002

Grayson BE, Levasseur PR, Williams SM, Smith MS, Marks DL, Grove KL (2010) Changes in melanocortin expression and inflammatory pathways in fetal offspring of nonhuman primates fed a high-fat diet. Endocrinology 151(4):1622–1632. doi:10.1210/en.2009-1019

Gregorio BM, Souza-Mello V, Carvalho JJ, Mandarim-de-Lacerda CA, Aguila MB (2010) Maternal high-fat intake predisposes nonalcoholic fatty liver disease in C57BL/6 offspring. Am J Obstet Gynecol 203(5):495 e491–e498. doi:10.1016/j.ajog.2010.06.042

Hales CN, Barker DJ (1992) Type 2 (non-insulin-dependent) diabetes mellitus: the thrifty phenotype hypothesis. Diabetologia 35(7):595–601

Han Z, Niu T, Chang J, Lei X, Zhao M, Wang Q, … Chai J (2010) Crystal structure of the FTO protein reveals basis for its substrate specificity. Nature 464(7292):1205–1209. doi:10.1038/nature08921

Harding J (2001) The nutritional basis of the fetal origins of adult disease. Int J Epidemiol 30(1):15–23. doi:10.1093/ije/30.1.15

Harpsoe MC, Basit S, Bager P, Wohlfahrt J, Benn CS, Nohr EA, … Jess T (2013) Maternal obesity, gestational weight gain, and risk of asthma and atopic disease in offspring: a study within the Danish National Birth Cohort. J Allergy Clin Immunol 131(4):1033–1040. doi:10.1016/j.jaci.2012.09.008

Hartil K, Vuguin PM, Kruse M, Schmuel E, Fiallo A, Vargas C, … Charron MJ (2009) Maternal substrate utilization programs the development of the metabolic syndrome in male mice exposed to high fat in utero. Pediatr Res 66(4):368–373. doi:10.1203/PDR.0b013e3181b33375

Hayes EK, Lechowicz A, Petrik JJ, Storozhuk Y, Paez-Parent S, Dai Q, … Raha S (2012) Adverse fetal and neonatal outcomes associated with a life-long high fat diet: role of altered development of the placental vasculature. PLoS One 7(3):e33370. doi:10.1371/journal.pone.0033370

Hayes EK, Tessier DR, Percival ME, Holloway AC, Petrik JJ, Gruslin A, Raha S (2014) Trophoblast invasion and blood vessel remodeling are altered in a rat model of lifelong maternal obesity. Reprod Sci 21(5):648–657. doi:10.1177/1933719113508815

Hirasawa A, Tsumaya K, Awaji T, Katsuma S, Adachi T, Yamada M, ... Tsujimoto G (2005) Free fatty acids regulate gut incretin glucagon-like peptide-1 secretion through GPR120. Nat Med 11(1):90–94. doi:10.1038/nm1168

Hotamisligil GS, Shargill NS, Spiegelman BM (1993) Adipose expression of tumor necrosis factor-alpha: direct role in obesity-linked insulin resistance. Science 259(5091):87–91

Hotamisligil GS, Arner P, Caro JF, Atkinson RL, Spiegelman BM (1995) Increased adipose tissue expression of tumor necrosis factor-alpha in human obesity and insulin resistance. J Clin Invest 95(5):2409–2415. doi:10.1172/JCI117936

Howie GJ, Sloboda DM, Kamal T, Vickers MH (2009) Maternal nutritional history predicts obesity in adult offspring independent of postnatal diet. J Physiol 587(Pt 4):905–915. doi:10.1113/jphysiol.2008.163477

Howie GJ, Sloboda DM, Reynolds CM, Vickers MH (2013) Timing of maternal exposure to a high fat diet and development of obesity and hyperinsulinemia in male rat offspring: same metabolic phenotype, different developmental pathways? J Nutr Metab 2013:517384. doi:10.1155/2013/517384

Huszar D, Lynch CA, Fairchild-Huntress V, Dunmore JH, Fang Q, Berkemeier LR, ... Lee F (1997) Targeted disruption of the melanocortin-4 receptor results in obesity in mice. Cell 88(1):131–141

Ichimura A, Hirasawa A, Poulain-Godefroy O, Bonnefond A, Hara T, Yengo L, ... Froguel P (2012) Dysfunction of lipid sensor GPR120 leads to obesity in both mouse and human. Nature 483(7389):350–354. doi:10.1038/nature10798

Ingvorsen C, Thysen AH, Fernandez-Twinn D, Nordby P, Nielsen KF, Ozanne SE, ... Hellgren LI (2014) Effects of pregnancy on obesity-induced inflammation in a mouse model of fetal programming. Int J Obes (Lond) 38(10):1282–1289. doi:10.1038/ijo.2014.69

Jansson N, Rosario FJ, Gaccioli F, Lager S, Jones HN, Roos S, ... Powell TL (2013) Activation of placental mTOR signaling and amino acid transporters in obese women giving birth to large babies. J Clin Endocrinol Metab 98(1):105–113. doi:10.1210/jc.2012-2667

Janzen C, Lei MY, Cho J, Sullivan P, Shin BC, Devaskar SU (2013) Placental glucose transporter 3 (GLUT3) is up-regulated in human pregnancies complicated by late-onset intrauterine growth restriction. Placenta 34(11):1072–1078. doi:10.1016/j.placenta.2013.08.010

Jensen MD, Ryan DH, Apovian CM, Ard JD, Comuzzie AG, Donato KA, ... Obesity S (2014) 2013 AHA/ACC/TOS guideline for the management of overweight and obesity in adults: a report of the American College of Cardiology/American Heart Association Task Force on Practice Guidelines and The Obesity Society. Circulation 129(25 Suppl 2):S102–S138. doi:10.1161/01.cir.0000437739.71477.ee

Jones HN, Woollett LA, Barbour N, Prasad PD, Powell TL, Jansson T (2009) High-fat diet before and during pregnancy causes marked up-regulation of placental nutrient transport and fetal overgrowth in C57/BL6 mice. FASEB J 23(1):271–278. doi:10.1096/fj.08-116889

Kappen C, Kruger C, MacGowan J, Salbaum JM (2012) Maternal diet modulates placenta growth and gene expression in a mouse model of diabetic pregnancy. PLoS One 7(6):e38445. doi:10.1371/journal.pone.0038445

Kepczynska MA, Wargent ET, Cawthorne MA, Arch JR, O'Dowd JF, Stocker CJ (2013) Circulating levels of the cytokines IL10, IFN gamma and resistin in an obese mouse model of developmental programming. J Dev Orig Health Dis 4(6):491–498. doi:10.1017/S2040174413000263

Kirk SF, Penney TL, McHugh TL, Sharma AM (2012) Effective weight management practice: a review of the lifestyle intervention evidence. Int J Obes (Lond) 36(2):178–185. doi:10.1038/ijo.2011.80

Kopp-Hoolihan LE, van Loan MD, Wong WW, King JC (1999) Fat mass deposition during pregnancy using a four-component model. J Appl Physiol 87:196–202

Koren O, Goodrich JK, Cullender TC, Spor A, Laitinen K, Kling Bäckhed H, ... Ley RE (2012) Host remodeling of the gut microbiome and metabolic changes during pregnancy. Cell 150(3):470–480. doi:http://dx.doi.org/10.1016/j.cell.2012.07.008

Lau DC, Douketis JD, Morrison KM, Hramiak IM, Sharma AM, Ur E, Obesity Canada Clinical Practice Guidelines Expert P (2007) 2006 Canadian clinical practice guidelines on the management and prevention of obesity in adults and children [summary]. CMAJ 176(8):S1-S13. doi:10.1503/cmaj.061409

Lau EY, Liu J, Archer E, McDonald SM, Liu J (2014) Maternal weight gain in pregnancy and risk of obesity among offspring: a systematic review. J Obes 2014:524939. doi:10.1155/2014/524939

Lee YS, Kim JW, Osborne O, da Oh Y, Sasik R, Schenk S, … Olefsky JM (2014) Increased adipocyte O_2 consumption triggers HIF-1alpha, causing inflammation and insulin resistance in obesity. Cell 157(6):1339–1352. doi:10.1016/j.cell.2014.05.012

Lee K-I, Chiang C-W, Lin H-C, Zhao J-F, Li C-T, Shyue S-K, Lee T-S (2015) Maternal exposure to di-(2-ethylhexyl) phthalate exposure deregulates blood pressure, adiposity, cholesterol metabolism and social interaction in mouse offspring. Arch Toxicol 1–14. doi:10.1007/s00204-015-1539-0

Leung TY, Leung TN, Sahota DS, Chan OK, Chan LW, Fung TY, Lau TK (2008) Trends in maternal obesity and associated risks of adverse pregnancy outcomes in a population of Chinese women. BJOG 115(12):1529–1537. doi:10.1111/j.1471-0528.2008.01931.x

Li C, Zhang BB (2000) Insulin signaling and action: glucose, lipids, protein. In: De Groot LJ, Beck-Peccoz P, Chrousos G, Dungan K, Grossman A, Hershman JM, Koch C, McLachlan R, New M, Rebar R, Singer F, Vinik A, Weickert MO (eds) Endotext. MDText.com, Inc, South Dartmouth

Loos RJ, Lindgren CM, Li S, Wheeler E, Zhao JH, Prokopenko I, … Mohlke KL (2008) Common variants near MC4R are associated with fat mass, weight and risk of obesity. Nat Genet 40(6):768–775. doi:10.1038/ng.140

Lowe A, Braback L, Ekeus C, Hjern A, Forsberg B (2011) Maternal obesity during pregnancy as a risk for early-life asthma. J Allergy Clin Immunol 128(5):1107–1109 e1101–e1102. doi:10.1016/j.jaci.2011.08.025

Lubis AR, Widia F, Soegondo S, Setiawati A (2008) The role of SOCS-3 protein in leptin resistance and obesity. Acta Med Indones 40(2):89–95

Lutsiv O, Mah J, Beyene J, McDonald SD (2015) The effects of morbid obesity on maternal and neonatal health outcomes: a systematic review and meta-analyses. Obes Rev. doi:10.1111/obr.12283

Mamun AA, Mannan M, Doi SA (2014) Gestational weight gain in relation to offspring obesity over the life course: a systematic review and bias-adjusted meta-analysis. Obes Rev 15(4):338–347. doi:10.1111/obr.12132

Mao J, Zhang X, Sieli PT, Falduto MT, Torres KE, Rosenfeld CS (2010) Contrasting effects of different maternal diets on sexually dimorphic gene expression in the murine placenta. Proc Natl Acad Sci U S A 107(12):5557–5562. doi:10.1073/pnas.1000440107

Mark P, Sisala C, Connor K, Patel R, Vickers MH (2011) A maternal high-fat diet in rat pregnancy reduces growth of the fetus and the placental junctional zone, but not placental labyrinth zone growth. J Dev Origins Health Dis 2(1):8. doi:http://dx.doi.org/10.1017/S2040174410000681

Masuyama H, Hiramatsu Y (2012) Effects of a high-fat diet exposure in utero on the metabolic syndrome-like phenomenon in mouse offspring through epigenetic changes in adipocytokine gene expression. Endocrinology 153(6):2823–2830. doi:10.1210/en.2011-2161

Matherly SC, Puri P (2012) Mechanisms of simple hepatic steatosis: not so simple after all. Clin Liver Dis 16(3):505–524. doi:10.1016/j.cld.2012.05.005

McCurdy CE, Bishop JM, Williams SM, Grayson BE, Smith MS, Friedman JE, Grove KL (2009) Maternal high-fat diet triggers lipotoxicity in the fetal livers of nonhuman primates. J Clin Invest 119(2):323–335. doi:10.1172/JCI32661

Metwally M, Ong KJ, Ledger WL, Li TC (2008) Does high body mass index increase the risk of miscarriage after spontaneous and assisted conception? A meta-analysis of the evidence. Fertil Steril 90(3):714–726. doi:10.1016/j.fertnstert.2007.07.1290

Murabayashi N, Sugiyama T, Zhang L, Kamimoto Y, Umekawa T, Ma N, Sagawa N (2013) Maternal high-fat diets cause insulin resistance through inflammatory changes in fetal adipose tissue. Eur J Obstet Gynecol Reprod Biol 169(1):39–44. doi:10.1016/j.ejogrb.2013.02.003

Musso G, Gambino R, Cassader M (2009) Recent insights into hepatic lipid metabolism in non-alcoholic fatty liver disease (NAFLD). Prog Lipid Res 48(1):1–26. doi:10.1016/j.plipres.2008.08.001

Nivoit P, Morens C, Van Assche FA, Jansen E, Poston L, Remacle C, Reusens B (2009) Established diet-induced obesity in female rats leads to offspring hyperphagia, adiposity and insulin resistance. Diabetologia 52(6):1133–1142. doi:10.1007/s00125-009-1316-9

Ogden CL, Carroll MD, Kit BK, Flegal KM (2014) Prevalence of childhood and adult obesity in the united states, 2011–2012. JAMA 311(8):806–814. doi:10.1001/jama.2014.732

Organization WH (2015) BMI classification. from World Health Organization http://apps.who.int/bmi/index.jsp?introPage=intro_3.html

Pate RR, O'Neill JR, Liese AD, Janz KF, Granberg EM, Colabianchi N, ... Taverno Ross SE (2013) Factors associated with development of excessive fatness in children and adolescents: a review of prospective studies. Obes Rev 14(8):645–658. doi:10.1111/obr.12035

Pelleymounter MA, Cullen MJ, Baker MB, Hecht R, Winters D, Boone T, Collins F (1995) Effects of the obese gene product on body weight regulation in ob/ob mice. Science 269(5223):540–543

Punnonen R (1977) The relationship between serum oestradiol levels and serum triglyceride, cholesterol and phospholipid levels in normal human pregnancy. Br J Obstet Gynaecol 84(11):838–845

Rajasingam D, Seed PT, Briley AL, Shennan AH, Poston L (2009) A prospective study of pregnancy outcome and biomarkers of oxidative stress in nulliparous obese women. Am J Obstet Gynecol 200(4):395 e391–399. doi:10.1016/j.ajog.2008.10.047

Ramsay JE, Ferrell WR, Crawford L, Wallace AM, Greer IA, Sattar N (2002) Maternal obesity is associated with dysregulation of metabolic, vascular, and inflammatory pathways. J Clin Endocrinol Metab 87(9):4231–4237. doi:10.1210/jc.2002-020311

Rasouli N, Molavi B, Elbein SC, Kern PA (2007) Ectopic fat accumulation and metabolic syndrome. Diabetes Obes Metab 9(1):1–10. doi:10.1111/j.1463-1326.2006.00590.x

Reynolds RM, Allan KM, Raja EA, Bhattacharya S, McNeill G, Hannaford PC, ... Norman JE (2013) Maternal obesity during pregnancy and premature mortality from cardiovascular event in adult offspring: follow-up of 1 323 275 person years. BMJ 347:f4539. doi:10.1136/bmj.f4539

Rooney K, Ozanne SE (2011) Maternal over-nutrition and offspring obesity predisposition: targets for preventative interventions. Int J Obes (Lond) 35(7):883–890. doi:10.1038/ijo.2011.96

Ryan EA, O'Sullivan MJ, Skyler JS (1985) Insulin action during pregnancy. Studies with the euglycemic clamp technique. Diabetes 34(4):380–389

Saben J, Kang P, Zhong Y, Thakali KM, Gomez-Acevedo H, Borengasser SJ, ... Shankar K (2014a) RNA-seq analysis of the rat placentation site reveals maternal obesity-associated changes in placental and offspring thyroid hormone signaling. Placenta 35(12):1013–1020. doi:10.1016/j.placenta.2014.09.015

Saben J, Lindsey F, Zhong Y, Thakali K, Badger TM, Andres A, ... Shankar K (2014b) Maternal obesity is associated with a lipotoxic placental environment. Placenta 35(3):171–177. doi:10.1016/j.placenta.2014.01.003

Sabroe I, Parker LC, Dower SK, Whyte MKB (2008) The role of TLR activation in inflammation. J Pathol 214(2):126–135. doi:10.1002/path.2264

Samuelsson A-M, Matthews PA, Argenton M, Christie MR, McConnell JM, Jansen EHJM, ... Taylor PD (2008) Diet-induced obesity in female mice leads to offspring hyperphagia, adiposity, hypertension, and insulin resistance: a novel murine model of developmental programming. Hypertension 51(2):383–392. doi:10.1161/hypertensionaha.107.101477

Satoh N, Ogawa Y, Katsuura G, Hayase M, Tsuji T, Imagawa K, ... Nakao K (1997) The arcuate nucleus as a primary site of satiety effect of leptin in rats. Neurosci Lett 224(3):149–152. doi:10.1016/S0304-3940(97)00163-8

Schack-Nielsen L, Michaelsen KF, Gamborg M, Mortensen EL, Sorensen TI (2010) Gestational weight gain in relation to offspring body mass index and obesity from infancy through adulthood. Int J Obes (Lond) 34(1):67–74. doi:10.1038/ijo.2009.206

Sferruzzi-Perri AN, Vaughan OR, Haro M, Cooper WN, Musial B, Charalambous M, ... Fowden AL (2013) An obesogenic diet during mouse pregnancy modifies maternal nutrient partitioning and the fetal growth trajectory. FASEB J 27(10):3928–3937. doi:10.1096/fj.13-234823

Shi H, Kokoeva MV, Inouye K, Tzameli I, Yin H, Flier JS (2006) TLR4 links innate immunity and fatty acid-induced insulin resistance. J Clin Invest 116(11):3015–3025. doi:10.1172/JCI28898
Srinivasan M, Katewa SD, Palaniyappan A, Pandya JD, Patel MS (2006) Maternal high-fat diet consumption results in fetal malprogramming predisposing to the onset of metabolic syndrome-like phenotype in adulthood. Am J Physiol Endocrinol Metab 291(4):E792–E799. doi:10.1152/ajpendo.00078.2006
Starling AP, Brinton JT, Glueck DH, Shapiro AL, Harrod CS, Lynch AM, ... Dabelea D (2015) Associations of maternal BMI and gestational weight gain with neonatal adiposity in the Healthy Start study. Am J Clin Nutr 101(2):302–309. doi:10.3945/ajcn.114.094946
Steculorum SM, Bouret SG (2011) Maternal diabetes compromises the organization of hypothalamic feeding circuits and impairs leptin sensitivity in offspring. Endocrinology 152(11):4171–4179. doi:10.1210/en.2011-1279
Stupin JH, Arabin B (2014) Overweight and obesity before, during and after pregnancy: part 1: pathophysiology, molecular biology and epigenetic consequences. Geburtshilfe Frauenheilkd 74(7):639–645. doi:10.1055/s-0034-1368486
Summers SA, Garza LA, Zhou H, Birnbaum MJ (1998) Regulation of insulin-stimulated glucose transporter GLUT4 translocation and Akt kinase activity by ceramide. Mol Cell Biol 18(9):5457–5464
Suter M, Bocock P, Showalter L, Hu M, Shope C, McKnight R, ... Aagaard-Tillery K (2011) Epigenomics: maternal high-fat diet exposure in utero disrupts peripheral circadian gene expression in nonhuman primates. FASEB J 25(2):714–726. doi:10.1096/fj.10-172080
Suter MA, Sangi-Haghpeykar H, Showalter L, Shope C, Hu M, Brown K, ... Aagaard KM (2012) Maternal high-fat diet modulates the fetal thyroid axis and thyroid gene expression in a nonhuman primate model. Mol Endocrinol 26(12):2071–2080. doi:10.1210/me.2012-1214
Szendroedi J, Yoshimura T, Phielix E, Koliaki C, Marcucci M, Zhang D, ... Roden M (2014) Role of diacylglycerol activation of PKCtheta in lipid-induced muscle insulin resistance in humans. Proc Natl Acad Sci U S A 111(26):9597–9602. doi:10.1073/pnas.1409229111
Tenenbaum-Gavish K, Hod M (2013) Impact of maternal obesity on fetal health. Fetal Diagn Ther 34(1):1–7. doi:10.1159/000350170
Thorn SR, Baquero KC, Newsom SA, El Kasmi KC, Bergman BC, Shulman GI, ... Friedman J E (2014) Early life exposure to maternal insulin resistance has persistent effects on hepatic NAFLD in juvenile nonhuman primates. Diabetes 63(8):2702–2713. doi:10.2337/db14-0276
Tung YC, Ayuso E, Shan X, Bosch F, O'Rahilly S, Coll AP, Yeo GS (2010) Hypothalamic-specific manipulation of Fto, the ortholog of the human obesity gene FTO, affects food intake in rats. PLoS One 5(1):e8771. doi:10.1371/journal.pone.0008771
Vaisse C, Clement K, Guy-Grand B, Froguel P (1998) A frameshift mutation in human MC4R is associated with a dominant form of obesity. Nat Genet 20(2):113–114. doi:10.1038/2407
Vickers MH (2014) Developmental programming and transgenerational transmission of obesity. Ann Nutr Metab 64(suppl 1):26–34
Vickers MH, Breier BH, Cutfield WS, Hofman PL, Gluckman PD (2000) Fetal origins of hyperphagia, obesity, and hypertension and postnatal amplification by hypercaloric nutrition. Am J Physiol Endocrinol Metab 279(1):E83–E87
Volpato AM, Schultz A, Magalhaes-da-Costa E, Correia ML, Aguila MB, Mandarim-de-Lacerda CA (2012) Maternal high-fat diet programs for metabolic disturbances in offspring despite leptin sensitivity. Neuroendocrinology 96(4):272–284. doi:10.1159/000336377
Vuguin PM, Hartil K, Kruse M, Kaur H, Lin CL, Fiallo A, ... Charron MJ (2013) Shared effects of genetic and intrauterine and perinatal environment on the development of metabolic syndrome. PLoS One 8(5):e63021. doi:10.1371/journal.pone.0063021
Wang J, Campbell IL (2002) Cytokine signaling in the brain: putting a SOCS in it? J Neurosci Res 67(4):423–427
Wang P, Yang FJ, Du H, Guan YF, Xu TY, Xu XW, ... Miao CY (2011) Involvement of leptin receptor long isoform (LepRb)-STAT3 signaling pathway in brain fat mass- and obesity-associated (FTO) downregulation during energy restriction. Mol Med 17(5–6):523–532. doi:10.2119/molmed.2010.00134

Wellen KE, Hotamisligil GS (2005) Inflammation, stress, and diabetes. J Clin Invest 115(5):1111–1119. doi:10.1172/JCI25102

Whitaker RC (2004) Predicting preschooler obesity at birth: the role of maternal obesity in early pregnancy. Pediatrics 114(1):e29–e36

White CL, Purpera MN, Morrison CD (2009) Maternal obesity is necessary for programming effect of high-fat diet on offspring. Am J Physiol Regul Integr Comp Physiol 296(5):R1464–R1472. doi:10.1152/ajpregu.91015.2008

Wilson RM, Marshall NE, Jeske DR, Purnell JQ, Thornburg K, Messaoudi I (2015) Maternal Obesity alters immune cell frequencies and responses in umbilical cord blood samples. Pediatr Allergy Immunol. doi:10.1111/pai.12387

Wu Q, Saunders RA, Szkudlarek-Mikho M, Serna Ide L, Chin KV (2010) The obesity-associated Fto gene is a transcriptional coactivator. Biochem Biophys Res Commun 401(3):390–395. doi:10.1016/j.bbrc.2010.09.064

Yang Z, Huffman SL (2013) Nutrition in pregnancy and early childhood and associations with obesity in developing countries. Matern Child Nutr 9(Suppl 1):105–119. doi:10.1111/mcn.12010

Yeo GS, Farooqi IS, Aminian S, Halsall DJ, Stanhope RG, O'Rahilly S (1998) A frameshift mutation in MC4R associated with dominantly inherited human obesity. Nat Genet 20(2):111–112. doi:10.1038/2404

Yeo GS, Farooqi IS, Challis BG, Jackson RS, O'Rahilly S (2000) The role of melanocortin signalling in the control of body weight: evidence from human and murine genetic models. QJM 93(1):7–14

Yu C, Chen Y, Cline GW, Zhang D, Zong H, Wang Y, ... Shulman GI (2002) Mechanism by which fatty acids inhibit insulin activation of insulin receptor substrate-1 (IRS-1)-associated phosphatidylinositol 3-kinase activity in muscle. J Biol Chem 277(52):50230–50236. doi:10.1074/jbc.M200958200

Chapter 9
The Role of Environmental Exposures in Preterm Birth

Kelly K. Ferguson and John D. Meeker

Abstract Preterm birth is a significant yet poorly understood public health problem that may arise in part from maternal exposure to chemicals in the environment. This review explores the state of the knowledge on prematurity in relation to: (1) Organic pollutants, including persistent organic pollutants, such as dichlorodiphenyltrichloroethane, polychlorinated biphenyls, and perfluorinated compounds, disinfection byproducts, such as trihalomethanes, non-persistent pesticides, such as atrazine, and non-persistent organics of emerging concern, such as phthalates and bisphenol-A; (2) Metals and metalloids, including lead, cadmium, arsenic, and mercury; and (3) Air pollutants, including EPA criteria air contaminants, environmental tobacco smoke, and polycyclic aromatic hydrocarbons. We also highlight pervasive study limitations as well as important directions for future research.

Keywords Epidemiology • Pregnancy • Birth outcomes • Gestation • Environment

9.1 Introduction

Preterm birth, defined commonly as birth before 37 weeks completed gestation, is a complex and poorly understood disease that is highly prevalent in the US and elsewhere. Preterm newborns are at a much higher risk of mortality and various morbidities, and longitudinal studies have linked being born preterm to a range of morbidities in childhood and later in life, such as asthma and metabolic disorders as well as neurodevelopment delays. The combined cost of caring for preterm infants plus addressing subsequent complications was estimated at 26.2 billion dollars in the US in 2005 (Behrman and Butler 2007). Preventing preterm birth is a priority of the March of Dimes Foundation, the Surgeon General, and the Institute of Medicine. Despite concerted efforts, factors that are known to cause preterm birth, and their

K.K. Ferguson, Ph.D. (✉) • J.D. Meeker, Sc.D., C.I.H.
Department of Environmental Health Sciences, University of Michigan School of Public Health, Ann Arbor, MI, USA
e-mail: kellferg@umich.edu; meekerj@umich.edu

C.L. Hughes, M.D. Waters (eds.), *Translational Toxicology*, Molecular and Integrative Toxicology, DOI 10.1007/978-3-319-27449-2_9

underlying mechanistic pathways, are few, and current strategies are likely to decrease preterm birth rates minimally by 2015 (Chang et al. 2013).

Of emerging concern is the contribution of environmental exposures to preterm birth. Women are exposed to a chemical milieu throughout life, and the gestational time period is no exception. Furthermore, many chemicals are capable of infiltrating the placenta and accumulating in the fetus. These exposures may precipitate preterm birth through several hypothesized pathways, such as inducing inflammation, prostaglandin release, hormonal changes, or oxidative stress in the maternal-fetal compartment, or through mechanisms that remain unexplored.

Examining environmental contributors to prematurity in an animal model is complicated by the fact that rodents do not naturally deliver preterm, but only do so with specific gene knockouts or with high doses of lipopolysaccharides injection (Cha et al. 2013; Kaga et al. 1996). Thus, evidence linking environmental contaminants to preterm birth largely comes from epidemiologic investigations. These studies have been rigorously examined in a number of substantive reviews (Ferguson et al. 2013; Wigle et al. 2008) and meta-analyses (Nieuwenhuijsen et al. 2013). In this chapter we highlight some of the more robust and influential of these studies as well as more recently published results (summarized by pollutant category in Tables 9.1, 9.2 and 9.3).

9.2 Organic Pollutants

Of the organic pollutants, persistent compounds have historically received the most attention in the study of preterm birth. These include chemicals such as dichlorodiphenyltrichloroethane (DDT) and other persistent pesticides, polychlorinated biphenyls (PCBs) and perfluorinated compounds previously used in various industrial applications, and brominated flame retardants such as polybrominated diphenyl ethers (PBDEs) which are persistent in the environment and human tissue. Parent compounds and in some instances metabolites can be measured reliably in serum or plasma as well as the placenta, which has enabled study designs with accurate subject-specific exposure metrics. While use of some of these compounds continues, most have been restricted in the US and other highly developed countries; thus, attention to lower levels of exposure may be particularly important. However, overall, the body of evidence suggests that higher doses of these compounds are more clearly linked to prematurity.

DDT and its metabolites dichlorodiphenyldichloroethylene (DDE) and dichlorodiphenyldichloroethane (DDD) were examined in a number of small ($N < 60$) case-control studies published in the 1980s and 1990s in relation to preterm birth. Evidence was largely conflicting, but associations were clearly stronger in studies with higher exposure levels (Berkowitz et al. 1996; Procianoy and Schvartsman 1981; Saxena et al. 1981; Wassermann et al. 1982). The most rigorous study to address this relationship measured concentrations of DDT, DDE, and DDD in maternal serum collected in the 3rd trimester of pregnancy in the Collaborative

Table 9.1 Findings from studies examining organic pollutants in association with preterm birth

Exposure	References	Primary results
DDT	Saxena et al. (1981)	*Higher DDT metabolite concentrations in placenta and blood from cases compared to controls*
	Procianoy and Schvartsman (1981)	*Higher DDT metabolite concentrations in cord blood but not maternal serum in cases compared to controls*
	Wassermann et al. (1982)	*Higher DDT metabolites in 3rd trimester serum of cases compared to controls*
	Berkowitz et al. (1996)	*No significant differences in DDE from 1st trimester maternal serum in cases of spontaneous PTB compared to controls*
	Longnecker et al. (2001)	*Elevated odds of PTB in women with higher 3rd trimester serum concentrations of DDE*
	Ribas-Fitó et al. (2002)	*Higher DDE in cord serum from preterm vs. term newborns*
	Torres-Arreola et al. (2003)	*No significant differences in maternal serum DDE at delivery in cases vs. controls*
	Farhang et al. (2005)	*No significant differences in maternal serum DDT or DDE in mothers who went on to deliver preterm vs. term*
	Wood et al. (2007)	*No change in odds of PTB in association with maternal serum DDE concentrations at delivery*
	Pathak et al. (2009)	*No differences in DDT or DDE concentrations in maternal or cord blood taken at delivery in cases vs. controls*
	Wojtyniak et al. (2010)	*No significant but some suggestive associations between DDE measured in maternal serum from the second half of pregnancy in cases compared to controls*
	Bergonzi et al. (2011)	*Higher DDE in serum and higher DDT in adipose tissue in mothers who delivered preterm compared to term; no differences in cord serum or placental levels*
HCB	Saxena et al. (1981)	*No difference in placenta or blood HCB in cases compared to controls*
	Wassermann et al. (1982)	*No difference in HCB in 3rd trimester serum of cases compared to controls*
	Ribas-Fitó et al. (2002)	*Higher HCB in cord serum from preterm vs. term newborns*
	Torres-Arreola et al. (2003)	*No difference in HCB in maternal serum at delivery in cases vs. controls*
	Bergonzi et al. (2011)	*No differences in HCB from 1st trimester maternal serum in cases of spontaneous PTB compared to controls*
	Basterrechea et al. (2014)	*No significant associations between HCB and PTB*

(continued)

Table 9.1 (continued)

Exposure	References	Primary results
HCH/aldrin/dieldrin	Saxena et al. (1981)	*Higher HCH and aldrin levels in placenta and blood from cases compared to controls*
	Wasserman et al. (1982)	*Higher dieldrin in 3rd trimester serum of cases compared to controls*
	Ribas-Fitó et al. (2002)	*No difference in HCH in cord serum from preterm vs. term newborns*
	Torres-Arreola et al. (2003)	*Suggestively increased odds of PTB in association with HCH in maternal serum collected at delivery*
	Pathak et al. (2009)	*Higher HCH in maternal or cord blood taken at delivery in cases vs. controls*
Chlordecone	Kadhel et al. (2014)	*Levels in plasma collected at delivery associated with increased odds of PTB*
PCBs	Wassermann et al. (1982)	*Higher summed PCBs in maternal 3rd trimester serum in cases vs. controls*
	Berkowitz et al. (1996)	*No differences in 1st trimester serum concentrations of summed PCBs in cases compared to controls*
	Ribas-Fitó et al. (2002)	*No differences in summed PCBs in cord serum from cases vs. controls*
	Longnecker et al. (2005)	*No significant associations between summed PCBs and PTB*
	Wojtyniak et al. (2010)	*No associations between maternal serum levels of PCB-153 and PTB*
	Bergonzi et al. (2011)	*No differences in summed PCB concentrations in maternal or cord serum, placenta, or adipose in mothers who delivered preterm vs. term*
PFCs	Apelberg et al. (2007)	*No differences in concentrations of PFOA or PFOS in cord serum from preterm vs. term newborns*
	Fei et al. (2007)	*Suggestive associations between PFOS and PFOA measured in 1st trimester maternal serum and odds of PTB*
	Nolan et al. (2009)	*No significant differences in rates of PTB in PFOA contaminated vs. uncontaminated areas*
	Hamm et al. (2010)	*No significant associations between PFOA or PFOS measured in 2nd trimester maternal serum and PTB*
	Chen et al. (2012)	*Increased odds of PTB in association with cord blood PFOS, but not PFOA, levels measured in cord blood*
	Arbuckle et al. (2012)	*No differences in PFOS or PFOA in cord serum from preterm vs. term births*

(continued)

Table 9.1 (continued)

Exposure	References	Primary results
	Savitz et al. (2012a)	*No significant associations between PFOA levels measured in drinking water and self-reported history of PTB*
	Savitz et al. (2012b)	*No association between exposure to PFOA estimated by drinking water modeling and PTB*
	Whitworth et al. (2012)	*Reduced odds of PTB in subjects with higher concentrations of PFOA and PFOS in maternal plasma from the 2nd trimester*
	Wu et al. (2012)	*Higher PFOA concentrations in maternal serum collected at birth in mothers who delivered preterm compared to term*
	Darrow et al. (2013)	*No association between PFOA or PFOS concentrations in maternal serum and PTB*
Dioxin	Eskenazi et al. (2003)	*No association between serum TCDD levels and odds of PTB*
	Lin et al. (2006)	*Suggestive but non-significant increased odds of PTB in association with PCDD/F exposure estimates calculated from statistical models*
	Wesselink et al. (2014)	*No association between serum TCDD levels and odds of PTB*
PBDEs	Wu et al. (2010)	*Higher levels of summed PBDEs in cord blood from newborns with adverse birth outcomes compared to those from normal pregnancies*
DBPs	Bove et al. (1995)	*No association between TTHMs in drinking water during pregnancy and PTB*
	Kramer et al. (1992)	*No association between chloroform in drinking water during pregnancy and PTB*
	Savitz et al. (1995)	*No association between TTHMs in drinking water during 3rd trimester and PTB*
	Gallagher et al. (1998)	*No association between TTHMs in drinking water during 3rd trimester and PTB*
	Dodds et al. (1999)	*No association between TTHMs in drinking water during 3rd trimester and PTB*
	Wright et al. (2003)	*No association between TTHMs in drinking water during pregnancy and PTB*
	Wright et al. (2004)	*No association between TTHMs or HAAs in drinking water during pregnancy and PTB*
	Lewis et al. (2007)	*No association between TTHMs in drinking water during pregnancy and PTB*
	Yang et al. (2007)	*No association between TTHMs in drinking water during pregnancy and PTB*

(continued)

Table 9.1 (continued)

Exposure	References	Primary results
	Hoffman et al. (2008)	*No association between TTHMs, HAAS, or total organic halides in drinking water during 2nd trimester and PTB*
	Horton et al. (2011)	*Significant increase in odds of PTB in association with 2nd trimester drinking water total organic halide levels, but not TTHM or HAA alone*
	Patelarou et al. (2011)	*No association between preterm birth and drinking water concentrations of TTHM measured in 1st, 2nd, or 3rd trimester, or with entire pregnancy average*
	Villanueva et al. (2011)	*No association between preterm birth and drinking water concentrations of TTHM measured in 1st, 2nd, or 3rd trimester, or with entire pregnancy average*
	Costet et al. (2012)	*No association between preterm birth and either 1st trimester urinary concentrations of TCAA or 3rd trimester drinking water TTHM*
	Rivera-Núñez and Wright (2013)	*Significant crude and suggestive adjusted associations between DBPs measured in drinking water and PTB*
TCE/PCE	Bove et al. (1995)	*No association between TCE or PCE measured in drinking water during pregnancy and PTB*
	Sonnenfeld et al. (2001)	*Increased odds of PTB in association with PCE in drinking water during pregnancy in a contaminated area*
	Aschengrau et al. (2008)	*No association between pregnancy PCE exposure estimated by fate-transport model of drinking water concentrations*
Benzene	Llop et al. (2010)	*Significantly increased odds of PTB in association with elevated exposure assessed via ambient air monitoring across the duration of pregnancy*
	Wilhelm et al. (2011)	*Increased ambient air concentrations significantly associated with increased odds of PTB*
Formaldehyde	Maroziene et al. (2002)	*No significant association with PTB in association with ambient air monitoring exposure levels measured monthly or averaged over pregnancy*

(continued)

Table 9.1 (continued)

Exposure	References	Primary results
Atrazine	Forand et al. (2011)	*No association change in risk of PTB for women residing in TCE or PCE contaminated areas*
	Villanueva et al. (2005)	*No significant increase in odds of PTB in association with drinking water levels during pregnancy*
	Ochoa-Acuña et al. (2009)	*No significant increase in prevalence of PTB in association with elevated drinking water levels in the first or last months of pregnancy*
	Rinsky et al. (2012)	*Significantly increased odds of PTB in association with elevated drinking water levels during pregnancy*
OP pesticides	Eskenazi et al. (2004)	*No association between OP pesticides and PTB, but significantly increased odds of PTB in association with lower cholinesterase activity during pregnancy*
	Sathyanarayana et al. (2010)	*No association between self-reported OP pesticide use and PTB*
Phthalates	Adibi et al. (2009)	*Reduced odds of PTB in association with 3rd trimester urinary DEHP metabolite concentrations*
	Meeker et al. (2009)	*Increased odds of PTB in association with 3rd trimester urinary DEHP and other phthalate metabolite concentrations*
	Ferguson et al. (2014b)	*Increased odds of PTB in association with average of DEHP and other phthalate metabolite levels measured in urine at up to 3 visits per subject during pregnancy*
	Huang et al. (2014)	*Increased odds of PTB in association with phthalates measured in cord blood*
BPA	Cantonwine et al. (2010b)	*Suggestively increased odds of PTB in association with 3rd trimester urinary BPA concentrations*

Abbreviations: *PTB* preterm birth, *DDT* dichlorodiphenyltrichloroethane, *DDE* dichlorodiphenyldichloroethylene, *HCB* hexachlorobenzene, *HCH* hexachlorocyclohexane, *PCBs* polychlorinated biphenyls, *PFCs* perfluorinated compounds, *PFOA* perfluorooctanoic acid, *PFOS* perfluorooctane sulfonic acid, *TCDD* 2,3,7,8-tetrachlorodibenzo-p-dioxin, *PCDD/F* polychlorinated dibenzo-p-dioxins and dibenzofurans, *PBDEs* polybrominated diphenyl ethers, *DBPs* disinfection byproducts, *TTHMs* total trihalomethands, *HAAs* total haloacetic acids, *TCAA* trichloroacetic acid, *OP* organophosphate, *DEHP* di-2-ethylhexyl phthalate, *BPA* bisphenol-A

Table 9.2 Findings from studies examining toxic metals and metalloids in association with PTB

Exposure	References	Primary results
Lead	Torres-Sánchez et al. (1999)	*Significant associations between umbilical cord blood lead levels and PTB among primiparous women in somewhat highly exposed population*
	Sowers et al. (2002)	*No associations between maternal blood lead levels measured at four time points during pregnancy and PTB in a moderately exposed population*
	Falcon et al. (2003)	*Significantly elevated levels of lead in placenta taken from pregnancies ending in preterm rupture of membranes or delivery compared to normal pregnancies*
	Jelliffe-Pawlowski et al. (2006)	*Significant association between elevated (>10 μg/dL) maternal blood lead levels during pregnancy and PTB*
	Cantonwine et al. (2010a)	*Second trimester (but not 1st or 3rd trimester) maternal blood lead levels associated with increased odds of PTB; No associations with cord blood levels*
	Zhu et al. (2010)	*No association between maternal blood lead levels before or at delivery and PTB*
	Vigeh et al. (2011)	*Significantly increased odds of PTB in association with first trimester maternal blood lead levels*
	Perkins et al. (2014)	*Significant association between maternal blood lead levels during pregnancy and PTB in male infants only; Exposure levels notably low across population*
	Taylor et al. (2014)	*No association between maternal blood lead levels mid-pregnancy and PTB*
Cadmium	Landgren (1996)	*Mothers residing in an area of cadmium contamination showed no greater risk of delivering preterm compared to mothers in non-contaminated areas*
	Fagher et al. (1993)	*Elevated blood cadmium levels in mothers who delivered preterm compared to term in a small case-control study*
	Nishijo et al. (2002)	*Significant association between gestational urinary cadmium levels and PTB in an area of high contamination in Japan*
	Zhang et al. (2004)	*No association between maternal blood, cord blood, or placental cadmium concentrations and PTB*
Arsenic	Ahmad et al. (2001)	*Significantly elevated PTB rates to mothers residing in areas with high well water concentrations of arsenic in Bangladesh*
	Yang et al. (2003)	*No association between residence in areas of high drinking water arsenic contamination and PTB in Taiwan*
	Mukherjee et al. (2005)	*No significant difference in PTB rates in mothers with exposure to elevated levels of arsenic in drinking water*
	Myers et al. (2010)	*No significant change in odds of PTB for mothers residing in villages with elevated well water arsenic concentrations in China*
Mercury	Xue et al. (2007)	*Increased odds of PTB in mothers with higher hair mercury levels between 15 and 27 weeks of pregnancy*
	Burch et al. (2014)	*Higher rates of PTB observed in African American mothers residing in areas with elevated mercury contamination of fish in South Carolina*
	Bashore et al. (2014)	*No association between umbilical cord blood mercury levels and PTB*

Abbreviations: *PTB*, preterm birth

Table 9.3 Findings from studies examining air pollution exposures in association with PTB

Exposure	References	Primary results
Criteria air pollutants	Sram et al. (2005)	*Insufficient evidence to demonstrate causal associations with PTB and any individual criteria air pollutant in a review of the literature*
	Stillerman et al. (2008)	*Associations with these pollutants and PTB tend to be small in magnitude but statistically significant in a review of the literature*
	Bonzini et al. (2010)	*Evidence for a relationship between exposure to criteria air pollutant exposure in the first trimester and PTB in a review of the literature*
	Shah et al. (2011)	*Maternal exposure to sulfur dioxide and* $PM_{2.5}$ *associated with PTB in a review of the literature. Inconclusive evidence for other criteria pollutants*
	Nieuwenhuijsen et al. (2013)	*In a summary of recent meta-analyses,* $PM_{2.5}$ *most clearly associated with PTB*
	Stieb et al. (2012)	*Meta-analysis demonstrated association between carbon monoxide and* PM_{10} *exposure from third trimester of pregnancy and PTB, but evidence was less strong for ozone or sulfur dioxide*
	Lai et al. (2013)	*Significant association between sulfur dioxide and PTB in a meta-analysis of Chinese populations*
ETS	Leonardi-Bee et al. (2008)	*No association between ETS and gestational age in a meta-analysis*
	Salmasi et al. (2010)	*No association between ETS exposure and length of gestation or preterm birth in a meta-analysis.*
PAH	Vassilev et al. (2001)	*Maternal ambient air concentrations of PAH, based on census tract, significantly associated with increased odds of PTB*
	Choi et al. (2008)	*Increased total PAH exposure levels measured using personal air monitors in the third trimester were associated with increased odds of PTB in African American, but not Dominican, mothers*
	Singh et al. (2008)	*Higher individual PAH concentrations in placenta from preterm compared to term pregnancies*
	Wilhelm et al. (2011)	*Increased odds of PTB in association with total and some individual PAH levels measured using ambient air monitors in Los Angeles County*
	Guo et al. (2012)	*Higher PAH concentrations in cord blood from pregnancies with adverse birth outcomes (including PTB, low birth weight, congenital malformations, and still birth) compared to others in a highly exposed region of China*
	Padula et al. (2014)	*Association between ambient air concentrations during the last 6 weeks of pregnancy and increased odds of early PTB but some protective associations between exposure and PTB as well*

Note: For criteria air pollutants and environmental tobacco smoke only meta-analyses and reviews from the past 10 years are included as there are numerous studies examining the association with preterm birth. Criteria air pollutants included in this table are: ozone, particulate matter ($PM_{2.5}$ and PM_{10}), carbon monoxide, nitrogen oxides, and sulfur dioxide. *Abbreviations*: *PTB* preterm birth, *ETS* environmental tobacco smoke, *PAH* polycyclic aromatic hydrocarbons

Perinatal Project in the US (N = 2380) (Longnecker et al. 2001). These women were recruited from 1959 to 1965 and consequently had higher levels of exposure compared to those observed today. A strong increase in odds of delivering preterm was observed for mothers with high (>60 μg/L) levels of serum DDE. However, a smaller study (N = 455) with slightly higher median exposure levels failed to detect any effects (Farhang et al. 2005). More recent studies with lower exposure levels in various countries have shown conflicting results, but in general more associations have been detected in populations with higher exposure levels (Bergonzi et al. 2011; Pathak et al. 2009; Ribas-Fitó et al. 2002; Torres-Arreola et al. 2003; Wojtyniak et al. 2010; Wood et al. 2007).

Other persistent pesticides examined in relation to preterm birth include hexachlorobenzene (HCB), hexachlorocyclohexane (HCH) or lindane, heptachlor or heptachlor epoxide and aldrin or dieldrin. These studies showed some suggestive associations but were largely inconclusive (Bergonzi et al. 2011; Fenster et al. 2006; Pathak et al. 2009; Ribas-Fitó et al. 2002; Saxena et al. 1981; Torres-Arreola et al. 2003; Wassermann et al. 1982). A large study in Spain (N = 1568) conducted recently also failed to detect any significant association between 1st trimester maternal HCB levels and preterm birth (Basterrechea et al. 2014). Interestingly, however, a small gene-environment interaction study (N = 156 cases, 151 controls) examining polymorphisms in the genes responsible for organochlorine pesticide metabolism and detoxification found significant interactions between a number of these compounds, particularly HCH, and risk of delivering preterm (Mustafa et al. 2013), suggesting some women may be more susceptible to these effects than others. Additionally, high levels of exposure may be more likely to have an effect. A study of mothers from the French West Indies, where chlordecone use is common, plasma levels measured at delivery were associated with increased odds of prematurity (Kadhel et al. 2014).

In the Collaborative Perinatal Project described above, Longnecker and colleagues also examined the association between exposure the industrial lubricants and insulators PCBs and preterm birth (Longnecker et al. 2005). Although this was one of the largest (N = 1034) and best designed studies to examine this relationship no associations were detected, and any suggestive associations were diminished after models were adjusted for DDT exposure levels. As with DDT, a number of smaller case control studies have shown some evidence for associations between prenatal PCB exposures and preterm birth; however, most of the studies to date have shown no associations (Bergonzi et al. 2011; Berkowitz et al. 1996; Govarts et al. 2012; Ribas-Fitó et al. 2002; Wassermann et al. 1982).

Perfluorinated compounds (PFCs), including perfluorooctanoic acid (PFOA) or perfluorooctane sulfonic acid (PFOS) were used until recently for industrial and some consumer product applications as repellants of oil, grease, and water. Human exposure occurs primarily through consumption of contaminated food and drinking water or through inhalation of dusts in the US and other populations. Several studies have examined PFOA and/or PFOS concentrations in cord or maternal serum in association with preterm birth in populations with exposure levels consistent with those observed currently in the US. Results from these studies have been conflicting, with most reporting null or even protective associations (Apelberg et al. 2007; Fei et al. 2007; Hamm et al. 2010; Whitworth et al. 2012), yet some evidence for increased odds of preterm birth or reduced gestational age at delivery in relation to PFOS concentrations specifically (Arbuckle et al. 2012; Chen et al. 2012).

Incidences of drinking water contamination have resulted in higher than average PFC exposures to some populations. In the US, industry contamination of water sources in the Mid-Ohio valley resulted in PFOA human exposure levels 5 times higher than in subjects from the National Health and Nutrition Examination Survey (NHANES), a nationally representative sample (Frisbee et al. 2009). The C8 Health Project was designed to investigate exposure levels in this population and to conduct epidemiologic studies of potential health consequences. Despite large sample sizes (N > 1000) and a variety of sophisticated modeling techniques, no associations were detected between PFOA or PFOS exposures and preterm birth in that study population (Nolan et al. 2009; Savitz et al. 2012a, b; Stein et al. 2009). In a more recent study in the same population utilizing biomarkers of exposure measured somewhat proximally to pregnancy, no associations between PFOA or PFOS and preterm birth were detected either (Darrow et al. 2013). However, in a Chinese population with elevated exposures due to environmental contamination by electronic waste, PFOA in maternal serum was associated with preterm birth (Wu et al. 2012).

Finally, in the realm of persistent organic pollutants, few studies exist measuring associations between preterm birth and either dioxin or PBDEs. Suggestive but generally null relationships have been observed with dioxin (Le and Johansson 2001; Lin et al. 2006; Revich et al. 2001), even in populations with high exposures resulting from a chemical explosion in Seveso, Italy (Eskenazi et al. 2003; Wesselink et al. 2014). One study observed higher cord blood levels of PBDE in newborns with an adverse birth outcome but that definition was not specific to preterm birth (Wu et al. 2010).

Disinfection byproducts (DBPs), formed from chlorine used to treat drinking water, have been examined in a number of studies in relation to preterm birth. These include trihalomethanes (THMs), such as chloroform, as well as haloacetic acids (HAAs) such as trichloroacetic acid (TCAA). A recent review and meta-analysis of 9 studies by Grellier and colleagues concluded that insufficient evidence exists for a relationship between THMs or HAAs on preterm birth (Grellier et al. 2010; Dodds et al. 1999; Gallagher et al. 1998; Hoffman et al. 2008; Kramer et al. 1992; Lewis et al. 2007; Savitz et al. 1995; Wright et al. 2003, 2004; Yang et al. 2007). Since the publication of these findings several additional studies have been published, primarily with null findings. In areas of drinking water contamination, two large (N > 1000) studies assigning exposure based on THM levels measured in drinking water paired with questionnaire data on drinking water use found no association with preterm birth (Patelarou et al. 2011; Villanueva et al. 2011). Another study with similar exposure assessment methods found an association between total organic halide exposure and preterm birth in an area of brominated disinfection byproduct contamination (Horton et al. 2011). One study utilizing urinary biomarkers of exposure to TCAA also failed to detect any associations with PTB, although detection in samples was quite low (Costet et al. 2012). However, the most recent study to examine the relationship between DBPs and preterm birth observed significantly increased crude odds ratios across all exposure categories for most DBPs measured in public drinking water systems in the Boston area (N=712,394) (Rivera-Núñez and Wright 2013). Most associations were attenuated with adjustment for pertinent covariates; relationships remained statistically significant or suggestive for chloroform, TCAA, and summed DBPs.

Trichloroethylene (TCE) and tetrachloroethylene (PCE) are also common, though unintentional, organic drinking water contaminants. Three studies examining associations between these compounds and preterm birth have utilized drinking water concentrations in areas of contamination and paired these data with information on maternal residence to assess exposures. None found significant associations despite large sample sizes (Aschengrau et al. 2008; Bove et al. 2002; Sonnenfeld et al. 2001). Additionally, one study examined air concentrations in an area with contaminated soil and found no association with preterm birth (Forand et al. 2011). Benzene, like TCE and PCE, is a volatile organic compound, but exposure occurs more typically through inhalation. Two studies examining the relationship between ambient air monitoring levels of benzene observed significantly increased odds of preterm birth in association with exposure levels (Llop et al. 2010; Wilhelm et al. 2011). Finally, one study examined the relationship between maternal formaldehyde exposure during pregnancy and prematurity and did not observe any statistically significant associations (Maroziene and Grazuleviciene 2002).

Non-persistent pesticides are also a drinking water contaminant of concern. Although not bioaccumulative or persistent in the environment, these compounds have longer half-lives in drinking water which can results in population-wide exposures, particularly in agricultural areas. Atrazine is one such pesticide that has been examined in a number of studies in association with preterm birth. Most studies have utilized drinking water measurements paired with information on maternal residence and in some instances drinking water use. However, detection in drinking water sources is typically low, and while one study observed a slight increase in odds of preterm birth (Villanueva et al. 2005) other findings have been null (Ochoa-Acuña et al. 2009; Rinsky et al. 2012). Two other studies have examined the non-persistent organophosphate pesticides in association with preterm birth in agricultural populations. One utilized biomarkers of parent compounds or their metabolites in urine samples collected during pregnancy and found no evidence of an association; although the authors did observe that decreased maternal cholinesterase activity, which is related to an overall increase in exposure to organophosphate pesticides, was associated with increased odds of preterm birth (Eskenazi et al. 2004). However, within the Agricultural Health Study, self-reported pesticide use was not associated with preterm birth (Sathyanarayana et al. 2010).

Finally, the non-persistent endocrine disrupting compounds phthalates and bisphenol-a (BPA) have received recent attention for their potential contribution to preterm birth. As with the persistent organic pollutants, most studies investigating these compounds have utilized personal exposure measurements; however, as these are metabolized rapidly in the human body, levels are measured in urine samples and are less stable over time. The first study to examine phthalate exposure in relation to length of gestation measured di-2-ethylhexyl phthalate (DEHP) and its primary metabolite mono-2-ethylhexyl phthalate in cord blood of preterm and term newborns and found that concentrations were associated with earlier delivery (Latini et al. 2003). Another study measuring a panel of 9 phthalates in cord blood also observed an association with increased odds of preterm birth (Huang et al. 2014). Most other studies have utilized a single spot urine sample, in most instances collected from mothers in the third trimester of pregnancy. While some of these studies reported increased odds of preterm birth or decreased gestational age at delivery in association with several phthalate metabolites, particularly those of DEHP (Meeker et al. 2009; Whyatt et al. 2009; Latini et al. 2003), others reported null (Suzuki et al. 2010) or even protective effects (Adibi et al. 2009; Wolff et al. 2008). A more recent study examined associations with average levels of up to three phthalate metabolite measurements per subject over the course of pregnancy and observed significantly elevated odds of preterm birth, and even stronger associations when spontaneous preterm births, defined as deliveries following spontaneous preterm labor and/or preterm premature rupture of the membranes, were examined alone (Ferguson et al. 2014b). When associations were examined with phthalates measured at each individual time point during pregnancy, relationships for spontaneous preterm birth were stronger when concentrations were measured later in pregnancy (Ferguson et al. 2014a). Only one study has examined the relationship between BPA exposure and preterm birth. Concentrations in 3rd trimester urine samples were associated with a suggestive increase in odds of preterm birth, and the relationship was stronger when the case definition was restricted to deliveries at 36 weeks or less (Cantonwine et al. 2010b). This relationship remained after adjustment for urinary phthalate metabolite concentrations.

9.3 Metals and Metalloids

Some of the most convincing evidence for a relationship between an environmental exposure and preterm birth comes from the literature on lead. Early findings, primarily from smaller case-control studies, are strongly indicative of a relationship between prenatal lead exposure and preterm birth, and suggest a dose-dependent effect (Andrews et al. 1994). Most of these studies also examined effects of high levels of exposure. More recent research on highly exposed populations has demonstrated associations with preterm birth as well (Torres-Sánchez et al. 1999; Jelliffe-Pawlowski et al. 2006). Following the removal of lead from gasoline the general population in the US has experienced much lower levels of exposure (Pirkle et al. 1994), thus heightening the interest in studying levels at lower concentrations (e.g., blood levels <10 μg/dL). Studies examining lower levels of lead during pregnancy have been less consistent in establishing a relationship with preterm birth. No associations were observed in studies in populations with median maternal blood lead levels of approximately 1–2 μg/dL (Sowers et al. 2002; Zhu et al. 2010), while three other studies measuring levels in 1st trimester maternal blood (Vigeh et al. 2011), mid pregnancy red blood cells (Taylor et al. 2014), and placenta at delivery (Falcon et al. 2003), also with low exposure levels, did detect statistically significant associations with preterm birth.

Some recent studies of preterm birth in association with lower levels of lead exposure illustrate that to detect more subtle effects additional attention must be paid to timing of exposure and specific characteristics of the pregnancy. One study in Mexico City showed significant associations between lead levels measured in the 2nd trimester, but no associations with levels measured in the 1st or 3rd trimesters (Cantonwine et al. 2010a). Another study in eastern Massachusetts found that low maternal blood lead levels during pregnancy were strongly associated with preterm birth of male infants but not females (Perkins et al. 2014). Thus, a relationship may exist at lower levels of exposure but more care must be paid in statistical analysis in order to detect effects.

Other toxic metals have been studied much less intensively in relation to preterm birth. An early ecologic study suggested that mothers residing in an area with elevated cadmium contamination during pregnancy were not more likely to have a preterm birth compared to women in non-contaminated areas (Landgren 1996). A small case-control study measuring cadmium levels in maternal blood, cord blood, and placenta likewise observed no significant associations between cadmium levels and preterm birth (Zhang et al. 2004). Two studies, however, suggest that a relationship may exist. Fagher and colleagues observed elevated maternal blood cadmium concentrations in mothers who delivered preterm compared to term in a small case-control study (Fagher et al. 1993). Additionally, a study that assessed exposure using urine biomarkers demonstrated a significant association between maternal gestational cadmium exposure and preterm birth in an area of high contamination in Japan (Nishijo et al. 2002). Thus, for cadmium there may be a threshold effect or, as with lower levels of lead exposure, the relationship with preterm birth may be more subtle in magnitude requiring more careful exploration in epidemiologic studies with attention to fetal gender, etiology, and timing of exposure.

Arsenic exposure has been examined in association with prematurity in several studies, all of which have assigned exposure based on drinking water contamination levels in areas of relatively high exposure. Most of these studies showed no association between maternal drinking water levels and preterm birth (Mukherjee et al. 2005; Myers et al. 2010; Yang et al. 2003), although small study in Bangladesh observed a significant association (Ahmad et al. 2001).

Finally, some evidence exists for a relationship between mercury exposure during pregnancy and preterm birth. One study demonstrated an association with elevated hair mercury concentrations representative of exposure during gestation (Xue et al. 2007). A second ecologic study observed higher rates of preterm birth among African American mothers residing in areas with high levels of mercury contamination in fish in South Carolina (Burch et al. 2014). However, a third study utilizing maternal urinary mercury concentrations during pregnancy as well as umbilical cord blood mercury levels did not detect any significant associations with prematurity in a population from Brooklyn, New York (Bashore et al. 2014). Associations with mercury may be particularly difficult to assess, as one of the major exposure sources is through consumption of contaminated fish, which have other characteristics (e.g., high concentration of omega-3 fatty acids) which may be beneficial to pregnancy. Additional studies with attention to adjustment for this confounding may be necessary to clarify this relationship.

9.4 Air Pollutants

A large number of studies have examined the relationship between air pollutant exposures and preterm birth, and likewise there have been many systematic reviews summarizing this information (Glinianaia et al. 2004; Shah and Balkhair 2011; Sram et al. 2005; Stillerman et al. 2008; Stieb et al. 2012; Bonzini et al. 2010; Nieuwenhuijsen et al. 2013; Lai et al. 2013). Most of the literature in this area has focused on EPA criteria air pollutants, identified by the Clean Air Act as common US exposures with need for regulation. These include ozone, particulate matter (including $PM_{2.5}$ and PM_{10}), carbon monoxide, nitrogen oxides, and sulfur dioxide. (Lead is also an EPA criteria air pollutant, as exposure prior to elimination from gasoline was common via inhalation routes.) The primary results from these reviews and meta-analyses conclude that there is strong evidence for a relationship between sulfur dioxide and $PM_{2.5}$ and preterm birth, but other exposures have only weak to moderate evidence for an association. The most recent of these published by Stieb et al. in 2012 calculated pooled estimates of preterm birth risk in association with each of these exposures in 62 studies meeting their inclusion criteria. The authors found that associations with exposures assessed in the third trimester of pregnancy were most precise, but were generally small in magnitude (odds ratios 0.80–1.15). Significantly increased odds of preterm birth were found in association with carbon monoxide and PM_{10} exposures. The limitations from these and other criteria air pollutant studies have been described in the aforementioned reviews, and also in more depth in a report from the International Workshop on Air Pollution and Human Reproduction (Slama et al. 2008). Generally, recommendations for research in this area include utilization of prospective study designs, consideration of a uniform set of confounders across studies (importantly, including season of exposure and delivery and maternal diet), promotion of subject-specific exposure assessment methods (e.g., biomarkers, personal air monitoring), and investigation of mechanism of air pollutant action using molecular epidemiology (Ferguson et al. 2013; Slama et al. 2008).

There are a number of non-criteria pollutants with primarily inhalation exposure routes that have been measured more frequently in epidemiologic studies on the individual level. These exposures include environmental tobacco smoke (ETS) and polycyclic aromatic hydrocarbons (PAH), and will be examined in more detail here. Studies on ETS likely stemmed from the strong evidence that active maternal smoking during pregnancy increases odds of delivering preterm (Ion and Bernal 2015). Results from research on ETS effects are less conclusive. A recent meta-analysis demonstrated that maternal ETS exposure was associated with only a slight increase in risk of preterm birth, and associations were not significant in adjusted models (Salmasi et al. 2010). Likewise, another meta-analysis concluded no significant association between ETS and gestational age at delivery (Leonardi-Bee et al. 2008). However, systematic reviews by other groups conclude that there is an association (Stillerman et al. 2008; Wigle et al. 2008).

PAH exposure can result inhalation exposure in ambient (industrial combustion, automobile emissions, etc.) as well as indoor (from heating or cooking emissions) air contamination. Additionally individuals can be exposed through consumption of certain foods, particularly grilled and smoked meats (ATSDR 1995). Several studies have examined the relationship between PAH exposure and preterm birth using personal exposure methods, including personal air monitoring or measurement of biomarkers. In New York City, one study observed that increased total PAH exposure measured using personal air monitors in the third trimester of pregnancy were associated with significantly increased odds of preterm birth in African American, but not Dominican mothers (Choi et al. 2008). Studies using biomarkers suggest an association as well. One small case-control study measured PAH concentrations in placental tissue and observed significantly higher levels of individual PAH compounds in cases compared to controls (Singh et al. 2008). In another study where PAH were measured in cord blood in a highly contaminated region of China, increased levels were also significantly associated with adverse birth outcomes, including preterm birth, low birth weight, congenital malformations, and stillbirth (Guo et al. 2012). Looking at length of gestation more specifically, there was an association between PAH exposure and decreased gestational age at delivery as well. Studies using ambient air monitoring for assessment purposes also suggest that maternal PAH exposure is associated with a significantly increased risk of having a preterm birth (Padula et al. 2014; Vassilev et al. 2001; Wilhelm et al. 2011).

9.5 Conclusions and Research Needs

Overall, these data suggest an important role of environmental chemical exposures in the etiology of preterm birth. Of the organic pollutants studied, the strongest evidence for a relationship with prematurity exists for DDT. Notably, however, most of the evidence for this association comes from studies where exposure levels were relatively high. This may be relevant to countries that continue to utilize DDT as an insecticide but less so in developed nations. Additionally, although research is nascent, there is strongly suggestive evidence for a relationship between phthalate exposure during pregnancy and preterm birth. While lead exposure, particularly at high concentrations, is strongly associated with preterm birth, other metals have received very little attention in this area of research. This may be an important area to expand upon in future studies, particularly as biomarkers of metals exposures are available and maternal exposures at relatively low levels have been linked to other adverse reproductive and developmental outcomes. Finally, reviews and meta-analyses of air pollution effects show somewhat conflicting evidence for relationships between criteria air pollutants and ETS on preterm birth. As previously mentioned, this may be due in large part to use of ambient air monitors to assess effects and poor ability to examine exposure levels on an individual basis. PAH exposure during pregnancy, which has been studied using ambient and personal air monitors and biomarkers, shows strong evidence for an effect, despite the fact that research in this area, as with phthalates, has been limited to the last decade.

Absence of statistically significant associations in many of these studies may be due to a true lack of relationship with preterm birth, or, alternatively, to issues with study design. Preterm birth is a complex disease that is defined typically, especially in studies of environmental contributors, by a cutoff point of 37 weeks gestation, rather than by diagnostic markers or by etiology. This broad diagnosis may be problematic in several ways. Attention to 37 weeks as a cutoff point, which is typical because of the association that this dichotomy has with adverse effects on the infant, may be inappropriate. Recent evidence shows that early term (e.g., 37–39 weeks gestation) delivery is also associated with increased risk of neonatal mortality and various morbidities (Boyle et al. 2012). Additionally, while cutoffs by time in gestation (e.g., <37 weeks, or early preterm birth, <32 weeks) are commonly examined in relation to exposures, these do not separate preterm cases by etiology. Many causes contribute to early delivery (e.g., spontaneous preterm labor, preterm premature rupture of the membranes, intrauterine growth restriction, preeclampsia, etc.) and, though there may be some overlap, these conditions arise through different mechanisms (McElrath et al. 2008; Savitz 2008). While research in the field of obstetrics generally characterizes preterm birth into categories based on etiology (e.g., considering only spontaneous preterm births in a given analysis), few studies in the realm of environmental health sciences examine this distinction and thus may be observing diluted effects.

Additional study design issues may contribute to the inconsistency of some effects observed in this literature. First, many studies examining the relationships with preterm birth were small or were designed with other objectives in mind and consequently have low power to detect effects. Studies with case-control designs, particularly with attention to the aforementioned subtypes of preterm birth, may be more appropriate for this area of research. Second, most studies fail to examine multiple exposure windows during pregnancy. Depending on the type of preterm birth, exposures early or late in pregnancy may be more relevant to this outcome. For example, if preterm birth originates from impaired placentation, exposures early in pregnancy may be more relevant. Conversely, if preterm birth originates from spontaneous preterm labor, exposures late in pregnancy may be the most significant predictors. Finally, some of the exposure assessment methods employed in this research may be inadequate to detect effects. For example, BPA is a rapidly metabolized and excreted compound and concentrations in urine show low reliability over time (Fisher et al. 2014). Spot urine samples may not fully characterize a mother's exposure either at the time point or over the course of gestation. Thus, greater attention to exposure assessment methods must be paid in this area of research as well.

Future research in the field of environmental health should address these study design limitations, their characterization of preterm birth, but additionally the effects of combined exposures to multiple pollutants during pregnancy and mechanism of toxicant action. Mothers are exposed to a complex milieu of environmental toxicants during gestation (Woodruff et al. 2011), and it is highly plausible that some of these compounds may act through similar pathways or through mechanisms that exacerbate one another to cause prematurity. Thus, study of effects of combined exposures should be a research priority.

In addition to improving and expanding epidemiologic studies, translation of research findings to public health practice is essential, especially for those compounds with well-demonstrated associations with preterm birth. There are two routes to this end. First, with a better understanding of mechanism of toxicant, interventions may be developed to stop the pathway from exposure to prematurity. For example, if it is clear that a compound is leading to preterm birth by inducing maternal oxidative stress during pregnancy, supplementation with antioxidants could be a useful intervention. Unfortunately, understanding mechanisms by which exposures lead to preterm birth is difficult to assess. Animal models of preterm birth are limited, as rodents rarely deliver prematurely except with high doses of lipopolysaccharide injection or with gene knock-outs (Cha et al. 2013; Kaga et al. 1996). However, such models may be useful in some circumstances if the limitations are fully acknowledged (Elovitz and Mrinalini 2004) or if specific mechanisms known to be relevant to human prematurity, such as inflammation, are the outcomes of interest. Additionally, mechanisms of action can be examined in human studies using biomarkers of intermediate effect.

Nevertheless, attempting to stop pathological processes connecting exposure and preterm birth may not be practical, as decades of research devoted to understanding mechanisms of prematurity have been relatively ineffective in developing successful interventions (Chang et al. 2013). A second and potentially much more feasible strategy for ameliorating effects of environmental chemicals is preventing maternal exposures during pregnancy. For compounds that have clearly demonstrated links with preterm birth, development of clinician recommendations and studies examining effectiveness for reducing exposure will be an important next step.

In summary, there is much evidence to suggest a relationship between environmental contaminant exposures and preterm birth, although additional work is necessary to fully assess the effect of individual chemicals. Nevertheless, this is an important area of future research, as maternal exposure to many chemicals may be modifiable, as opposed to other factors simultaneous under investigation, such as genetic polymorphisms. Also, identifying exposures that may be prevented prior to or during pregnancy may be more effective than interventions that can be implemented at delivery, which have shown low potential for reduction of preterm prevalence thus far (Chang et al. 2013). Thus, further investigation of the role of environmental exposures in the etiology of prematurity is a promising line of research in the initiative to prevent this significant public health problem.

References

Adibi JJ, Hauser R, Williams PL, Whyatt RM, Calafat AM, Nelson H, Herrick R, Swan SH (2009) Maternal urinary metabolites of di-(2-ethylhexyl) phthalate in relation to the timing of labor in a US multicenter pregnancy cohort study. Am J Epidemiol 169(8):1015–1024

Ahmad SA, Sayed MH, Barua S, Khan MH, Faruquee MH, Jalil A, Hadi SA, Talukder HK (2001) Arsenic in drinking water and pregnancy outcomes. Environ Health Perspect 109(6):629–631

Andrews KW, Savitz DA, Hertz-Picciotto I (1994) Prenatal lead exposure in relation to gestational age and birth weight: a review of epidemiologic studies. Am J Ind Med 26(1):13–32

Apelberg BJ, Witter FR, Herbstman JB, Calafat AM, Halden RU, Needham LL, Goldman LR (2007) Cord serum concentrations of perfluorooctane sulfonate (PFOS) and perfluorooctanoate (PFOA) in relation to weight and size at birth. Environ Health Perspect 115(11):1670–1676

Arbuckle TE, Kubwabo C, Walker M, Davis K, Lalonde K, Kosarac I, Wen SW, Arnold DL (2012) Umbilical cord blood levels of perfluoroalkyl acids and polybrominated flame retardants. Int J Hyg Environ Health 216(2):184–194

Aschengrau A, Weinberg J, Rogers S, Gallagher L, Winter M, Vieira V, Webster T, Ozonoff D (2008) Prenatal exposure to tetrachloroethylene-contaminated drinking water and the risk of adverse birth outcomes. Environ Health Perspect 116(6):814–820

ATSDR (1995) Toxicological profile for polycyclic aromatic hydrocarbons. Agency for Toxic Substances and Disease Registry. http://www.atsdr.cdc.gov/toxprofiles/tp.asp?id=122&tid=25. Accessed 26 Sept 2014

Bashore CJ, Geer LA, He X, Puett R, Parsons PJ, Palmer CD, Steuerwald AJ, Abulafia O, Dalloul M, Sapkota A (2014) Maternal mercury exposure, season of conception and adverse birth outcomes in an urban immigrant community in Brooklyn, New York, U.S.A. Int J Environ Res Public Health 11(8):8414–8442

Basterrechea M, Lertxundi A, Iñiguez C, Mendez M, Murcia M, Mozo I, Goñi F, Grimalt J, Fernández M, Guxens M (2014) Prenatal exposure to hexachlorobenzene (HCB) and reproductive effects in a multicentre birth cohort in Spain. Sci Total Environ 466:770–776

Behrman RE, Butler AS (eds) (2007) Preterm birth: causes, consequences, and prevention. The National Academies Press, Washington, DC

Bergonzi R, De Palma G, Specchia C, Dinolfo M, Tomasi C, Frusca T, Apostoli P (2011) Persistent organochlorine compounds in fetal and maternal tissues: evaluation of their potential influence on several indicators of fetal growth and health. Sci Total Environ 409(15):2888–2893

Berkowitz GS, Lapinski RH, Wolff MS (1996) The role of DDE and polychlorinated biphenyl levels in preterm birth. Arch Environ Contam Toxicol 30(1):139–141

Bonzini M, Carugno M, Grillo P, Mensi C, Bertazzi PA, Pesatori AC (2010) Impact of ambient air pollution on birth outcomes: systematic review of the current evidences. Med Lav 101(5):341–363

Bove FJ, Fulcomer MC, Klotz JB, Esmart J, Dufficy EM, Savrin JE (1995) Public drinking water contamination and birth outcomes. Am J Epidemiol 141:850–862

Bove F, Shim Y, Zeitz P (2002) Drinking water contaminants and adverse pregnancy outcomes: a review. Environ Health Perspect 110(Suppl 1):61–74

Boyle EM, Poulsen G, Field DJ, Kurinczuk JJ, Wolke D, Alfirevic Z, Quigley MA (2012) Effects of gestational age at birth on health outcomes at 3 and 5 years of age: population based cohort study. BMJ 344:e896

Burch JB, Wagner Robb S, Puett R, Cai B, Wilkerson R, Karmaus W, Vena J, Svendsen E (2014) Mercury in fish and adverse reproductive outcomes: results from South Carolina. Int J Health Geogr 13:30

Cantonwine D, Hu H, Sanchez BN, Lamadrid-Figueroa H, Smith D, Ettinger AS, Mercado-Garcia A, Hernandez-Avila M, Wright RO, Tellez-Rojo MM (2010a) Critical windows of fetal lead exposure: adverse impacts on length of gestation and risk of premature delivery. J Occup Environ Med 52(11):1106–1111

Cantonwine D, Meeker JD, Hu H, Sanchez BN, Lamadrid-Figueroa H, Mercado-Garcia A, Fortenberry GZ, Calafat AM, Tellez-Rojo MM (2010b) Bisphenol a exposure in Mexico City and risk of prematurity: a pilot nested case control study. Environ Health 9:62

Cha J, Bartos A, Egashira M, Haraguchi H, Saito-Fujita T, Leishman E, Bradshaw H, Dey SK, Hirota Y (2013) Combinatory approaches prevent preterm birth profoundly exacerbated by gene-environment interactions. J Clin Invest 123(9):4063–4075

Chang HH, Larson J, Blencowe H, Spong CY, Howson CP, Cairns-Smith S, Lackritz EM, Lee SK, Mason E, Serazin AC, Walani S, Simpson JL, Lawn JE, Born Too Soon Preterm Prevention Analysis Group (2013) Preventing preterm births: analysis of trends and potential reductions with interventions in 39 countries with very high human development index. Lancet 381(9862):223–234

Chen MH, Ha EH, Wen TW, Su YN, Lien GW, Chen CY, Chen PC, Hsieh WS (2012) Perfluorinated compounds in umbilical cord blood and adverse birth outcomes. PLoS One 7(8):e42474

Choi H, Rauh V, Garfinkel R, Tu Y, Perera FP (2008) Prenatal exposure to airborne polycyclic aromatic hydrocarbons and risk of intrauterine growth restriction. Environ Health Perspect 116(5):658–665

Costet N, Garlantezec R, Monfort C, Rouget F, Gagniere B, Chevrier C, Cordier S (2012) Environmental and urinary markers of prenatal exposure to drinking water disinfection by-products, fetal growth, and duration of gestation in the PELAGIE birth cohort (Brittany, France, 2002–2006). Am J Epidemiol 175(4):263–275

Darrow LA, Stein CR, Steenland K (2013) Serum perfluorooctanoic acid and perfluorooctane sulfonate concentrations in relation to birth outcomes in the Mid-Ohio Valley, 2005–2010. Environ Health Perspect 121(10):1207

Dodds L, King W, Woolcott C, Pole J (1999) Trihalomethanes in public water supplies and adverse birth outcomes. Epidemiology 10(3):233–237

Elovitz MA, Mrinalini C (2004) Animal models of preterm birth. Trends Endocrinol Metab TEM 15(10):479–487

Eskenazi B, Mocarelli P, Warner M, Chee WY, Gerthoux PM, Samuels S, Needham LL, Patterson DG Jr (2003) Maternal serum dioxin levels and birth outcomes in women of Seveso, Italy. Environ Health Perspect 111(7):947–953

Eskenazi B, Harley K, Bradman A, Weltzien E, Jewell NP, Barr DB, Furlong CE, Holland NT (2004) Association of in utero organophosphate pesticide exposure and fetal growth and length of gestation in an agricultural population. Environ Health Perspect 112(10):1116–1124

Fagher U, Laudanski T, Schutz A, Sipowicz M, Akerlund M (1993) The relationship between cadmium and lead burdens and preterm labor. Int J Gynaecol Obstet 40(2):109–114

Falcon M, Vinas P, Luna A (2003) Placental lead and outcome of pregnancy. Toxicology 185(1–2):59–66

Farhang L, Weintraub JM, Petreas M, Eskenazi B, Bhatia R (2005) Association of DDT and DDE with birth weight and length of gestation in the Child Health and Development Studies, 1959–1967. Am J Epidemiol 162(8):717–725

Fei C, McLaughlin JK, Tarone RE, Olsen J (2007) Perfluorinated chemicals and fetal growth: a study within the Danish National Birth Cohort. Environ Health Perspect 115(11):1677–1682

Fenster L, Eskenazi B, Anderson M, Bradman A, Harley K, Hernandez H, Hubbard A, Barr DB (2006) Association of in utero organochlorine pesticide exposure and fetal growth and length of gestation in an agricultural population. Environ Health Perspect 114(4):597–602

Ferguson KK, O'Neill MS, Meeker JD (2013) Environmental contaminant exposures and preterm birth: a comprehensive review. J Toxicol Environ Health B Crit Rev 16(2):69–113

Ferguson KK, McElrath TF, Ko YA, Mukherjee B, Meeker JD (2014a) Variability in urinary phthalate metabolite levels across pregnancy and sensitive windows of exposure for the risk of preterm birth. Environ Int 70:118–124

Ferguson KK, McElrath TF, Meeker JD (2014b) Environmental phthalate exposure and preterm birth. JAMA Pediatr 168(1):61–67

Fisher M, Arbuckle TE, Mallick R, LeBlanc A, Hauser R, Feeley M, Koniecki D, Ramsay T, Provencher G, Berube R, Walker M (2014) Bisphenol a and phthalate metabolite urinary concentrations: daily and across pregnancy variability. J Expo Sci Environ Epidemiol 25(3):231–239

Forand SP, Lewis-Michl EL, Gomez MI (2011) Maternal exposure to tetrachloroethylene and trichloroethylene through soil vapor intrusion and adverse birth outcomes in New York State. Environ Health Perspect 120:616–621

Frisbee SJ, Brooks AP Jr, Maher A, Flensborg P, Arnold S, Fletcher T, Steenland K, Shankar A, Knox SS, Pollard C (2009) The C8 health project: design, methods, and participants. Environ Health Perspect 117(12):1873–1882

Gallagher MD, Nuckols JR, Stallones L, Savitz DA (1998) Exposure to trihalomethanes and adverse pregnancy outcomes. Epidemiology 9(5):484–489

Glinianaia SV, Rankin J, Bell R, Pless-Mulloli T, Howel D (2004) Particulate air pollution and fetal health: a systematic review of the epidemiologic evidence. Epidemiology 15(1):36–45

Govarts E, Nieuwenhuijsen M, Schoeters G, Ballester F, Bloemen K, de Boer M, Chevrier C, Eggesbo M, Guxens M, Kramer U, Legler J, Martinez D, Palkovicova L, Patelarou E, Ranft U, Rautio A, Petersen MS, Slama R, Stigum H, Toft G, Trnovec T, Vandentorren S, Weihe P, Kuperus NW, Wilhelm M, Wittsiepe J, Bonde JP, Obelix E (2012) Birth weight and prenatal exposure to polychlorinated biphenyls (PCBs) and dichlorodiphenyldichloroethylene (DDE): a meta-analysis within 12 European Birth Cohorts. Environ Health Perspect 120(2):162–170

Grellier J, Bennett J, Patelarou E, Smith RB, Toledano MB, Rushton L, Briggs DJ, Nieuwenhuijsen MJ (2010) Exposure to disinfection by-products, fetal growth, and prematurity: a systematic review and meta-analysis. Epidemiology 21(3):300–313

Guo Y, Huo X, Wu K, Liu J, Zhang Y, Xu X (2012) Carcinogenic polycyclic aromatic hydrocarbons in umbilical cord blood of human neonates from Guiyu, China. Sci Total Environ 427–428:35–40

Hamm MP, Cherry NM, Chan E, Martin JW, Burstyn I (2010) Maternal exposure to perfluorinated acids and fetal growth. J Expo Sci Environ Epidemiol 20(7):589–597

Hoffman CS, Mendola P, Savitz DA, Herring AH, Loomis D, Hartmann KE, Singer PC, Weinberg HS, Olshan AF (2008) Drinking water disinfection by-product exposure and duration of gestation. Epidemiology 19(5):738–746

Horton BJ, Luben TJ, Herring AH, Savitz DA, Singer PC, Weinberg HS, Hartmann KE (2011) The effect of water disinfection by-products on pregnancy outcomes in two southeastern US communities. J Occup Environ Med 53(10):1172–1178

Huang Y, Li J, Garcia JM, Lin H, Wang Y, Yan P, Wang L, Tan Y, Luo J, Qiu Z, Chen JA, Shu W (2014) Phthalate levels in cord blood are associated with preterm delivery and fetal growth parameters in Chinese women. PLoS One 9(2):e87430

Ion R, Bernal AL (2015) Smoking and Preterm birth. Reprod Sci 22(8):918–926

Jelliffe-Pawlowski LL, Miles SQ, Courtney JG, Materna B, Charlton V (2006) Effect of magnitude and timing of maternal pregnancy blood lead (Pb) levels on birth outcomes. J Perinatol 26(3):154–162

Kadhel P, Monfort C, Costet N, Rouget F, Thome JP, Multigner L, Cordier S (2014) Chlordecone exposure, length of gestation, and risk of preterm birth. Am J Epidemiol 179(5):536–544

Kaga N, Katsuki Y, Obata M, Shibutani Y (1996) Repeated administration of low-dose lipopolysaccharide induces preterm delivery in mice: a model for human preterm parturition and for assessment of the therapeutic ability of drugs against preterm delivery. Am J Obstet Gynecol 174(2):754–759

Kramer MD, Lynch CF, Isacson P, Hanson JW (1992) The association of waterborne chloroform with intrauterine growth retardation. Epidemiology 3(5):407–413

Lai H-K, Tsang H, Wong C-M (2013) Meta-analysis of adverse health effects due to air pollution in Chinese populations. BMC Public Health 13(1):360

Landgren O (1996) Environmental pollution and delivery outcome in southern Sweden: a study with central registries. Acta Paediatr 85(11):1361–1364

Latini G, De Felice C, Presta G, Del Vecchio A, Paris I, Ruggieri F, Mazzeo P (2003) In utero exposure to di-(2-ethylhexyl)phthalate and duration of human pregnancy. Environ Health Perspect 111(14):1783–1785

Le TN, Johansson A (2001) Impact of chemical warfare with agent orange on women's reproductive lives in Vietnam: a pilot study. Reprod Health Matters 9(18):156–164

Leonardi-Bee J, Smyth A, Britton J, Coleman T (2008) Environmental tobacco smoke and fetal health: systematic review and meta-analysis. Arch Dis Child Fetal Neonatal Ed 93(5):F351–F361

Lewis C, Suffet IH, Hoggatt K, Ritz B (2007) Estimated effects of disinfection by-products on preterm birth in a population served by a single water utility. Environ Health Perspect 115(2):290–295

Lin CM, Li CY, Mao IF (2006) Birth outcomes of infants born in areas with elevated ambient exposure to incinerator generated PCDD/Fs. Environ Int 32(5):624–629
Llop S, Ballester F, Estarlich M, Esplugues A, Rebagliato M, Iniguez C (2010) Preterm birth and exposure to air pollutants during pregnancy. Environ Res 110(8):778–785
Longnecker MP, Klebanoff MA, Zhou H, Brock JW (2001) Association between maternal serum concentration of the DDT metabolite DDE and preterm and small-for-gestational-age babies at birth. Lancet 358(9276):110–114
Longnecker MP, Klebanoff MA, Brock JW, Guo X (2005) Maternal levels of polychlorinated biphenyls in relation to preterm and small-for-gestational-age birth. Epidemiology 16(5):641–647
Maroziene L, Grazuleviciene R (2002) Maternal exposure to low-level air pollution and pregnancy outcomes: a population-based study. Environ Health 1(1):6
McElrath TF, Hecht JL, Dammann O, Boggess K, Onderdonk A, Markenson G, Harper M, Delpapa E, Allred EN, Leviton A, Investigators ES (2008) Pregnancy disorders that lead to delivery before the 28th week of gestation: an epidemiologic approach to classification. Am J Epidemiol 168(9):980–989
Meeker JD, Hu H, Cantonwine DE, Lamadrid-Figueroa H, Calafat AM, Ettinger AS, Hernandez-Avila M, Loch-Caruso R, Tellez-Rojo MM (2009) Urinary phthalate metabolites in relation to preterm birth in Mexico city. Environ Health Perspect 117(10):1587–1592
Mukherjee SC, Saha KC, Pati S, Dutta RN, Rahman MM, Sengupta MK, Ahamed S, Lodh D, Das B, Hossain MA, Nayak B, Mukherjee A, Chakraborti D, Dulta SK, Palit SK, Kaies I, Barua AK, Asad KA (2005) Murshidabad–one of the nine groundwater arsenic-affected districts of West Bengal, India. Part II: dermatological, neurological, and obstetric findings. Clin Toxicol 43(7):835–848
Mustafa M, Banerjee B, Ahmed RS, Tripathi A, Guleria K (2013) Gene–environment interaction in preterm delivery with special reference to organochlorine pesticides. Mol Hum Reprod 19(1):35–42
Myers SL, Lobdell DT, Liu Z, Xia Y, Ren H, Li Y, Kwok RK, Mumford JL, Mendola P (2010) Maternal drinking water arsenic exposure and perinatal outcomes in inner Mongolia, China. J Epidemiol Community Health 64(4):325–329
Nieuwenhuijsen MJ, Dadvand P, Grellier J, Martinez D, Vrijheid M (2013) Environmental risk factors of pregnancy outcomes: a summary of recent meta-analyses of epidemiological studies. Environ Health 12(1):6
Nishijo M, Nakagawa H, Honda R, Tanebe K, Saito S, Teranishi H, Tawara K (2002) Effects of maternal exposure to cadmium on pregnancy outcome and breast milk. Occup Environ Med 59(6):394–396; discussion 397
Nolan LA, Nolan JM, Shofer FS, Rodway NV, Emmett EA (2009) The relationship between birth weight, gestational age and perfluorooctanoic acid (PFOA)-contaminated public drinking water. Reprod Toxicol 27(3–4):231–238
Ochoa-Acuña H, Frankenberger J, Hahn L, Carbajo C (2009) Drinking-water herbicide exposure in Indiana and prevalence of small-for-gestational-age and preterm delivery. Environ Health Perspect 117(10):1619–1624
Padula AM, Noth EM, Hammond SK, Lurmann FW, Yang W, Tager IB, Shaw GM (2014) Exposure to airborne polycyclic aromatic hydrocarbons during pregnancy and risk of preterm birth. Environ Res 135:221–226
Patelarou E, Kargaki S, Stephanou EG, Nieuwenhuijsen M, Sourtzi P, Gracia E, Chatzi L, Koutis A, Kogevinas M (2011) Exposure to brominated trihalomethanes in drinking water and reproductive outcomes. Occup Environ Med 68(6):438–445
Pathak R, Ahmed RS, Tripathi AK, Guleria K, Sharma CS, Makhijani SD, Banerjee BD (2009) Maternal and cord blood levels of organochlorine pesticides: association with preterm labor. Clin Biochem 42(7–8):746–749
Perkins M, Wright RO, Amarasiriwardena CJ, Jayawardene I, Rifas-Shiman SL, Oken E (2014) Very low maternal lead level in pregnancy and birth outcomes in an Eastern Massachusetts population. Ann Epidemiol 24(12):915–919

Pirkle JL, Brody DJ, Gunter EW, Kramer RA, Paschal DC, Flegal KM, Matte TD (1994) The decline in blood lead levels in the United States. The National Health and Nutrition Examination Surveys (NHANES). JAMA 272(4):284–291

Procianoy RS, Schvartsman S (1981) Blood pesticide concentration in mothers and their newborn infants. Relation to prematurity. Acta Paediatr Scand 70(6):925–928

Revich B, Aksel E, Ushakova T, Ivanova I, Zhuchenko N, Klyuev N, Brodsky B, Sotskov Y (2001) Dioxin exposure and public health in Chapaevsk, Russia. Chemosphere 43(4–7):951–966

Ribas-Fitó N, Sala M, Cardo E, Mazon C, De Muga ME, Verdu A, Marco E, Grimalt JO, Sunyer J (2002) Association of hexachlorobenzene and other organochlorine compounds with anthropometric measures at birth. Pediatr Res 52(2):163–167

Rinsky JL, Hopenhayn C, Golla V, Browning S, Bush HM (2012) Atrazine exposure in public drinking water and preterm birth. Public Health Rep 127(1):72–80

Rivera-Núñez Z, Wright JM (2013) Association of brominated trihalomethane and haloacetic acid exposure with fetal growth and preterm delivery in Massachusetts. J Occup Environ Med 55(10):1125–1134

Salmasi G, Grady R, Jones J, McDonald SD (2010) Environmental tobacco smoke exposure and perinatal outcomes: a systematic review and meta-analyses. Acta Obstet Gynecol Scand 89(4):423–441

Sathyanarayana S, Basso O, Karr CJ, Lozano P, Alavanja M, Sandler DP, Hoppin JA (2010) Maternal pesticide use and birth weight in the agricultural health study. J Agromedicine 15(2):127–136

Savitz DA (2008) Invited commentary: disaggregating preterm birth to determine etiology. Am J Epidemiol 168(9):990–992; discussion 993–994

Savitz DA, Andrews KW, Pastore LM (1995) Drinking water and pregnancy outcome in central North Carolina: source, amount, and trihalomethane levels. Environ Health Perspect 103(6):592–596

Savitz DA, Stein CR, Bartell SM, Elston B, Gong J, Shin HM, Wellenius GA (2012a) Perfluorooctanoic acid exposure and pregnancy outcome in a highly exposed community. Epidemiology 23(3):386–392

Savitz DA, Stein CR, Elston B, Wellenius GA, Bartell SM, Shin HM, Vieira VM, Fletcher T (2012b) Relationship of perfluorooctanoic acid exposure to pregnancy outcome based on birth records in the Mid-Ohio Valley. Environ Health Perspect 120:1201–1207

Saxena MC, Siddiqui MK, Seth TD, Krishna Murti CR, Bhargava AK, Kutty D (1981) Organochlorine pesticides in specimens from women undergoing spontaneous abortion, premature of full-term delivery. J Anal Toxicol 5(1):6–9

Shah PS, Balkhair T (2011) Air pollution and birth outcomes: a systematic review. Environ Int 37(2):498–516

Singh VK, Singh J, Anand M, Kumar P, Patel DK, Krishna Reddy MM, Javed Siddiqui MK (2008) Comparison of polycyclic aromatic hydrocarbon levels in placental tissues of Indian women with full- and preterm deliveries. Int J Hyg Environ Health 211(5–6):639–647

Slama R, Darrow L, Parker J, Woodruff TJ, Strickland M, Nieuwenhuijsen M, Glinianaia S, Hoggatt KJ, Kannan S, Hurley F, Kalinka J, Sram R, Brauer M, Wilhelm M, Heinrich J, Ritz B (2008) Meeting report: atmospheric pollution and human reproduction. Environ Health Perspect 116(6):791–798

Sonnenfeld N, Hertz-Picciotto I, Kaye WE (2001) Tetrachloroethylene in drinking water and birth outcomes at the US Marine Corps Base at Camp Lejeune, North Carolina. Am J Epidemiol 154(10):902–908

Sowers M, Jannausch M, Scholl T, Li W, Kemp FW, Bogden JD (2002) Blood lead concentrations and pregnancy outcomes. Arch Environ Health 57(5):489–495

Sram RJ, Binkova B, Dejmek J, Bobak M (2005) Ambient air pollution and pregnancy outcomes: a review of the literature. Environ Health Perspect 113(4):375–382

Stein CR, Savitz DA, Dougan M (2009) Serum levels of perfluorooctanoic acid and perfluorooctane sulfonate and pregnancy outcome. Am J Epidemiol 170(7):837–846

Stieb DM, Chen L, Eshoul M, Judek S (2012) Ambient air pollution, birth weight and preterm birth: a systematic review and meta-analysis. Environ Res 117:100–111

Stillerman KP, Mattison DR, Giudice LC, Woodruff TJ (2008) Environmental exposures and adverse pregnancy outcomes: a review of the science. Reprod Sci 15(7):631–650
Suzuki Y, Niwa M, Yoshinaga J, Mizumoto Y, Serizawa S, Shiraishi H (2010) Prenatal exposure to phthalate esters and PAHs and birth outcomes. Environ Int 36(7):699–704
Taylor C, Golding J, Emond A (2014) Adverse effects of maternal lead levels on birth outcomes in the ALSPAC study: a prospective birth cohort study. BJOG 122(3):322–328
Torres-Arreola L, Berkowitz G, Torres-Sanchez L, Lopez-Cervantes M, Cebrian ME, Uribe M, Lopez-Carrillo L (2003) Preterm birth in relation to maternal organochlorine serum levels. Ann Epidemiol 13(3):158–162
Torres-Sánchez LE, Berkowitz G, Lopez-Carrillo L, Torres-Arreola L, Rios C, Lopez-Cervantes M (1999) Intrauterine lead exposure and preterm birth. Environ Res 81(4):297–301
Vassilev ZP, Robson MG, Klotz JB (2001) Outdoor exposure to airborne polycyclic organic matter and adverse reproductive outcomes: a pilot study. Am J Ind Med 40(3):255–262
Vigeh M, Yokoyama K, Seyedaghamiri Z, Shinohara A, Matsukawa T, Chiba M, Yunesian M (2011) Blood lead at currently acceptable levels may cause preterm labour. Occup Environ Med 68(3):231–234
Villanueva CM, Durand G, Coutte MB, Chevrier C, Cordier S (2005) Atrazine in municipal drinking water and risk of low birth weight, preterm delivery, and small-for-gestational-age status. Occup Environ Med 62(6):400–405
Villanueva CM, Gracia-Lavedan E, Ibarluzea J, Santa Marina L, Ballester F, Llop S, Tardon A, Fernandez MF, Freire C, Goni F, Basagana X, Kogevinas M, Grimalt JO, Sunyer J (2011) Exposure to trihalomethanes through different water uses and birth weight, small for gestational age, and preterm delivery in Spain. Environ Health Perspect 119(12):1824–1830
Wassermann M, Ron M, Bercovici B, Wassermann D, Cucos S, Pines A (1982) Premature delivery and organochlorine compounds: polychlorinated biphenyls and some organochlorine insecticides. Environ Res 28(1):106–112
Wesselink A, Warner M, Samuels S, Parigi A, Brambilla P, Mocarelli P, Eskenazi B (2014) Maternal dioxin exposure and pregnancy outcomes over 30 years of follow-up in Seveso. Environ Int 63:143–148
Whitworth KW, Haug LS, Baird DD, Becher G, Hoppin JA, Skjaerven R, Thomsen C, Eggesbo M, Travlos G, Wilson R, Cupul-Uicab LA, Brantsaeter AL, Longnecker MP (2012) Perfluorinated compounds in relation to birth weight in the Norwegian mother and child cohort study. Am J Epidemiol 175:1209–1216
Whyatt RM, Adibi JJ, Calafat AM, Camann DE, Rauh V, Bhat HK, Perera FP, Andrews H, Just AC, Hoepner L, Tang D, Hauser R (2009) Prenatal di(2-ethylhexyl)phthalate exposure and length of gestation among an inner-city cohort. Pediatrics 124(6):e1213–e1220
Wigle DT, Arbuckle TE, Turner MC, Berube A, Yang Q, Liu S, Krewski D (2008) Epidemiologic evidence of relationships between reproductive and child health outcomes and environmental chemical contaminants. J Toxicol Environ Health B 11(5–6):373–517
Wilhelm M, Ghosh JK, Su J, Cockburn M, Jerrett M, Ritz B (2011) Traffic-related air toxics and preterm birth: a population-based case-control study in Los Angeles County, California. Environ Health 10:89
Wojtyniak BJ, Rabczenko D, Jonsson BA, Zvezday V, Pedersen HS, Rylander L, Toft G, Ludwicki JK, Goralczyk K, Lesovaya A, Hagmar L, Bonde JP (2010) Association of maternal serum concentrations of 2,2′,4,4′5,5′-hexachlorobiphenyl (CB-153) and 1,1-dichloro-2,2-bis (p-chlorophenyl)-ethylene (p,p′-DDE) levels with birth weight, gestational age and preterm births in Inuit and European populations. Environ Health 9:56
Wolff MS, Engel SM, Berkowitz GS, Ye X, Silva MJ, Zhu C, Wetmur J, Calafat AM (2008) Prenatal phenol and phthalate exposures and birth outcomes. Environ Health Perspect 116(8):1092–1097
Wood SL, Jarrell JJ, Swaby C, Chan S (2007) Endocrine disruptors and spontaneous premature labor: a case control study. Environ Health 6:35

Woodruff TJ, Zota AR, Schwartz JM (2011) Environmental chemicals in pregnant women in the United States: NHANES 2003–2004. Environ Health Perspect 119(6):878–885

Wright JM, Schwartz J, Dockery DW (2003) Effect of trihalomethane exposure on fetal development. Occup Environ Med 60(3):173–180

Wright JM, Schwartz J, Dockery DW (2004) The effect of disinfection by-products and mutagenic activity on birth weight and gestational duration. Environ Health Perspect 112(8):920–925

Wu K, Xu X, Liu J, Guo Y, Li Y, Huo X (2010) Polybrominated diphenyl ethers in umbilical cord blood and relevant factors in neonates from Guiyu, China. Environ Sci Technol 44(2):813–819

Wu K, Xu X, Peng L, Liu J, Guo Y, Huo X (2012) Association between maternal exposure to perfluorooctanoic acid (PFOA) from electronic waste recycling and neonatal health outcomes. Environ Int 48:1–8

Xue F, Holzman C, Rahbar MH, Trosko K, Fischer L (2007) Maternal fish consumption, mercury levels, and risk of preterm delivery. Environ Health Perspect 115(1):42–47

Yang CY, Chang CC, Tsai SS, Chuang HY, Ho CK, Wu TN (2003) Arsenic in drinking water and adverse pregnancy outcome in an arseniasis-endemic area in northeastern Taiwan. Environ Res 91(1):29–34

Yang CY, Xiao ZP, Ho SC, Wu TN, Tsai SS (2007) Association between trihalomethane concentrations in drinking water and adverse pregnancy outcome in Taiwan. Environ Res 104(3):390–395

Zhang YL, Zhao YC, Wang JX, Zhu HD, Liu QF, Fan YG, Wang NF, Zhao JH, Liu HS, Ou-Yang L, Liu AP, Fan TQ (2004) Effect of environmental exposure to cadmium on pregnancy outcome and fetal growth: a study on healthy pregnant women in China. J Environ Sci Health A Tox Hazard Subst Environ Eng 39(9):2507–2515

Zhu M, Fitzgerald EF, Gelberg KH, Lin S, Druschel CM (2010) Maternal low-level lead exposure and fetal growth. Environ Health Perspect 118(10):1471–1475

Chapter 10
The Impact of Environmental Stressors on DNA Methylation, Neurobehavioral Development, and Chronic Physical Aggression: Prospects for Early Protective Interventions

Richard E. Tremblay, Linda Booij, Nadine Provençal, and Moshe Szyf

Abstract There is now convincing evidence from prospective and retrospective epidemiological studies that prenatal and early post-natal stressors have long term impacts on life span health and well-being. Unraveling the mechanisms by which early environmental stressors have an impact on DNA methylation and neurobehavioral development should provide the foundation for creating effective early protective interventions. We review the recent convergence of four research domains to explain the mechanisms leading to chronic physical aggression (behavior development, epigenetics, serotonin neurotransmission and immunology) and we discuss the next generation of studies that are needed to identify effective pre and post natal preventive interventions.

Keywords Environmental stressors • Aggression • Neurodevelopment • Epigenetics • Prevention

---

R.E. Tremblay, Ph.D. (✉)
School of Public Health, University College Dublin, Belfield, Dublin, Ireland

Université de Montréal, 3050, Édouard-Montpetit Blvd, Montréal, QC H3T 1J7, Canada
e-mail: richard.tremblay@ucd.ie; richard.ernest.tremblay@umontreal.ca

L. Booij, Ph.D.
Department of Psychology, Concordia University,
7141 Sherbrooke St. West, Montreal, QC H4B 1R6, Canada

CHU Sainte-Justine, 3175 Chemin de la Côte Sainte-Catherine, Montréal, QC, Canada
e-mail: linda.booij@concordia.ca; linda.booij@umontreal.ca

N. Provençal, Ph.D.
Max Planck Institute of Psychiatry, Kraepelinstr. 2-10, 80804 Munich, Germany
e-mail: nadine_provencal@psych.mpg.de

M. Szyf, Ph.D.
Department of Pharmacology and Therapeutics, McGill Faculty of Medicine,
McIntyre Medical Building, 3655 Prom. Sir-William-Osler, Room 1309/1310,
Montreal, QC H3G 1Y6, Canada
e-mail: moshe.szyf@mcgill.ca

C.L. Hughes, M.D. Waters (eds.), *Translational Toxicology*, Molecular and Integrative Toxicology, DOI 10.1007/978-3-319-27449-2_10

10.1 The Developmental Origins of Chronic Physical Aggression

The development of physical aggression throughout early childhood has only very recently been studied with large population samples. This lack of attention to the developmental origins of physical aggression during early childhood appears to be the result of a long-held belief that humans start to use physical aggression after early childhood as a result of exposure to violent behavior (e.g., Bandura 1977; Lefkowitz et al. 1977; Zimbardo 2007). Indeed the 2002 World Health Organization report on violence concluded: "The majority of young people who become violent are adolescent-limited offenders who, in fact, show little or no evidence of high levels of aggression or other problem behaviours during their childhood." (Krug et al. 2002). The putative developmental mechanisms leading to this phenomenon had been described a decade earlier in the 1993 report of the US Academy of Science Panel on Understanding Violent Behavior: "Modern Psychological perspectives emphasize that aggressive and violent behaviors are learned responses to frustration, that they can also be learned as instruments for achieving goals, and that the learning occurs by observing models of such behavior. Such models may be observed in the family, among peers, elsewhere in the neighborhood, through the mass media …" (Reiss and Roth 1993).

However, from an evolutionary perspective there are good reasons to doubt that humans have to learn to aggress. Physical aggression is a crucial component of human's behavioural heritage. Our ancestors needed to be skilled in the art of physical aggression to eat, to defend themselves against predators, to compete for mating, to protect their brood, and to acquire resources. However, like all other social animals, humans need to learn to use aggression sparingly because physically aggressive encounters can be fatal, and lack of self-control among social animals leads to social exclusion (Boivin et al. 2005; Suomi 2005). The relative absence of fear from being murdered is recent. Historical analyses of homicide rates indicate that physical violence has systematically and substantially decreased among European citizens over the past 500 years (Eisner 2003). Homicides in European cities decreased from 40 to 1 per 100,000 citizens per year. Looking further back from an evolutionary perspective, the estimated rate of "homicide" among our closest non-human relatives, chimpanzees, is 261 per 100,000 (Wrangham et al. 2006).

10.2 Development of Physical Aggression During Early Childhood

Studies of physical aggression during infancy have clearly shown that humans start to use physical aggression towards the end of the first year after birth when they have acquired the motor coordination to push, pull, hit, kick, etc. (Alink et al. 2006; Hay et al. 2011; Naerde et al. 2014; Tremblay et al. 1999). For example, analyses of physical aggression developmental trajectorie from 17 to 60 months with a population birth cohort (Côté et al. 2007) showed that all children increased the frequency of their physical aggression from 17 to 42 months of age and then decresed their frequency until 60 months. thatA third of the children were on a low trajectory of physical aggression, half were on a middle trajectory, while 17 % were on a high trajectory.

These analyses are based on prospective repeated assessments of physical aggressions reported by mothers over 4 years. The prospective studies of physical aggression during early childhood indicate that the peak frequency in use of physical aggression for most humans is somewhere between 2 and 4 years of age (Tremblay and Côté 2009; NICHD Early Child Care Research Network 2004)

The developmental trajectories of physical aggression after early childhood have now been studied in many different cultures. From these studies we can expect between 7 and 11 % of elementary school children on a CPA trajectory (Broidy et al. 2003; Campbell et al. 2010; Nagin and Tremblay 1999). That percentage tends to be higher for preschool children (Côté et al. 2007; Tremblay et al. 2004) and lower for adolescents (Brame et al. 2001). This decrease in CPA cases with age corresponds to the general decrease in frequency of physical aggression with age, after the peak in early childhood.

Most children use physical aggression during the preschool years, but most children also learn to use alternatives to physical aggression with age, and this applies to a number of chronic cases during early childhood and preadolescence (Nagin and Tremblay 1999). In fact there is good evidence that the learning process to gain control over physical aggression continues throughout adulthood. A longitudinal study of juvenile delinquents up to old age showed that the number of their violent offenses decreased with age (Sampson and Laub 2003; see also Sweeten et al. 2013).

Crime records from the middle ages to modern times suggest that this phenomenon is not new. The likelihood of committing a homicide and most other crimes has always decreased from late adolescence and early adulthood to old age (Eisner 2003; Quetelet 1984). Trajectories of physical aggression covering different age periods (early childhood to childhood, childhood to adolescence, adolescence to adulthood) also indicate that CPA very rarely onsets after early childhood (Barker et al. 2007; NICHD Early Child Care Research Network 2004; van Lier et al. 2009).

Longitudinal studies of physical aggression trajectories during childhood have been used to study how well the trajectories predict future outcomes such as school performance, social adjustment, mental health and violent behavior. The first longitudinal study to describe developmental trajectories of physical aggression from school entry to adolescence (Nagin and Tremblay 1999) reported that boys on a teacher-rated trajectory of frequent physical aggression from 6 to 15 years of age were at highest risk of self-reported violence as well as other forms of delinquency at 17 years of age, even after having controlled for hyperactivity and oppositional behavior. The chronically aggressive boys were also at highest risk of school dropout. A study which used 6 longitudinal cohorts from Canada, New Zealand and the US (Broidy et al. 2003) reached the same conclusion for male adolescent violent delinquency, but not for female adolescent violent delinquency. The authors attributed the sex difference in prediction to the fact that the prevalence of female adolescent violent delinquency was too low. However, a later analysis of one of the female samples (Fontaine et al. 2008) reported that elementary school girls who were on a chronic physical aggression trajectory combined with a chronic hyperactivity trajectory were more likely than others to report physical and psychological aggression towards intimate partners by age 21 years. They were also more likely to report early pregnancy, welfare assistance, nicotine use problems and low educational attainment. A more recent analysis of a population sample of males and females (Pingault et al. 2013) reported that the 9.5 % of children on a high physical aggression trajectory between 6 and 12 years, according to mother and teacher rating, represented 28.2 % of all those who had a criminal record by age 24 years. In addition, they represented 45.9 % of all recorded criminal charges and 57.4 % of the violence charges. Therefore, children on a high trajectory of physical aggression during elementary school are not only more likely to have a criminal record but also to have more criminal charges. There is evidence that the criminal outcomes of childhood physical aggression during adolescence and adulthood are preceded by a large range of negative social and academic outcomes by the end of elementary school for boys and girls (Campbell et al. 2010).

10.3 Early Risk Factors of Chronic Physical Aggression and Putative Mechanisms

Sex of the individual is one of the most important risk factor for chronic physical aggression. When children start using physical aggression at the end of the first year after birth there are no significant difference in frequency of physical aggressions between boys and girls (Hay et al. 2011), however the differences appear soon after and increase until adolescence (e.g., Baillargeon et al. 2007; Côté 2007). Males between 10 and 15 years of age are close to 20 times (OR = 18.84) more at risk than females of being on a chronic physical aggression trajectory (van Lier et al. 2009).

Twin studies have become important tools to understand the contributions of environmental and genetic factors in the development of human characteristics, including aggression. However, to date there appears to be only one longitudinal study that used a large sample of twins from infancy onwards to study the contributions of genetic and environmental factors in the development of physical aggression. The study reported that 19 months after birth 58 % of the variance in frequency of physical aggression rated by mothers could be attributed to genetic contributions and 42 % to common environmental contributions (Dionne et al. 2003), furthermore, a large part of the variance in frequency of aggression change over time was attributed to genetic factors (Lacourse et al. 2014).

Although these results suggest a substantial contribution to frequency of physical aggression by genetic factors from infancy to school entry, this development does not happen in a vacuum, environmental factors are also very important. The developmental trajectories of physical aggression described above indicate that the environmental conditions are essential to learn alternatives to physical aggression during early childhood. Studies of physical aggression trajectories during early childhood with singletons have identified the following types of environmental risk factors: (a) Maternal characteristics, including life style and mental health, (b) family characteristics, (c) maternal parenting, (d) child characteristics (Campbell et al. 2010; Côté et al. 2006; Hay et al. 2011; NICHD Early Child Care Research Network 2004; Tremblay et al. 2004).

Maternal and family characteristics are key for planning preventive interventions because they can be used to identify pregnant women at risk of having children on a CPA trajectory (e.g., Olds et al. 1998). The maternal characteristics associated to chronic physical aggression include mothers' young age at birth of their child, mothers' smoking during pregnancy, mothers' antisocial behaviour during adolescence, mothers' depression, and mothers' low level of education. Family characteristics included low income, family dysfunction and the presence of siblings. High risk maternal parenting behaviour includes mother's hostile-coercive-harsh parenting and lack of sensitivity.

It is important to note that these studies of environmental risk factors were not done in the context of genetically informative designs (e.g., twin studies or sibling studies), hence we do not know to what extent the significant environmental risk factors are correlated or interact with genetic factors (Plomin 1994; Szyf et al. 2009). Nonetheless, the environmental risk factors identified by these studies can be used to identify at risk groups for preventive experiments. Such experiments are useful to test the effectiveness of the interventions as well as test causal hypotheses (Schwartz et al. 1981; Tremblay 2003). Maternal and family characteristics are especially key for early preventive interventions because they can be used to identify at risk pregnant women (e.g., Olds et al. 1998).

10.4 Physical Aggression, Brain Development, and the Role of Serotonin

Although the neurobiological substrates of physical aggression are many, one of the most consistent biological correlates of physical aggression is altered serotonin (5-HT) neurotransmission (Siever 2008). Following the early observation that low levels of the 5-HT metabolite 5-hydroxyindoleacetic acid (5-HIAA) in the cerebrospinal fluid (CSF) is associated with impulsive aggression (Brown et al. 1979), a number of studies have further confirmed this association, using other indices of 5-HT neurotransmission, such as neuro-endocrine challenge methods (e.g., Coccaro et al. 1989) and tryptophan depletion procedures (Cleare and Bond 1995; Moeller et al. 1996).

Neuro-imaging studies provided evidence that aggression problems might result from top-down suppression deficits in (cortical regions (Orbitofrontal (OBFC), anterior cingulate) in combination with excessive *bottom up* signalling by the amygdala (Booij et al. 2010; Davidson et al. 2000; Siever 2008). All of these brain regions are densily innervated with 5-HT neurons and 5-HT receptors (Azmitia and Gannon 1986; O'Rourke and Fudge 2006; Steinbusch 1981). As 5-HT in the Prefrontal cortex (PFC) facilitates inhibition of the amygdala, impaired 5-HT function in these regions could result in an increased amygdala response to social threat, which could trigger an aggressive response (Davidson et al. 2000; Siever 2008). Consistent with this, neuro-imaging studies in individuals with high levels of aggression found reduced metabolic response in the frontal-limbic regions in response to administration of 5-HT agonists (New et al. 2002; Siever et al. 1999). In addition, Positron Emission Tomography (PET) studies in combination with 5-HT radioligands showed that individuals with high levels of impulsive aggressive behaviors indicated lower 5-HT transporter (SLC6A4) density (Frankle et al. 2005) and lower 5-HT synthesis capacity, in the same regions (Leyton et al. 2001; Leyton et al. 2006), relative to controls. Taken these findings together suggests that altered 5-HT function in the frontal-limbic circuits are associated with aggression problems.

Of relevance is our study, investigating 5-HT synthesis in a community sample of healthy 27 year adult males who were rated regularly between age 6 and 15 by their teachers on measures of physical aggression as well as on a variety of psychosocial and psychological factors. Subjective measures of aggression confirmed that individuals in the high physical aggression developmental trajectory exhibited more aggressive behaviors during childhood and adolescence, but not in adulthood. However, males with high physical aggression in childhood compared to those with low physical aggression during childhood and adolescence had lower brain 5-HT synthesis bilaterally and specifically in the OBFC (Booij et al. 2010). Though 5-HT function was not measured in childhood and adolescence in this sample, the results are in line with studies in nonhuman primates, showing that 5-HT metabolism remains stable for at least a decade (Howell et al. 2007). Hence, diminished 5-HT neurotransmission in the frontal-limbic circuitry may be a persisting trait rather than state characteristic. Notably, our cohort members did not differ in terms of current aggression levels, impulsivity, mood, working memory or psychosocial functioning; Although other biological factors cannot be ruled out, taken these findings together suggest that 5-HT alterations may be an important etiological predisposing biological factor in aggression. However, whether actual aggressive behaviors are expressed, is likely to depend on other biological factors, experiences and environmental support during development. In line with this so-called "diathesis-stress" model are studies of gene-environment interactions, showing associations between specific genotypes of 5-HT genes (i.e., monoamine oxidase A enzyme (MAOA) gene, SLC6A4 gene) and aggression, but only in the context of early life adversity (Aslund et al. 2011; Caspi et al. 2002; Reif et al. 2007; Suomi 2006).

An important question in this context is *how* 5-HT alterations develop, and how these alterations could lead to physical aggression. In addition to being a neurotransmitter, animal and human studies have shown that 5-HT is important for the development and maturation of the human brain (see Booij et al. 2014, review). For instance, in humans, the first 5-HT neurons in the raphe neurons appear already by 5 weeks of gestation (Booij et al. 2014; Whitaker-Azmitia 2001, review). Many aspects of 5-HT system development, including innervation, fiber density and synthesis appear to be largely matured in early childhood. However, 5-HT receptors, enzymes and proteins appear to have a unique developmental pattern, with some them fluctuating up to adulthood, while other 5-HT proteins and enzymes stabilizing already early in life (Booij et al. 2014). The expression of specific receptors, enzymes and generally relevant proteins appears in turn to modulate specific brain developmental processes (Gaspar et al. 2003).

Of particular interest in the context of physical aggression is the gene coding MAOA; an enzyme responsible for breaking down monoamines, and thus an important regulator of the levels of monoamines throughout the brain, including 5-HT. A deficiency of this enzyme has been associated with aggression in animal models (Cases et al. 1995) and humans (Brunner et al. 1993). In terms of development, research has shown that activity of the MAOA enzyme in the human frontal cortex increases up to birth, then diminishes during the first year of life, and stabilizes around age 2 (Kornhuber et al. 1989); a pattern quite consistent with what has been observed in rodents (see Booij et al. 2014). Notably, this developmental pattern (early peak, stabilization in early life) seems to precede, and is quite similar to the developmental pattern of physical aggression (see above).

Although the specific role of SLC6A4 in aggression is less clear, with studies both positive and negative associations, it has now widely been demonstrated that SLC6A4 in particular plays a major role in brain development (Booij et al. 2013, 2014; Gaspar et al. 2003; Lesch and Mössner 2006). Studies have shown that SLC6A4 is one of the first 5-HT molecules to develop during pregnancy (Booij et al. 2014, review). When it is stabilized in development depends on brain region. For instance, studies in rodents have shown that SLC6A4 density in the forebrain increases up to adulthood, while densities remain steady after P25 (late adolescence) in the striatum, midbrain and brain stem (Booij et al. 2014).

Hence, both MAOA and SLC6A4 are important regulators of 5-HT neurotransmission, and also have been implicated in physical aggression. Although it is clear that no specific 5-HT molecule is unequivocally associated with any specific behaviour, it would be of interest for longitudinal studies to further examine developmental changes in 5-HT expression levels over time, and how these changes in 5-HT molecules are associated with developmental changes in aggression. Given the observed sex differences in chronic aggression (see above), as well as the observed sex differences in (adult) serotonin functioning (Nishizawa et al. 1997), it would further be of interest to study how sex differences in physical aggression would relate to sex differences in 5-HT development.

Given the role of 5-HT in brain development and in aggression, it is reasonable to assume that a disruption in the very early stage in the system could interfere with normal brain development, with consequences for risk of physical aggression. Most clear evidence supporting this notion comes from genetic knockout studies. For example, SLC6A4 KO mice show functional and structural cortical brain alterations (e.g., Ansorge et al. 2004). In addition, in a sample of transgenetic mice, deletion in the gene encoding MAOA led to increased levels of aggressive behaviors and several changes in brain structure, some of them persisting in adulthood (Cases et al. 1995; Shih et al. 1999). The increased aggression levels in MAOA KO mice can be reversed by SSRIs, suggesting that the behavioural effects possibly underlie developmental alterations of the SLC6A4 (Godar et al. 2014), thereby further supporting the important role of SLC6A4 in brain development.

In humans, the role of 5-HT in brain development is supported by molecular imaging studies (see Booij et al. 2014, review). For instance, a number of studies now have shown that carriers of so called "risk" alleles (i.e., the s allele of the serotonin transporter, the low expression variant of the MAOA gene), show increased amygdala activity in response to emotional stimuli and diminished activation in the PFC, including the OFBC and the cingulate; limbic volume reductions; and altered connectivity between the prefrontal cortex and amygdala (e.g., Buckholtz et al. 2008; Canli et al. 2006; Heinz et al. 2005; Meyer-Lindenberg et al. 2006; Munafo et al. 2008).

Psychosocial stressors in early life can also alter 5-HT function. For instance, early maternal separation in rodents alters 5-HT neurotransmission in the frontal cortex (Jezierski et al. 2006; Matthews et al. 2001). In primates, rhesus macaques that have previously experienced high rates of rejection from their mothers after birth have lower levels of the 5-HT metabolite 5-HIAA in adulthood (Higley et al. 1996). The influence of early psychosocial adversity on the development of the 5-HT system is further supported by in vivo PET studies of the SLC6A4, showing that adolescent monkeys who were raised by their peers had, relative to mother-raised monkeys, decreased SLC6A4 and 5-HT_{1A} binding in the raphe and frontal-limbic regions (Ichise et al. 2006; Stevens et al. 2009). In vivo measures of reduced SLC6A4 density have also been associated with higher levels of impulsive aggression in humans (Frankle et al. 2005). Evidence for 5-HT alterations as a result of adversity also exists in humans. For instance, adults in low socioeconomic environments or who reported high rates of parental neglect in childhood showed blunted prolactin response to fenfluramine challenge (Manuck et al. 2005), as an indicator of impaired 5-HT function, and lower 5-HT metabolite levels in the cerebrospinal fluid (Roy et al. 2002). In a study of males followed since childhood, we observed that lower 5-HT synthesis in the frontal-limbic circuitry (OBFC and hippocampus) was associated with maternal smoking in pregnancy, lower birth weight and birth complications. All of these three in utero and early postnatal adversities have previously been shown to be predictive of physical aggression (Arseneault et al. 2002; Huijbregts et al. 2008). Hence it is tempting to speculate that 5-HT alterations in these regions may be an underlying mechanism of how such early adversities can translate into risk for chronic physical aggression.

10.5 Epigenetic Effects of Early Stressors and Pathways to Chronic Physical Aggression

10.5.1 The Serotonergic System

An emerging number of studies provide evidence for the role of epigenetic processes as physiological mechanism through which genetic and environmental factors may interact, and consequently brain and behavioural alterations may develop (Booij et al. 2014). Following the initial observation of how early stress affect methylation in rodents and in the post-mortem human brain (McGowan et al. 2009; Weaver et al. 2004), a number of studies have shown that early stress was associated with altered levels of methylation in 5-HT genes like MAOA and SLC6A4 in the living human brain, using peripheral cells (see Booij et al. 2013, review). For example, Beach et al. (2011) identified an association between women victims of child sexual abuse and overall hypermethylation of the SLC6A4 (5HTT) promoter region (Beach et al. 2011). In addition, they observed a significant association between DNA methylation in SLC6A4 promoter with symptoms of ASPD in women that was also partly mediated by 5-HTTLPR polymorphism (Beach et al. 2011). Thus, child sexual abuse may create long-lasting epigenetic changes in the SLC6A4 gene promoter and lead to female antisocial behavior. With regard to physical aggression, we observed that SLC6A4 promoter methylation in white blood cells of adults was associated with higher childhood physical aggression and with lower in vivo measures of brain 5-HT synthesis in adult males (Wang et al. 2012). Although causality cannot be established in any of these studies, it can be hypothesized that early stress may lead to altered methylation levels of the SLC6A4 promoter, with consequences here for brain chemistry and behaviour. This was further supported by a very recent report that greater SLC6A4 methylation assessed in whole blood DNA, was associated with lower hippocampal volume, a brain region with rich 5-HT innervation and important for stress regulation (Booij et al. 2015). Similarly, we recently observed greater methylation in the MAOA promoter in individuals with Antisocial Personality disorders relative to controls (Checknita et al. 2015); while another research group reported a correlation between MAOA methylation levels and in vivo PET measures of MAOA (Shumay et al. 2012).

Hence, taken together these findings suggest that, in addition to genetic factors, the in utero or early postnatal environment influence 5-HT homeostasis. A disruption in 5-HT homeostasis, could alter the trophic properties of 5-HT at play during the critical periods of brain development, in many different ways, affecting the brain at various levels, through various mechanisms. It may alter 5-HT innervation, expression levels of certain proteins, enzymes, receptor functioning, synthesis, all of which may interact. This may in turn predispose the individual to structural and functional alterations in brain circuits such as the frontal cortex and the amygdala, previously identified as key regions in the modulation of aggression; and increase vulnerability to enduring patterns of aggressive behaviors, in the context of further adversity.

It is clear that 5-HT neurotransmission and associated proteins are also influenced by multiple factors other than the trophic factors considered here; those include, among others, age and sex, other 5-HT molecules, and, importantly, the impact of other biological systems. The specific consequences on behaviour on disrupting 5-HT neurotransmission, if any, ultimately depend on the complex interplay between these various factors.

10.5.2 The Immune System

Using peripheral blood cells DNA from monocytes and T cells, we recently reported an association between childhood CPA in men and differential DNA methylation in regulatory regions of cytokine and transcription factor genes (Provencal et al. 2013b). Moreover, these cytokines were also shown to be repressed in men with CPA compared to men on normative developmental trajectory of aggressive behaviour (Provençal et al. 2013a). Interestingly, one of these downregulated cytokine in men with CPA, Interleukin-6 (IL-6), was previously shown to be involved in aggressive behaviour in mice since it's knockout (IL-6 (–/–)) resulted in increased aggressive behaviour phenotype in these mice (Alleva et al. 1998). In humans, a growing body of research also suggests that inflammatory cytokines might have systemic effects in addition to their traditional roles in the immune response. Indeed, recent studies have shown that cytokines are associated with various behavioural disorders such as anxiety, depression, suicide, childhood mood disorder and PTSD (Dowlati et al. 2010; Groer and Morgan 2007; Hoge et al. 2009; Janelidze et al. 2010; Koo and Duman 2008; Smith et al. 2011; von Kanel et al. 2007) as well as aggression (Marsland et al. 2008; Suarez et al. 2002). Moreover, early life stress such as social isolation and prenatal anxiety has been found to alter the immune system (Barreau et al. 2004; Danese et al. 2007; Powell et al. 2013; Sloan et al. 2007). Previous studies from our group and others that have examined associations of genome-wide DNA methylation profiles with adverse exposures have pointed to immune pathways, both in the brain and in the periphery. Maternal deprivation in rhesus macaques (Provençal et al. 2012), early life socioeconomic position (Borgho l et al. 2012; Kieseppa et al. 2004), child abuse (Suderman et al. 2014) and PTSD (Mehta et al. 2013; Smith et al. 2011; Uddin et al. 2010), all found DNA methylation associations in promoters regulating genes in the immune response pathways. Together these results suggest that immunoregulators are responsive to early life stress and might be involved in aggression where DNA methylation could be one of the mechanisms that mediate this association. These immonuregulators could influence brain circuitry and behaviour through a wide variety of other biological systems, including serotonin and the HPA-axis. For instance, with regard to 5-HT, cytokines have been shown to influence 5-HT synthesis and transporter expression (Capuron and Miller 2011). The interaction between cytokine and 5-HT has been hypothesized as a mechanism for the etiology of depression (Myint and Kim 2003). It could be speculated that it may also hold for chronic aggression.

Importantly, effects of immuneregulators also occur through their action on the hypothalamic-pituitary-adrenal (HPA) axis previously shown to play a role in aggression.

The HPA axis is considered to be the most important system in stress regulation. Upon its activation corticotrophin releasing hormone (CRH) and vasopressin (AVP) are released from the hypothalamus and stimulate adrenocorticotropic hormone (ACTH) release from the pituitary into the blood. This results in cortisol secretion from the adrenal cortex. The cellular actions of the cortisol are mediated by its binding to the glucocorticoid receptor (GR) and the mineralocorticoid receptor that act as transcription factors and are expressed in most tissues. Once activated, GR and MR translocate into the nucleus where they can exert their function as transcription factors regulating adaptive responses to stress, including metabolism, immune activation and cell proliferation and differentiation. At multiple levels of the HPA axis, the activation of the GR will initiate a negative feedback loop that is responsible for terminating the stress response and therefore the secretion of cortisol. A decrease in GR expression/activation is generally associated with an increase in the response to stress due to an impaired negative feedback. In addition, there is strong evidence of a crosstalk between the immune system and the brain through the HPA axis. It is well known that increases in glucocorticoid levels in response to activation of the HPA axis results in a profound silencing of gene expression of pro-inflammatory proteins and cytokines. Also, it was shown that early life social class can affect the expression of genes bearing response elements to transcription factors regulating immune genes such as CREB/ATF, NFKB and GR. Thus, the effects observed on the immune system in relation to aggression could be due to a dysregulated HPA axis and therefore alterations in the cortisol release and actions.

In general, correlations have been found between reduced cortisol levels and increased aggression levels in adolescents and young men (Loney et al. 2006; Popma et al. 2007; Shirtcliff et al. 2005). In contrast, boys with conduct disorder (CD) from the same cohort as the one studied here had elevated salivary cortisol levels compared to those without CD. Moreover, boys with an aggressive form of CD had even higher cortisol levels. A strong correlation was also observed between reactive aggression and elevated cortisol (van Bokhoven et al. 2005). It is important to note that maltreatment in childhood also leads to low basal cortisol in association with conduct and aggressive disorders (Tarullo and Gunnar 2006). Together, these results indicate that both hyper- and hypoactive HPA axis might explain children's aggression, where hyperactivity may be involved in reactive aggression and hypoactivity may be involved in proactive aggression.

Prenatal stress exposure to high levels of glucocorticoids, were also shown to promote aggressive behaviour (Glover 2011). In chicken, *in ovo* injection of high dose of cortisol during embryonic development was shown to increase aggressive behaviours through alteration of the HPA axis and serotonin system (Ahmed et al. 2014). Reduced hypothalamic levels of GR protein and CRH mRNA levels accompanied by increase in DNA methylation in the GR and CRH gene promoters were observed in the chicks. Here, prenatal cortisol exposure caused epigenetic reprogramming of critical genes that in turn, altered the HPA axis and enhanced aggressive behaviour.

In rats, exposure to early adverse life experiences was shown to induce high and sustained rates of increased aggressive behaviour in adulthood. In their model, Marquez et al. found that peripubertal exposure to stress (fear-induction experiences) induces pathological aggression in male rats (Marquez et al. 2013). These peripubertal stressed rat also exhibited hyperactivity in the amygdala and hypoactivity in the medial orbitofrontal cortex after exposure to social challenge. Interestingly, these neuroimaging brain activity data were accompanied by a sustained increase in *MAOA* expression in the PFC of stressed animals that is likely to be explained by epigenetic modulation. Indeed, they found an increase in histone 3, but not histone 4, acetylation levels in the promoter of the *MAOA* gene. Histones acetylation are known to promote gene transcription by increasing the accessibility to active transcription regulators binding (Kuo and Allis 1998) and especially histone 3 acetylation have been shown to play a role in regulating long-term changes in gene expression (Tsankova et al. 2007). Together these finding with the previous work on *MAOA* gene support the hypothesis that either MAOA hypo- or hyperactivity contribute to pathological aggression (Nelson and Trainor 2007) possibly through epigenetic programming.

The epigenetic association studies presented above mainly focused on candidate genes that either were suspected or previously shown to be involved in aggression. Another approach that was successfully used in the past to find significant association with diseases and behaviour is an unbiased genome-wide approach (Mehta et al. 2013). Using this approach to analyze men T cell epigenomes, we previously identified significant associations of DNA methylation levels with childhood CPA in distinct gene promoters (n=448) involved in biological pathways related to behaviour and immune function, and their colocalisation in genomic clusters (Provençal et al. 2014). Interestingly, some of these differentially methylated genes, such as the *AVP receptor* 1A (*AVPR1A*), *SLC6A3* (dopamine transporter) and *serotonin receptor 1D* (*HTR1D*), were previously associated with aggressive phenotype in humans (Guo et al. 2007; Vage et al. 2010; Vaughn et al. 2009) and animals (Ferris et al. 2006; Hammock et al. 2005). As anticipated from our previous study in cytokine genes (Provencal et al. 2013b), the inflammatory and immune biological function with specific signalling pathway such as cytokines signalling between immune cells, IL-6 and IL-10 signalling were found enriched with genes differentially methylated in men with CPA. Specific cytokines and receptors involved in these pathways were previously shown to be involved in aggression and human mood disorders such as IL1R1 and IL1RN (Pesce et al. 2011). Together, these findings suggest a well-defined, genome-wide epigenetic pattern associated with chronic physical aggression in men.

Not only aggressive behaviour in men but also in women were studied for DNA methylation associations. Indeed, in another study performed by our group, we observed similar DNA methylation signatures associated with childhood CPA in women (N=430 promoters) as seen in men where 31 gene promoters were significantly associated in both sexes (Guillemin et al. 2014). Interestingly, a significant portion of this overlap is due to identical genomic sites being differentially methylated in a sex-independent fashion. The almost perfect overlap between functional categories represented by both men and women signatures provides further evidence for these signatures to be, at least in part, associated with aggression rather than confounding factors. Here also, specific genes involved in serotonin metabolism and HPA axis regulation, previously shown to be involved in aggression, were found differentially methylated in women with childhood CPA. These HPA regulating genes (*NR3C1* and *CRHBP*) were only found differentially methylated in women with CPA. This may be explained in part by the fact that the HPA axis negative feedback control have been shown to be more sensitive in females than in males (Keck et al. 2002). These sex-specific and sex-independent components of the epigenetic signature are consistent with the existence of sex differences and similarities observed in human physical aggression.

10.6 Prenatal and Early Postnatal Prevention of CPA

The observation of the long-term influence of adverse factors occurring during the prenatal or early postnatal period on development of biological systems (brain, immune, HPA-axis), is consistent with the notion that early interventions in at risk populations should start as close as possible to conception, and continue supporting the family and the child as long as needed (Petitclerc and Tremblay 2009; Tremblay 2010). Preventive interventions during pregnancy for at risk pregnant women and family members are in fact corrective interventions for women who have a long history of social and mental adjustment problems. Many of them started to have children during adolescence and choose mates that also have a long history of behavior problems, thus increasing the likelihood of numerous forms of adversity during and after pregnancy. Interestingly, this was one of the main conclusions of the Swiss child psychiatrist Lucien Bovet in the first report the World Health Organization commissioned after its creation (Bovet 1951).

Preventive interventions with at risk women during and after pregnancy will not modify the genetic code, but they are likely to impact DNA expression. Epigenetic studies are giving tools to assess the epigenetic effects of preventive interventions during pregnancy and infancy. Because short term effects of adversity during pregnancy on DNA expression may be good markers of long term effects, they should be used to compare the effectiveness of different forms of interventions. These experimental preventive interventions can also be used to test environmental effects on the cascade of biological effects that follow gene expression and lead to the behavior problems which we described above. For example, we can test the effects of smoking cessation experimental interventions during pregnancy on DNA expression at birth and eventually on CPA (Caporaso et al. 2009; Markunas et al. 2014; Nielsen et al. 2014).

However, interventions that only target specific risk factors such as smoking during pregnancy are unlikely to have long term impacts on children's development since most of the at risk women have a large array of risk factors besides smoking during pregnancy (ex. stressful marital relations, inadequate nutrition, depression, impulsive and antisocial behaviour). Preventive interventions need to target all the risk factors. To our knowledge no experimental prenatal or early postnatal interventions have specifically targeted the reduction of CPA in the offspring. However, good models of preventive interventions targeting women with many risk factors during the prenatal and the early post natal period were implemented three decades ago and have shown long term impacts on numerous aspects of children's development that are related to CPA. For example, in the Elmira Home Visitation prevention experiment (Donelan-McCall et al. 2009; Olds et al. 1997, 1998) participants were mostly low-income, unmarried, pregnant adolescents. Other pregnant women were included in the study to prevent stigmatization. Three experimental groups were created by random allocation. Women in the first group were visited weekly by a nurse for the first month after enrolment in the study, twice a month until birth, weekly for the first 6 weeks after birth, twice a month until the baby reached 21 months, and monthly until the child reached the end of the second year. Women in the second group received home visits only during pregnancy, while women in the third group had a screening interview after birth and free transport to the health clinic between the child's birth and the end of the second year. Mothers and children have been followed up to the children's early adulthood and significant positive impacts were observed from early childhood (Olds et al. 1986) to early adulthood (Eckenrode et al. 2010). We do not know to what extent the intervention had an impact of CPA because this range of behavior was not systematically assessed, but the population they targeted includes those that are at high risk of having children with CPA, while the content and intensity of the intervention would be expected to impact the cascade of biological processes described above. Interestingly, the assessment 19 years after the children's birth indicated that the female children had benefited more from the intervention than the males. The girls from the intervention group had been significantly less involved in deviant behavior. Also the girls born to unmarried mothers and low-income mothers had significantly fewer children and less Medicaid use (Eckenrode et al. 2010). These results suggest that intensive support to at risk pregnant mothers has significant impacts, not only on their female offspring, but also on the third generation. Much progress has been made in the past decade in terms of refinements in techniques to study DNA methylation, as well as immune system and 5-HT system development. Such techniques allow the evaluation of the impact of preventive experiments on a large spectrum of putative mechanisms.

10.7 Summary and Conclusion

The present chapter reviewed recent research on the mechanisms leading to chronic physical aggression during childhood and adolescence, with an emphasis on familial characteristics, epigenetics, immunology and the serotonin system. While most of the studies on the biological mechanisms leading to chronic physical aggression investigated solely associations between one biological variable (e.g., neurotransmitters, neuro-immunoregulators, hormones) and behavioral outcome (i.e., chronic aggression), an emerging number of studies now have provided support for a more refined hypothesis on mechanisms of chronic physical aggression, taking into account the importance of the developmental role of these factors on brain developmental processes; and how early environmental factors could alter the functioning of these biological systems, with consequences for risk for chronic physical aggression. Specifically, there is now emerging evidence that early social-familial adversity leads to long lasting epigenetic alterations. These alterations may influence brain development, and, consequently, the ability to learn to regulate and control aggressive behaviour. In the context of early adversity and chronic physical aggression, epigenetic alterations in genes regulating the serotonin system, the immune system and the HPA axis, are especially of interest. These systems may also interact with other biological factors, thereby further complicating neurobiological theories.

In addition, the finding that many of these biological factors involved in chronic physical aggression develop very early (i.e., before birth), highlights the need to not only study the impact of the early postnatal environment on physical aggression, but also to take into account what is happening between conception and birth.

Indeed, a number of studies have now identified perinatal risk factors for a wide variety of behavioral problems (e.g., hyperactivity, opposition, rule breaking) that are highly related to maternal characteristics and maternal behavior during pregnancy. The effects of the perinatal environment on gene expression is presently the most interesting hypothesis. However, the reader needs to realize that we are only starting to explore this avenue. It is tempting, in a leap of faith, to jump on a band wagon, but many recent cases of unbridled enthusiasm should lead to restrained optimism (Risch et al. 2009).

The benefits of prevention starting as close as possible to conception seems obvious. If the earlier the risks the more wide spread the negative effects, then the corollary is the earlier the preventive intervention the more wide spread the benefits. The later we intervene, the less chance we have to impact the basic weaknesses of the organism. Preventive interventions should help at risk pregnant women live a life style that will facilitate "normal" gene expression. If we wait 3 or 4 years after birth (still worse if we wait until adolescence) we are dealing with a disruptive child who, as a figure of speech, may have only half of his gene expression potential to gain control over himself, and we are dealing with a mother who has a life long experience of failure now facing the probability of the same life-course for her child.

Although the focus of this chapter was on early risk factors for chronic aggression, early prevention and their underlying biological mechanisms related to serotonin, the immune system and the HPA-axis, since they have been most widely studied in the context of aggression, it is important to take into account the role of *protective* factors that may mitigate the impact of an adverse environment. One factor of particular interest is breastfeeding. Breastfeeding has been shown to be associated with many positive behaviors, including increased maternal sensitivity and increased attachment security (e.g., Tharner et al. 2012). One possible mechanism underlying these associations is the stimulating effect of breastfeeding on the production of the hormone oxytocin (see e.g., Febo et al. 2005). Oxytocin has also been shown to have a positive influence on child development, (Carter 2003; Insel 2010) and experimental placebo controlled studies using intranasal oxytocin has shown a positive effect of oxytocin on prosocial behaviors (MacDonald and MacDonald 2010). Two important research questions can be formulated from this perspective: (a) is to what extent can breastfeeding enhance the efficacy of preventive interventions for aggression problems in at risk populations? (b) what would be the consequences for the serotonin system, immune regulation and HPA-axis functioning?

The focus of this review was chronic physical aggression, but the basic principles of the epigenetic process suggest that intensive prenatal interventions will have an impact on numerous aspects of physical and mental health as well as social adjustment, including the major modern health problems: low birth weight, obesity, cardiovascular problems, cancer, hyperactivity, mood disorders and substance abuse.

Finally, one of the major conclusions we can draw from five decades of longitudinal studies on chronic physical aggression is that the time is ripe for investments in large collaborative early experimental preventive interventions. Randomized control trials are the best tools to test causal hypotheses while testing effective interventions (Schwartz et al. 1981; Tremblay 2003). It is hard to believevthat there have been so few bio-psycho-social experimental interventions with pregnant women at risk of intergenerational transmission of the psychiatric problems that start during early childhood. Good models were implemented two decades ago (Donelan-McCall et al. 2009; Olds et al. 1998). We need to use these interventions that are well tested and study carefully the development of their potential bio-psycho-social effects from the prenatal period to at least the third generations' prenatal period.

This chapter is an adaptation of the following papers:

- Booij, L., Tremblay, R. E., Szyf, M., Benkelfat, C. (2014). Genetic and early environmental influences on the serotonin system: Consequences for brain development and risk for psychopathology. *Journal of Psychiatry and Neuroscience*. Oct 7. doi: 10.1503/jpn.140099. [Epub ahead of print]
- Booij, L., Wang, D., Lévesque, M. L., Tremblay, R. E, & Szyf, M. (2013). Looking beyond the DNA Sequence: the relevance of DNA methylation processes for the stress-diathesis model of depression. *Philosophical Transactions of the Royal Society B-Biological Sciences, 368*(1615), 1–16. doi: 10.1098/rstb.2012.0251
- Provençal, N., Booij, L., & Tremblay, R. E. (2015). The developmental origins of chronic physical aggression: Early life adversity, epigenetics and impact on other biological systems. *Journal of Experimental Biology*
- Tremblay, R. E. (2010). Developmental origins of disruptive behaviour problems: The 'original sin' hypothesis, epigenetics and their consequences for prevention. *Journal of Child Psychology and Psychiatry, 51*(4), 341–367. doi: 10.1111/j.1469-7610.2010.02211.x

References

Ahmed AA, Ma WQ, Ni YD, Zhou Q, Zhao RQ (2014) Embryonic exposure to corticosterone modifies aggressive behavior through alterations of the hypothalamic pituitary adrenal axis and the serotonergic system in the chicken. Horm Behav 65(2):97–105. doi:10.1016/j.yhbeh.2013.12.002

Alink LR, Mesman J, van Zeijl J, Stolk MN, Juffer F, Koot HM, Bakersman-Kranenburg MJ, van Ijzendoorn MH (2006) The early childhood aggression curve: development of physical aggression in 10- to 50-months-old children. Child Dev 77(4):954–966

Alleva E, Cirulli F, Bianchi M, Bondiolotti GP, Chiarotti F, De Acetis L, Panerai AE (1998) Behavioural characterization of interleukin-6 overexpressing or deficient mice during agonistic encounters. Eur J Neurosci 10(12):3664–3672. doi:10.1046/j.1460-9568.1998.00377.x

Ansorge MS, Zhou MM, Lira A, Hen R, Gingrich JA (2004) Early-life blockade of the 5-HT transporter alters emotional behavior in adult mice. Science 306(5697):879–881. doi:10.1126/science.1101678

Arseneault L, Tremblay RE, Boulerice B, Saucier J-F (2002) Obstetrical complications and violent delinquency: testing two developmental pathways. Child Dev 73(2):496–508

Aslund C, Nordquist N, Comasco E, Leppert J, Oreland L, Nilsson K (2011) Maltreatment, MAOA, and delinquency: sex differences in gene-environment interaction in a large population-based cohort of adolescents. Behav Genet 41(2):262–272. doi:10.1007/s10519-010-9356-y

Azmitia EC, Gannon PJ (1986) The primate serotonergic system: a review of human and animal studies and a report on Macaca fascicularis. Adv Neurol 43:407–468

Baillargeon RH, Zoccolillo M, Keenan K, Côté S, Pérusse D, Wu H-X, Boivin M, Tremblay RE (2007) Gender differences in physical aggression: a prospective population-based survey of children before and after 2 years of age. Dev Psychol 43(1):13–26

Bandura A (1977) Social learning theory. Prentice Hall Inc., Englewood Cliffs

Barker ED, Séguin JR, White HR, Bates ME, Lacourse E, Carbonneau R, Tremblay RE (2007) Developmental trajectories of male physical violence and theft: relations to neurocognitive performance. Arch Gen Psychiatry 64(5):592–599. doi:10.1001/archpsyc.64.5.592

Barreau F, Cartier C, Ferrier L, Fioramonti J, Bueno L (2004) Nerve growth factor mediates alterations of colonic sensitivity and mucosal barrier induced by neonatal stress in rats. Gastroenterology 127(2):524–534. doi:10.1053/s0016-5085(04)00865-0

Beach SRH, Brody GH, Todorov AA, Gunter TD, Philibert RA (2011) Methylation at 5HTT mediates the impact of child sex abuse on women's antisocial behavior: an examination of the Iowa adoptee sample. Psychosom Med 73(1):83–87. doi:10.1097/PSY.0b013e3181fdd074

Boivin M, Vitaro F, Poulin F (2005) Peer relationships and the development of aggressive behavior in early childhood. In: Tremblay RE, Hartup WW, Archer J (eds) Developmental origins of aggression. Guilford Press, New York, pp 376–397

Booij L, Szyf M, Carballedo A, Frey NN, Morris D, Ly V, Fahey C, Meaney J, Gill M, Frodi T (submitted) The role of SLC6A4 DNA methylation in stress-related changes in hippocampal volume: a study in depressed patients and healthy controls

Booij L, Tremblay RE, Leyton M, Seguin JR, Vitaro F, Gravel P, Perreau-Linck E, Levesque ML, Durand F, Diksic M, Turecki G, Benkelfat C (2010) Brain serotonin synthesis in adult males characterized by physical aggression during childhood: a 21-year longitudinal study. PloS One 5(6). doi:10.1371/journal.pone.0011255

Booij L, Wang DS, Levesque ML, Tremblay RE, Szyf M (2013) Looking beyond the DNA sequence: the relevance of DNA methylation processes for the stress-diathesis model of depression. Philos Trans R Soc B Biol Sci 368(1615):1–16. doi:10.1098/rstb.2012.0251

Booij L, Tremblay RE, Szyf M, Benkelfat C (2014) Genetic and early environmental influences on the serotonin system: consequences for brain development and risk for psychopathology. J Psychiatry Neurosci 39(5):1400–1499. doi:10.1503/jpn.140099

Booij L, Szyf M, Carballedo A, Frey M, Morris D, Dymov S, Vaisheva F, Ly V, Fahey C, Meaney J, Gill M, Frodl T (2015) DNA methylation of the serotonin transporter gene in peripheral cells and stress-related changes in hippocampal volume: a study in depressed patients and healthy controls. PLoS One 10(3), e0119061

Borghol N, Suderman M, McArdle W, Racine A, Hallett M, Pembrey M, Hertzman C, Power C, Szyf M (2012) Associations with early-life socio-economic position in adult DNA methylation. Int J Epidemiol 41(1):62–74. doi:10.1093/ije/dyr147

Bovet L (1951) Psychiatric aspects of juvenile delinquency. World Health Organization, Geneva

Brame B, Nagin DS, Tremblay RE (2001) Developmental trajectories of physical aggression from school entry to late adolescence. J Child Psychol Psychiatry 42(4):503–512

Broidy LM, Nagin DS, Tremblay RE, Bates JE, Brame B, Dodge KA, Fergusson D, Horwood JL, Loeber R, Laird R, Lynam DR, Moffitt TE, Pettit GS, Vitaro F (2003) Developmental trajectories of childhood disruptive behaviors and adolescent delinquency: a six site, cross national study. Dev Psychol 39(2):222–245

Brown GL, Goodwin FK, Ballenger JC, Gover PF, Major LF (1979) Aggression in humans correlates with cerebrospinal fluid amine metabolites. Psychiatr Res 1(2):131–139

Brunner HG, Nelen MR, Van Zandvoort P, Abeling NG, Van Gennip AH, Wolters EC, Kuiper MA, Ropers HH, Van Oost BA (1993) X-linked borderline mental retardation with prominent behavioral disturbance: phenotype, genetic localization, and evidence for disturbed monoamine metabolism. Am J Hum Genet 52(6):1032–1039

Buckholtz JW, Callicott JH, Kolachana B, Hariri AR, Goldberg TE, Genderson M, Egan MF, Mattay VS, Weinberger DR, Meyer-Lindenberg A (2008) Genetic variation in MAOA modulates ventromedial prefrontal circuitry mediating individual differences in human personality. Mol Psychiatry 13(3):313–324. doi:10.1038/sj.mp.4002020

Campbell SB, Spieker S, Vandergrift N, Belsky J, Burchinal M, NICHD Early Child Care Research Network (2010) Predictors and sequelae of trajectories of physical aggression in school-age boys and girls. Dev Psychopathol 22(1):133–150. doi:10.1017/s0954579409990319

Canli T, Qiu M, Omura K, Congdon E, Haas BW, Amin Z, Herrmann MJ, Constable RT, Lesch KP (2006) Neural correlates of epigenesis. Proc Natl Acad Sci U S A 103(43):16033–16038. doi:10.1073/pnas.0601674103

Caporaso N, Gu F, Chatterjee N, Sheng-Chih J, Yu K, Yeager M, Chen C, Jacobs K, Wheeler W, Landi MT, Ziegler RG, Hunter DJ, Chanock S, Hankinson S, Kraft P, Bergen AW (2009) Genome-wide and candidate gene association study of cigarette smoking behaviors. PLoS One 4(2):e4653

Capuron L, Miller AH (2011) Immune system to brain signaling: neuropsychopharmacological implications. Pharmacol Ther 130(2):226–238. doi:10.1016/j.pharmthera.2011.01.014

Carter CS (2003) Developmental consequences of oxytocin. Physiol Behav 79(3):383–397

Cases O, Seif I, Grimsby J, Gaspar P, Chen K, Pournin S, Muller U, Aguet M, Babinet C, Shih JC, Demaeyer E (1995) Aggressive-behavior and altered amounts of brain-serotonin and norepinephrine in mice lacking MAOA. Science 268(5218):1763–1766. doi:10.1126/science.7792602

Caspi A, McClay J, Moffitt TE, Mill J, Martin J, Craig IW, Taylor A, Poulton R (2002) Role of genotype in the cycle of violence in maltreated children. Science 297(5582):851–854. doi:10.1126/science.1072290

Checknita D, Maussion G, Labonté B, Comai S, Tremblay RE, Vitaro F, Turecki N, Bertazzo A, Gobbi G, Côté G, Turecki G (2015) Monoamine oxidase A gene promoter methylation and transcriptional downregulation in an offender population with antisocial personality disorder. Br J Psychiatr 206(3):216–222

Cleare AJ, Bond AJ (1995) The effect of tryptophan depletion and enhancement on subjective and behavioral aggression in normal-male subjects. Psychopharmacology (Berl) 118(1):72–81. doi:10.1007/bf02245252

Coccaro EF, Siever LJ, Klar HM, Maurer G, Cochrane K, Cooper TB, Mohs RC, Davis KL (1989) Serotonergic studies in patients with affective and personality-disorders: correlates with suicidal and impulsive aggressive-behavior. Arch Gen Psychiatry 46(7):587–599

Côté SM (2007) Sex differences in physical and indirect aggression: a developmental perspective. Eur J Crim Policy Res 13(3–4):183–200

Côté SM, Vaillancourt T, LeBlanc JC, Nagin DS, Tremblay RE (2006) The development of physical aggression from toddlerhood to pre-adolescence: a nation wide longitudinal study of Canadian children. J Abnorm Child Psychol 34(1):71–85

Côté SM, Boivin M, Nagin DS, Japel C, Xu Q, Zoccolillo M, Tremblay RE (2007) The role of maternal education and non-maternal care services in the prevention of children's physical aggression. Arch Gen Psychiatry 64(11):1305–1312

Danese A, Pariante CM, Caspi A, Taylor A, Poulton R (2007) Childhood maltreatment predicts adult inflammation in a life-course study. Proc Natl Acad Sci U S A 104(4):1319–1324. doi:10.1073/pnas.0610362104

Davidson RJ, Putnam KM, Larson CL (2000) Dysfunction in the neural circuitry of emotion regulation – a possible prelude to violence. Science 289(5479):591–594

Dionne G, Tremblay RE, Boivin M, Laplante D, Pérusse D (2003) Physical aggression and expressive vocabulary in 19 month-old twins. Dev Psychol 39(2):261–273

Donelan-McCall N, Eckenrode J, Olds DL (2009) Home visiting for the prevention of child maltreatment: lessons learned during the past 20 years. Pediatr Clin North Am 56(2):389–403. doi:10.1016/j.pcl.2009.01.002

Dowlati Y, Herrmann N, Swardfager W, Liu H, Sham L, Reim EK, Lanctot KL (2010) A meta-analysis of cytokines in major depression. Biol Psychiatry 67(5):446–457. doi:10.1016/j.biopsych.2009.09.033

Eckenrode J, Campa M, Luckey DW, Henderson CR, Cole R, Kitzman H, Anson E, Sidora-Arcoleo K, Power J, Olds D (2010) Long-term effects of prenatal and infancy nurse home visitation on the life course of youths: 19-year follow-up of a randomized trial. Arch Pediatr Adolesc Med 164(1):9–15

Eisner M (2003) Long-term historical trends in violent crime. In: Tonry M (ed) Crime and justice: a review of research, vol 30. University of Chicago Press, Chicago, pp 83–142

Febo M, Numan M, Ferris CF (2005) Functional Magnetic Resonance Imaging shows oxytocin activates brain regions associated with mother–pup bonding during suckling. J Neurosci 25(50):11637–11644

Ferris CF, Lu SF, Messenger T, Guillon CD, Heindel N, Miller M, Koppel G, Bruns FR, Simon NG (2006) Orally active vasopressin V1a receptor antagonist, SRX251, selectively blocks aggressive behavior. Pharmacol Biochem Behav 83(2):169–174. doi:10.1016/j.pbb.2006.01.001

Fontaine N, Carbonneau R, Barker ED, Vitaro F, Hébert M, Côté SM, Nagin DS, Zoccolillo M, Tremblay RE (2008) Girls' hyperactivity and physical aggression during childhood and adjustment problems in early adulthood. Arch Gen Psychiatry 65(3):320–328

Frankle WG, Lombardo I, New AS, Goodman M, Talbot PS, Huang YY, Hwang DR, Slifstein M, Curry S, Abi-Dargham A, Laruelle M, Siever LJ (2005) Brain serotonin transporter distribution in subjects with impulsive aggressivity: a positron emission study with [11C]McN 5652. Am J Psychiatry 162(5):915–923. doi:10.1176/appi.ajp.162.5.915

Gaspar P, Cases O, Maroteaux L (2003) The developmental role of serotonin: news from mouse molecular genetics. Nat Rev Neurosci 4(12):1002–1012. doi:10.1038/nrn1256

Glover V (2011) Prenatal stress and the origins of psychopathology: an evolutionary perspective. J Child Psychol Psychiatry 52(4):356–367. doi:10.1111/j.1469-7610.2011.02371.x

Godar SC, Bortolato M, Castelli MP, Casti A, Casu A, Chen K, Ennas MG, Tambaro S, Shih JC (2014) The aggression and behavioral abnormalities associated with monoamine oxidase A deficiency are rescued by acute inhibition of serotonin reuptake. J Psychiatr Res 56:1–9. doi:10.1016/j.jpsychires.2014.04.014

Groer MW, Morgan K (2007) Immune, health and endocrine characteristics of depressed postpartum mothers. Psychoneuroendocrinology 32(2):133–139. doi:10.1016/j.psyneuen.2006.11.007

Guillemin C, Provençal N, Suderman M, Côté SM, Vitaro F, Hallett M, Tremblay RE, Szyf M (2014) DNA methylation signature of childhood chronic physical aggression in T cells of both men and women. PLoS One 9(1):1–16 (e86822). doi:10.1371/journal.pone.0086822

Guo G, Roettger ME, Shih JC (2007) Contributions of the DAT1 and DRD2 genes to serious and violent delinquency among adolescents and young adults. Hum Genet 121(1):125–136. doi:10.1007/s00439-006-0244-8

Hammock EAD, Lim MM, Nair HP, Young LJ (2005) Association of vasopressin 1a receptor levels with a regulatory microsatellite and behavior. Genes Brain Behav 4(5):289–301. doi:10.1111/j.1601-183X.2005.00119.x

Hay DF, Mundy L, Roberts S, Carta R, Waters CS, Perra O, Jones R, Jones I, Goodyer I, Harold G, Thapar A, van Goozen S (2011) Known risk factors for violence predict 12-month-old infants' aggressiveness with peers. Psychol Sci 22(9):1205–1211. doi:10.1177/0956797611419303

Heinz A, Braus DF, Smolka MN, Wrase J, Puls I, Hermann D, Klein S, Grusser SM, Flor H, Schumann G, Mann K, Buchel C (2005) Amygdala-prefrontal coupling depends on a genetic variation of the serotonin transporter. Nat Neurosci 8(1):20–21. doi:10.1038/nn1366

Higley JD, Mehlman PT, Higley SB, Fernald B, Vickers J, Lindell SG, Taub DM, Suomi SJ, Linnoila M (1996) Excessive mortality in young free-ranging male nonhuman primates with low cerebrospinal fluid 5-hydroxyindoleacetic acid concentrations. Arch Gen Psychiatry 53(6):537–543

Hoge EA, Brandstetter K, Moshier S, Pollack MH, Wong KK, Simon NM (2009) Broad spectrum of cytokine abnormalities in panic disorder and posttraumatic stress disorder. Depress Anxiety 26(5):447–455. doi:10.1002/da.20564

Howell S, Westergaard G, Hoos B, Chavanne TJ, Shoaf SE, Cleveland A, Snoy PJ, Suomi SJ, Higley JD (2007) Serotonergic influences on life-history outcomes in free-ranging male rhesus Macaques. Am J Primatol 69(8):851–865. doi:10.1002/ajp.20369

Huijbregts SCJ, Séguin JR, Zoccolillo M, Boivin M, Tremblay RE (2008) Maternal prenatal smoking, parental antisocial behavior, and early childhood physical aggression. Dev Psychopathol 20(2):437–453. doi:10.1017/S0954579408000217

Ichise M, Vines DC, Gura T, Anderson GM, Suomi SJ, Higley JD, Innis RB (2006) Effects of early life stress on C-11 DASB positron emission tomography imaging of serotonin transporters in adolescent peer- and mother-reared rhesus monkeys. J Neurosci 26(17):4638–4643. doi:10.1523/jneurosci.5199-05.2006

Insel TR (2010) The challenge of translation in social neuroscience: a review of oxytocin, vasopressin, and affiliative behaviour. Neuron 65(6):383–397

Janelidze S, Mattei D, Westrin A, Träskman-Bendz L, Brundin L (2010) Cytokine levels in the blood may distinguish suicide attempters from depressed patients. Brain Behav Immun 25(2):335–339. doi:10.1016/j.bbi.2010.10.010

Jezierski G, Braun K, Gruss M (2006) Epigenetic modulation of the developing serotonergic neurotransmission in the semi-precocial rodent Octodon degus. Neurochem Int 48(5):350–357. doi:10.1016/j.nueint.2005.11.009

Keck ME, Wigger A, Welt T, Muller MB, Gesing A, Reul J, Holsboer F, Landgraf R, Neumann ID (2002) Vasopressin mediates the response of the combined dexamethasone/CRH test in hyper-anxious rats: implications for pathogenesis of affective disorders. Neuropsychopharmacology 26(1):94–105. doi:10.1016/s0893-133x(01)00351-7

Kieseppa T, Partonen T, Haukka J, Kaprio J, Lonnqvist J (2004) High concordance of bipolar I disorder in a nationwide sample of twins. Am J Psychiatry 161(10):1814–1821. doi:10.1176/appi.ajp.161.10.1814

Koo JW, Duman RS (2008) IL-1 beta is an essential mediator of the antineurogenic and anhedonic effects of stress. Proc Natl Acad Sci U S A 105(2):751–756. doi:10.1073/pnas.0708092105

Kornhuber J, Konradi C, Mackburkhardt F, Riederer P, Heinsen H, Beckmann H (1989) Ontogenesis of monoamine oxidase-A and oxidase-B in the human-brain frontal-cortex. Brain Res 499(1):81–86. doi:10.1016/0006-8993(89)91136-0

Krug EG, Dahlberg LL, Mercy JA, Zwi AB, Lozano R (2002) World report on violence and health, 2002, from www.who.int/violence_injury_prevention/violence/world_report/wrvh1/en

Kuo MH, Allis CD (1998) Roles of histone acetyltransferases and deacetylases in gene regulation. Bioessays20(8):615–626.doi:10.1002/(sici)1521-1878(199808)20:8<615::aid-bies4>3.0.co;2-h

Lacourse E, Boivin M, Brendgen M, Petitclerc A, Girard A, Vitaro F, Paquin S, Ouellet-Morin I, Dionne G, Tremblay RE (2014) A longitudinal twin study of physical aggression in early childhood: evidence for a developmentally dynamic genome. Psychol Med 44(12):2617–2627. doi:10.1017/S0033291713003218

Lefkowitz MM, Eron LD, Walder LO, Huesmann LR (1977) Growing up to be violent. A longitudinal study of the development of aggression. Pergamon Press, New York

Lesch KP, Mössner R (2006) Inactivation of 5HT transport in mice: modeling altered 5HT homeostasis implicated in emotional dysfunction, affective disorders, and somatic syndromes. Handb Exp Pharmacol (175):417–456

Leyton M, Okazawa H, Diksic M, Paris J, Rosa P, Mzengeza S, Young SN, Blier P, Benkelfat C (2001) Brain regional alpha- C-11 Methyl-(L)-Tryptophan trapping in impulsive subjects with borderline personality disorder. Am J Psychiatry 158(5):775–782. doi:10.1176/appi.ajp.158.5.775

Leyton M, Paquette V, Gravel P, Rosa-Neto P, Weston F, Diksic M, Benkelfat C (2006) alpha- C-11 Methyl-L-tryptophan trapping in the orbital and ventral medial prefrontal cortex of suicide attempters. Eur Neuropsychopharmacol 16(3):220–223. doi:10.1016/j.euroneuro.2005.09.006
Loney BR, Butler MA, Lima EN, Counts CA, Eckel LA (2006) The relation between salivary cortisol, callous-unemotional traits, and conduct problems in an adolescent non-referred sample. J Child Psychol Psychiatry 47(1):30–36. doi:10.1111/j.1469-7610.2005.01444.x
MacDonald K, MacDonald TM (2010) The peptide that binds: a systematic review of oxytocin and its prosocial effects. Harv Rev Psychiatry 18(1):18–21
Manuck SB, Bleil ME, Petersen KL, Flory JD, Mann JJ, Ferrell RE, Muldoon MF (2005) The socio-economic status of communities predicts variation in brain serotonergic responsivity. Psychol Med 35(4):519–528. doi:10.1017/s0033291704003757
Markunas CA, Xu ZL, Harlid S, Wade PA, Lie RT, Taylor JA, Wilcox AJ (2014) Identification of DNA methylation changes in newborns related to maternal smoking during pregnancy. Environ Health Perspect 122(10):1147–1153. doi:10.1289/ehp.1307892
Marquez C, Poirier GL, Cordero MI, Larsen MH, Groner A, Marquis J, Magistretti PJ, Trono D, Sandi C (2013) Peripuberty stress leads to abnormal aggression, altered amygdala and orbitofrontal reactivity and increased prefrontal MAOA gene expression. Translat Psychiatry 3:e216. doi:10.1038/tp.2012.144
Marsland AL, Prather AA, Petersen KL, Cohen S, Manuck SB (2008) Antagonistic characteristics are positively associated with inflammatory markers independently of trait negative emotionality. Brain Behav Immun 22(5):753–761. doi:10.1016/j.bbi.2007.11.008
Matthews K, Dalley JW, Matthews C, Tsai TH, Robbins TW (2001) Periodic maternal separation of neonatal rats produces region- and gender-specific effects on biogenic amine content in postmortem adult brain. Synapse 40(1):1–10. doi:10.1002/1098-2396(200104)40:1<1::aid-syn1020>3.0.co;2-e
McGowan PO, Sasaki A, D'Alessio AC, Dymov S, Labonte B, Szyf M, Turecki G, Meaney MJ (2009) Epigenetic regulation of the glucocorticoid receptor in human brain associates with childhood abuse. Nat Neurosci 12(3):342–348. doi:10.1038/nn.2270
Mehta D, Klengel T, Conneely KN, Smith AK, Altmann A, Pace TW, Rex-Haffner M, Loeschner A, Gonik M, Mercer KB, Bradley B, Muller-Myhsok B, Ressler KJ, Binder EB (2013) Childhood maltreatment is associated with distinct genomic and epigenetic profiles in posttraumatic stress disorder. Proc Natl Acad Sci U S A 110(20):8302–8307. doi:10.1073/pnas.1217750110
Meyer-Lindenberg A, Buckholtz JW, Kolachana B, Hariri RA, Pezawas L, Blasi G, Wabnitz A, Honea R, Verchinski B, Callicott JH, Egan M, Mattay V, Weinberger DR (2006) Neural mechanisms of genetic risk for impulsivity and violence in humans. Proc Natl Acad Sci U S A 103(16):6269–6274
Moeller FG, Dougherty DM, Swann AC, Collins D, Davis CM, Cherek DR (1996) Tryptophan depletion and aggressive responding in healthy males. Psychopharmacology (Berl) 126(2):97–103. doi:10.1007/bf02246343
Munafo MR, Brown SM, Hariri AR (2008) Serotonin transporter (5-HTTLPR) genotype and amygdala activation: a meta-analysis. Biol Psychiatry 63(9):852–857. doi:10.1016/j.biopsych.2007.08.016
Myint AM, Kim YK (2003) Cytokine-serotonin interaction through IDO: a neurodegeneration hypothesis of depression. Med Hypotheses 61(5–6):519–525. doi:10.1016/s0306-9877(03)00207-x
Naerde A, Ogden T, Janson H, Zachrisson HD (2014) Normative development of physical aggression from 8 to 26 months. Dev Psychol 50(6):1710–1720. doi:10.1037/a0036324
Nagin D, Tremblay RE (1999) Trajectories of boys' physical aggression, opposition, and hyperactivity on the path to physically violent and nonviolent juvenile delinquency. Child Dev 70(5):1181–1196
Nelson RJ, Trainor BC (2007) Neural mechanisms of aggression. Nat Rev Neurosci 8(7):536–546
New AS, Hazlett EA, Buchsbaum MS, Goodman M, Reynolds D, Mitropoulou V, Sprung L, Shaw RB, Koenigsberg H, Platholi J, Silverman J, Siever LJ (2002) Blunted prefrontal cortical (18)

fluorodeoxyglucose positron emission tomography response to meta-chlorophenylpiperazine in impulsive aggression. Arch Gen Psychiatry 59(7):621–629. doi:10.1001/archpsyc.59.7.621
NICHD Early Child Care Research Network (2004) Trajectories of physical aggression from toddlerhood to middle school: predictors, correlates, and outcomes. SRCD Monogr 69(4, 278):1–146
Nielsen CH, Larsen A, Nielsen AL (2014) DNA methylation alterations in response to prenatal exposure of maternal cigarette smoking: a persistent epigenetic impact on health from maternal lifestyle? Arch Toxicol Dec 6. [Epub ahead of print]. doi:10.1007/s00204-014-1426-0
Nishizawa S, Benkelfat C, Young SN, Leyton M, Mzengeza S, de Montigny C, Blier P, Diksic M (1997) Differences between males and females in rates of serotonin synthesis in human brain. Proc Natl Acad Sci U S A 94(10):5308–5313
Olds D, Henderson CR, Chamberlin R, Talelbaum R (1986) Preventing child abuse and neglect: a randomized trial of nurse home visitation. Pediatrics 78:65–78
Olds DL, Eckenrode J, Henderson CR Jr, Kitzman H, Powers J, Cole R, Sidora K, Morris P, Pettitt LM, Luckey D (1997) Long-term effects of home visitation on maternal life course and child abuse and neglect: fifteen-year follow-up of a randomized trial. JAMA 278:637–643
Olds D, Henderson CR, Cole R, Eckenrode J, Kitzman H, Luckey D, Pettitt L, Sidora K, Morris P, Powers J (1998) Long-term effects of nurse home visitation on children's criminal and antisocial behavior: 15-year follow-up of a randomized controlled trial. JAMA 280(14):1238–1244
O'Rourke H, Fudge JL (2006) Distribution of serotonin transporter labeled fibers in amygdaloid subregions: implications for mood disorders. Biol Psychiatry 60(5):479–490. doi:10.1016/j.biopsych.2005.09.020
Pesce M, Speranza L, Franceschelli S, Ialenti V, Patruno A, Febo MA, De Lutiis MA, Felaco M, Grilli A (2011) Biological role of Interleukin-1 beta in defensive aggressive behavior. J Biol Regul Homeost Agents 25(3):323–329
Petitclerc A, Tremblay RE (2009) Childhood disruptive behaviour disorders: review of their origin, development, and prevention. Can J Psychiatry Revue Canadienne De Psychiatrie 54(4):222–231
Pingault J-B, Côté SM, Lacourse E, Galéra C, Vitaro F, Tremblay RE (2013) Childhood hyperactivity, physical aggression and criminality: a 19 year prospective population-based study. PLoS One 8(5):1–7
Plomin R (1994) Genetics and experience: the interplay between nature and nurture. Sage, Thousand Oaks
Popma A, Vermeiren R, Geluk C, Rinne T, van den Brink W, Knol DL, Jansen LMC, van Engeland H, Doreleijers TAH (2007) Cortisol moderates the relationship between testosterone and aggression in delinquent male adolescents. Biol Psychiatry 61(3):405–411. doi:10.1016/j.biopsych.2006.06.006
Powell ND, Sloan EK, Bailey MT, Arevalo JMG, Miller GE, Chen E, Kobor MS, Reader BF, Sheridan JF, Cole SW (2013) Social stress up-regulates inflammatory gene expression in the leukocyte transcriptome via beta-adrenergic induction of myelopoiesis. Proc Natl Acad Sci U S A 110(41):16574–16579. doi:10.1073/pnas.1310655110
Provençal N, Suderman MJ, Guillemin C, Massart R, Ruggiero A, Wang D, Bennett AJ, Pierre PJ, Friedman DP, Côté SM, Hallett M, Tremblay RE, Suomi SJ, Szyf M (2012) The signature of maternal rearing in the methylome in Rhesus Macaque prefrontal cortex and T cells. J Neurosci 32(44):15626–15642. doi:10.1523/JNEUROSCI.1470-12.2012
Provençal N, Suderman MJ, Caramaschi D, Wang DS, Hallett M, Vitaro F, Tremblay RE, Szyf M (2013a) Differential DNA methylation regions in cytokine and transcription factor genomic Loci associate with childhood physical aggression. PLoS One 8(8). doi:10.1371/journal.pone.0071691
Provencal N, Suderman MJ, Vitaro F, Szyf M, Tremblay RE (2013b) Childhood chronic physical aggression associates with adult cytokine levels in plasma. PLoS One 8(7):1–7 (e69481). doi:10.1371/journal.pone.0069481

Provençal N, Suderman MJ, Guillemin C, Vitaro F, Côté SM, Hallett M, Tremblay RE, Szyf M (2014) Association of childhood chronic physical aggression with a DNA methylation signature in adult human T cells. PLoS One 9(4):1–10 (e89839). doi:10.1371/journal.pone.0089839

Provençal N, Booij L, Tremblay RE (2015) The developmental origins of chronic physical aggression: biological pathways triggered by early life adversity. J Exp Biol 218(Pt 1): 123–133

Quetelet A (1984) Research on the propensity for crime at different ages (trans: Sylvester SF). Anderson, Cincinnati (Original work published in 1833)

Reif A, Rosler M, Freitag CM, Schneider M, Eujen A, Kissling C, Wenzler D, Jacob CP, Retz-Junginger P, Thome J, Lesch KP, Retz W (2007) Nature and nurture predispose to violent behavior: serotonergic genes and adverse childhood environment. Neuropsychopharmacology 32(11):2375–2383. doi:10.1038/sj.npp.1301359

Reiss AJ, Roth JA (eds) (1993) Understanding and preventing violence. National Academy Press, Washington, DC

Risch N, Herrell R, Lehner T, Liang KY, Eaves L, Hoh J, Griem A, Kovacs M, Ott J, Merikangas KR (2009) Interaction between the serotonin transporter gene (5-HTTLPR), stressful life events, and risk of depression a meta-analysis. JAMA 301(23):2462–2471

Roy A, Berman J, Williams R, Kuhn C, Gonzalez B (2002) Higher levels of CSF homovanillic acid in recently abstinent cocaine-dependent patients. Am J Psychiatry 159(6):1053–1055. doi:10.1176/appi.ajp.159.6.1053

Sampson RJ, Laub JH (2003) Life-course desisters? Trajectories of crime among delinquent boys followed to age 70. Criminology 41(3):555–592

Schwartz D, Flamant R, Lelouch J (1981) Clinical trials. Academic, London

Shih JC, Chen K, Ridd MJ (1999) Monoamine oxidase: from genes to behavior. Annu Rev Neurosci 22:197–217. doi:10.1146/annurev.neuro.22.1.197

Shirtcliff EA, Granger DA, Booth A, Johnson D (2005) Low salivary cortisol levels and externalizing behavior problems in youth. Dev Psychopathol 17(1):167–184. doi:10.1017/s0954579405050091

Shumay E, Logan J, Volkow ND, Fowler JS (2012) Evidence that the methylation state of the monoamine oxidase A (MAOA) gene predicts brain activity of MAO A enzyme in healthy men. Epigenetics 7(10):1151–1160. doi:10.4161/epi.21976

Siever LJ (2008) Neurobiology of aggression and violence. Am J Psychiatry 165(4):429–442. doi:10.1176/appi.ajp.2008.07111774

Siever LJ, Buchsbaum MS, New AS, Spiegel-Cohen J, Wei T, Hazlett EA, Sevin E, Nunn M, Mitropoulou V (1999) d, l-fenfluramine response in impulsive personality disorder assessed with F-18 fluorodeoxyglucose positron emission tomography. Neuropsychopharmacology 20(5):413–423. doi:10.1016/s0893-133x(98)00111-0

Sloan EK, Capitanio JP, Tarara RP, Mendoza SP, Mason WA, Cole SW (2007) Social stress enhances sympathetic innervation of primate lymph nodes: mechanisms and implications for viral pathogenesis. J Neurosci 27(33):8857–8865. doi:10.1523/jneurosci.1247-07.2007

Smith AK, Conneely KN, Kilaru V, Mercer KB, Weiss TE, Bradley B, Tang YL, Gillespie CF, Cubells JF, Ressler KJ (2011) Differential immune system DNA methylation and cytokine regulation in post-traumatic stress disorder. Am J Med Genet Part B Neuropsychiatr Genet 156B(6):700–708. doi:10.1002/ajmg.b.31212

Steinbusch HWM (1981) Distribution of serotonin-immunoreactivity in the central nervous-system of the rat: cell-bodies and terminals. Neuroscience 6(4):557–618. doi:10.1016/0306-4522(81)90146-9

Stevens HE, Leckman JF, Coplan JMD, Suomi SJ (2009) Risk and resilience: early manipulation of Macaque social experience and persistent behavioral and neurophysiological outcomes. J Am Acad Child Adolesc Psychiatry 48(2):114–127. doi:10.1097/CHI.0b013e318193064c

Suarez EC, Lewis JG, Kuhn C (2002) The relation of aggression, hostility, and anger to lipopolysaccharide-stimulated tumor necrosis factor (TNF)-alpha by blood monocytes from normal men. Brain Behav Immun 16(6):675–684. doi:10.1016/s0889-1591(02)00019-3

Suderman M, Borghol N, Pappas JJ, Pereira SMP, Pembrey M, Hertzman C, Power C, Szyf M (2014) Childhood abuse is associated with methylation of multiple loci in adult DNA. BMC Med Genomics 7:1–12. doi:10.1186/1755-8794-7-13

Suomi SJ (2005) Genetic and environmental factors influencing the expression of impulsive aggression and serotonergic functioning in Rhesus monkeys. In: Tremblay RE, Hartup WW, Archer J (eds) Developmental origins of aggression. Guilford Press, New York, pp 63–82

Suomi SJ (2006) Risk, resilience, and gene x environment interactions in rhesus monkeys. In: Lester BM, Masten AS, McEwen B (eds) Resilience in children, vol 1094. Blackwell Publishing, Oxford, pp 52–62

Sweeten G, Piquero AR, Steinberg L (2013) Age and the explanation of crime. Revisited [Published online]. J Youth Adolesc 1–18. doi:10.1007/s10964-013-9926-4

Szyf M, Weaver ICG, Provençal N, McGowan P, Tremblay RE, Meaney MJ (2009) Epigenetics and behaviour. In: Tremblay RE, van Aken MAG, Koops W (eds) Development and prevention of behaviour problems: from genes to social policy. Psychology Press, Hove, pp 25–59

Tarullo AR, Gunnar MR (2006) Child maltreatment and the developing HPA axis. Horm Behav 50(4):632–639. doi:10.1016/j.yhbeh.2006.06.010

Tharner A, Luijk MP, Raat H, IJzendoorn MH, Bakermans-Kranenburg MJ, Moll HA, Jaddoe VW, Hofman A, Verhulst FC, Tiemeier H (2012) Breastfeeding and its relation to maternal sensitivity and infant attachment. J Dev Behav Pediatr 33(5):396–404

Tremblay RE (2003) Why socialization fails? The case of chronic physical aggression. In: Lahey BB, Moffitt TE, Caspi A (eds) Causes of conduct disorder and juvenile delinquency. Guilford Publications, New York, pp 182–224

Tremblay RE (2010) Developmental origins of disruptive behaviour problems: the 'original sin' hypothesis, epigenetics and their consequences for prevention. J Child Psychol Psychiatry 51(4):341–367. doi:10.1111/j.1469-7610.2010.02211.x

Tremblay RE, Côté SM (2009) Development of sex differences in physical aggression: the maternal link to epigenetic mechanisms [comment]. Behav Brain Sci 32(3–4):290–291. doi:10.1017/S0140525X09990951

Tremblay RE, Japel C, Pérusse D, McDuff P, Boivin M, Zoccolillo M, Montplaisir J (1999) The search for the age of "onset" of physical aggression: Rousseau and Bandura revisited. Crim Behav Mental Health 9(1):8–23

Tremblay RE, Nagin D, Séguin JR, Zoccolillo M, Zelazo PD, Boivin M, Pérusse D, Japel C (2004) Physical aggression during early childhood: trajectories and predictors. Pediatrics 114(1):e43–e50

Tsankova N, Renthal W, Kumar A, Nestler EJ (2007) Epigenetic regulation in psychiatric disorders. Nat Rev Neurosci 8(5):355–367. doi:10.1038/nrn2132

Uddin M, Aiello AE, Wildman DE, Koenen KC, Pawelec G, de los Santos R, Goldmann E, Galea S (2010) Epigenetic and immune function profiles associated with posttraumatic stress disorder. Proc Natl Acad Sci U S A 107(20):9470–9475. doi:10.1073/pnas.0910794107

Vage J, Wade C, Biagi T, Fatjo J, Amat M, Lindblad-Toh K, Lingaas F (2010) Association of dopamine- and serotonin-related genes with canine aggression. Genes Brain Behav 9(4):372–378. doi:10.1111/j.1601-183X.2010.00568.x

van Bokhoven I, Van Goozen SHM, van Engeland H, Schaal B, Arseneault L, Séguin JR, Nagin DS, Vitaro F, Tremblay RE (2005) Salivary cortisol and aggression in a population-based longitudinal study of adolescent males. J Neural Transm 112(8):1083–1096. doi:10.1007/s00702-004-0253-5

van Lier PAC, Vitaro F, Barker ED, Koot HM, Tremblay RE (2009) Developmental links between trajectories of physical violence, vandalism, theft, and alcohol-drug use from childhood to adolescence. J Abnorm Child Psychol 37(4):481–492. doi:10.1007/s10802-008-9289-6

Vaughn MG, Delisi M, Beaver KM, Wright JP (2009) DAT1 and 5HTT are associated with pathological criminal behavior in a nationally representative sample of youth. Crim Justice Behav 36(11):1113–1124. doi:10.1177/0093854809342839 von Kanel R, Hepp U, Kraemer B, Traber R, Keel M, Mica L, Schnyder U (2007) Evidence for low-grade systemic proinflammatory activity in patients with posttraumatic stress disorder. J Psychiatr Res 41(9):744–752. doi:10.1016/j.jpsychires.2006.06.009

Wang D, Szyf M, Benkelfat C, Provençal N, Caramaschi D, Côté SM, Vitaro F, Tremblay RE, Booij L (2012) Peripheral SLC6A4 DNA methylation is associated with in vivo measures of human brain serotonin synthesis and childhood physical aggression. PLoS One 7(6):1–8 (e39501). doi:10.1371/journal.pone.0039501

Weaver ICG, Cervoni N, Champagne FA, D'Alessio AC, Sharma S, Seckl JR, Dymov S, Szyf M, Meaney MJ (2004) Epigenetic programming by maternal behavior. Nat Neurosci 7(8):847–854. doi:10.1038/nn1276

Whitaker-Azmitia PM (2001) Serotonin and brain development: role in human developmental diseases. Brain Res Bull 56(5):479–485

Wrangham RW, Wilson ML, Muller MN (2006) Comparative rates of violence in chimpanzees and humans. Primates 47(1):14–26

Zimbardo PG (2007) The Lucifer effect: understanding how good people turn evil. Random House, New York

Chapter 11
Coffee Health Effects from Early Fetal Development Through Childhood and Adolescence

Roseane Maria M. Santos and Darcy Roberto A. Lima

Abstract Coffee is a complex mixture of bioactive compounds and the major source of caffeine in the adult diet. Caffeine is a psychoactive stimulant, which is widely used by adults and increasingly adopted by the youth in the form of coffee, sodas and energy drinks. Caffeine is a legal drug that is still the object and focus of research on possible toxic and teratogenic effects on human health. As a consequence, coffee consumption during pregnancy, possible effects on the fetus and early and later ages of human development, such as childhood and adolescence, are important areas of translational toxicology to be investigated and better understood. This chapter is a tentative effort to review the most important data over the last 20 years on coffee effects, in particular, about its caffeine content, in terms of what is available regarding harmful or beneficial effects of drinking coffee regularly during pregnancy and its possible consequences to the fetus, neonate, infant and adolescent. The great majority of the studies found were epidemiological studies where the frequency of coffee consumption was self-reported by a questionnaire. Some valuable animal studies are also included as well as human studies with healthy volunteers. Overall, the take home message is that coffee consumed in moderation, 3–4 cups a day in adults and 1–2 cups a day during pregnancy is safe for human health.

Keywords Coffee • Effects • Fetus • Child • Adolescence

R.M.M. Santos, BS, MS, Ph.D. (✉)
Department of Pharmaceutical Sciences, South University School of Pharmacy, Savannah, GA, USA
e-mail: rsantos@southuniversity.edu

D.R.A. Lima, M.D., Ph.D.
Instituto de Neurologia Deolindo Couto, Universidade Federal do Rio de Janeiro, Rio de Janeiro, Brazil

C.L. Hughes, M.D. Waters (eds.), *Translational Toxicology*, Molecular and Integrative Toxicology, DOI 10.1007/978-3-319-27449-2_11

11.1 Introduction

More than one billion people start their day by drinking a cup of coffee, making it the most popular drink worldwide. After oil, the coffee industry is second in the world economy (Santos and Lima 2007). Despite 20 years of reassuring research, many people still avoid coffee because they worry about its health effects, but those concerns are understandable (Medical 2004). Older studies had linked coffee drinking to a wide range of health problems from teeth discoloration (Kumar et al. 2012) to pancreatic cancer and heart disease (Medical 2004); from neural tube defects (Yoon et al. 2001) to risk of pre-term birth (Brent et al. 2011) and child hood acute leukemia (Cheng et al. 2014) to sleepiness and tiredness from caffeinated beverages in adolescents (Orbeta et al. 2006). Many of those studies focused on the caffeine content of coffee, since caffeine is a CNS stimulant known to produce harmful health effects when ingested in high concentrations (Lean and Crozier 2012).

Coffee is not only caffeine. Coffee is a complex mixture of bioactive compounds that actually can affect health positively such as, antioxidants (polyphenols), minerals (high content of potassium, low content of sodium, high content of iron and calcium) and vitamins (niacin) (Santos and Lima 1989). This chapter is devoted to the compilation of the most recent data on the effects of coffee on human development from the fetus to adolescence. It will be divided according to age ranges that share developmental and social characteristics that pose specific health related issues. The effects of coffee will be presented divided in 3 developmental periods: (a) coffee effects on the fetus and pregnant women; (b) coffee effects in infancy and early childhood; and (c) coffee effects in adolescence.

It is necessary to make a comment on how the effects of coffee were searched in the medical literature for the preparation of this chapter. Caffeine still remains as the main concern in terms of toxicity in children. Therefore, considering that coffee is the major source of caffeine in the American and worldwide diet, most of what is stated in this chapter was collected from studies on caffeine effects, as long as coffee was mentioned as part of the diet being studied.

11.2 Coffee Effects on the Fetus and Pregnant Women

Over the last 30 years, an enormous number of research papers were published on coffee consumption during pregnancy and possible risks for the pregnant woman and the unborn child. The association of cigarette smoking and coffee drinking habit is very common and many times a confounding factor, which most of the older studies did not take into account and led to controversial results. The majority of the studies utilized epidemiological or animal studies dealing with congenital malformations, miscarriage, pre-term birth or growth retardation. We will present a representative sample of those studies.

11.2.1 Coffee and Congenital Malformations

11.2.1.1 Orofacial Clefts

A study from the National Center on Birth Defects and Developmental Disabilities NCBD, Centers for Disease Control and Prevention, Atlanta, Georgia in collaboration with New York State Department of Health and Oak Ridge Institute for Science and Education, Tennessee (Collier et al. 2009) included 1531 infants with cleft lip with or without cleft palate (CL/P) and 813 with cleft palate only (CPO) and 5711 infants with no major birth defects (controls) born between October 1997 through December 2004. Mothers reported dietary caffeine intake from coffee, tea, sodas, and chocolate in the year before pregnancy and reported intake of medications containing caffeine during pregnancy. Eleven percent reported consuming at least 300 mg of caffeine per day (2–3 cups/day) and 17 % reported consuming less than 10 mg of caffeine per day. The results did not suggest an association between maternal dietary caffeine intake and orofacial clefts, but caffeine-containing medications merit further study.

Another study comes from Norway, Department of Nutrition at University of Oslo. In Norway, coffee consumption is relatively high, reporting an average coffee intake for adult population (40–60 years old) of half a liter a day (Johansen et al. 2009). This study included 573 cases, 377 with CL/P and 196 with CPO and 763 randomly selected controls of infants delivered between 1996 and 2001. Mothers completed a 32-page questionnaire covering coffee intake, reported as number of cups per day, considering 100 mg of caffeine per cup as well as any medications and smoking and alcohol consumption during the first trimester of pregnancy. They found no evidence of an association between maternal coffee consumption during the first trimester and the risk of CPO, but there was a dose-response relationship with risk of delivering an infant with CL/P. The mechanism by which coffee intake might increase the risk of CL/P is not known, although the effects on homocysteine could be a potential pathway. Folic acid supplements reduce homocysteine plasma levels and reduce the risk of CL/P. Coffee intake, on the other hand, increases homocysteine levels (Grubben et al. 2000; Urgert et al. 2000), as well as smoking (Little et al. 2004), which is a well-established risk factor for CL/P (Johansen et al. 2009).

11.2.1.2 Neural Tube Defects (NTD)

The National Birth Defects Prevention Study (NBDPS) is an ongoing multi-center population-based case control study of birth defects, which started on 1997 (Yoon et al. 2001). A subset of those participants, involving 306 mothers of NTD cases and 669 control infants and their parents were genotyped for CYP 1A2, the enzyme responsible for 95 % of caffeine metabolization. CYP 1A2 shows genetic polymorphism that reflects the bimodal distribution in the population (Sachse et al. 1999). The wild type, most common genotype is denoted by the presence of CYP 1A2*1A allele and the variant allele is when CYP 1A2*1 F allele is present. The variant allele confers a slow caffeine oxidative phenotype and the wild type is associated with fast caffeine metabolization (Butler et al. 1989, Gu et al. 1992, Han et al. 2001). The purpose was to explore the association between NTDs and maternal and infant gene variants involved in caffeine metabolism. They found that maternal caffeine and its metabolites may be associated with increased risk for NTD-affected pregnancies in genetically susceptible subgroups. Caffeine-consuming mothers who were CYP 1A2 fast oxidizers, which are the most common genotype in the normal population, had non-significantly increased risk for NTD-affected pregnancy (Schmidt et al. 2010).

11.2.1.3 Cardiovascular Malformations (CVM)

Another subset from NBDPS population studied the consumption of caffeinated beverages (coffee, tea, colas, etc.) reported from 4196 CVM case infants overall and 3957 control infants (Browne et al. 2007). No significant positive association was found between maternal caffeine consumption and CVMs. On the contrary, an inverse trend between coffee intake and risk of atrial septal defect was observed.

On the contrary, animal studies on pregnant mice assessing both the short-term effects on cardiac development and embryo growth and long-term effects on cardiac function of *in utero* caffeine (coffee) exposure found different results (Wendler et al. 2009). In this study, animals were exposed to hypoxia (10 % O_2) and treated with caffeine equivalent to the circulating levels in humans after 2 cups of coffee, from embryonic days 8.5–10.5. The hypothesis was that caffeine is an adenosine receptor antagonist and could disrupt adenosine protective action against hypoxia in utero, leading to acute effects on the embryo and long-term effects in adult mice. They found caffeine-induced effects on cardiac ventricle thickness in embryos and that effects on body composition (increased body fat) and cardiac function in adulthood were more greatly influenced by prenatal caffeine exposure.

Another recent study (Buscariollo et al. 2014) following the same methodology as the previous one, found that the observed altered cardiac function and morphology in adult mice exposed to caffeine (coffee) *in utero* was mediated by adenosine A1 receptors through DNA methylation. Differentially methylated regions within the genome were associated with cardiac hypertrophy.

11.2.1.4 Occurrence of Trisomy 21

A study involving a case-control of 997 liveborn infants with Trisomy 21 and 1007 liveborn controls without a birth defect between 1991 and 1993 in California was used to evaluate possible effects of maternal smoking and coffee (caffeine) consumption on the occurrence of a recognized pregnancy with Trisomy 21 (Torfs and Christianson 2000). An inverse association was found only for non-smoking mothers who drank ≥4 cups of coffee per day, suggesting that high coffee consumption is more likely to reduce the viability of a Trisomy 21 conceptus than that of a normal conceptus.

A search of the MEDLINE/PUBMED from 1966 through October 2004 was made for all epidemiologic studies with maternal intake of caffeine as an exposure and major congenital malformations (Browne 2006). Their conclusion was that there is no evidence of teratogenic effect of caffeine in humans.

11.2.2 Coffee, Pregnancy Loss and Birth Weight Reduction

11.2.2.1 Spontaneous Abortion (SA)

Studies from 1996 (Dlugosz et al. 1996) and 1997 (Fenster et al. 1997) investigated the use of caffeinated beverages (coffee, tea and soda) within 2967 and 5144 pregnant women respectively. Both studies concluded that as compared with abstention of caffeinated beverages, the adjusted odds ratios were increased for spontaneous abortion in association with three or more cups of coffee a day (>300 mg of caffeine). However, the first study could not identify the cause and attributed the effect to "some ingredient (or correlate) of coffee or tea that may account for the observed association" and the second study says that "we suspect that this association may be biased from relations among fetal viability, symptoms of pregnancy such as nausea, and consumption patterns during pregnancy."

A review article recently published (Brent et al. 2011) covering period between 2000 and 2010 found that out of the 17 epidemiological studies dealing with the risk of SA from exposure to caffeine, only one actually measured the serum levels of caffeine and metabolites to determine the actual exposure. These authors noted that among these studies the results were inconsistent, some referring to levels of 300 mg or more of caffeine increasing the risk of SA and others reporting exposures of 500–900 mg of caffeine as not associated with risk of SA. The review concludes that when the wide range of human exposures have been utilized in animal reproductive studies, increased pregnancy loss in mammalian reproductive studies have not been detected.

11.2.2.2 Small for Gestational Age (SGA) and Pre-term Birth

Caffeine is known to cross the placenta and reach the fetus and that the clearance of caffeine in the pregnant women is delayed (Chiaffarino et al. 2006). Studies of coffee drinking and pre-term births have produced conflicting results, had limited information about caffeine sources, and did not control for any confounders. Parazzini and Chiffarino (Chiaffarino et al. 2006) from University of Milan in Italy addressed this issue and found that 1966 women who gave birth at term (≤37 weeks) used as controls and 502 who delivered early (≤37 weeks) that there was an inverse association with coffee consumption in the 3rd trimester of pregnancy in the SGA cases compared with the normal gestational age. However, compared with the non-coffee drinkers, a low consumption of coffee during pregnancy may not have significant effects on pre-term birth. A systematic review from Brazil (Pacheco et al. 2007) on caffeine consumption and prevalence of low birth weight and prematurity concluded that an association between moderate caffeine consumption and fetal growth was not demonstrated. The latest review and caffeine dose-response meta-analysis have found that the risk for low birth weight increased with increasing levels of caffeine intake (Chen et al. 2014). They found that the risk for low birth weight was significantly higher even in the low and moderate caffeine intake groups, 50–149 mg/day and 150–349 mg/day respectively, as compared with the reference group. Nonetheless, they cannot exclude the possibility for potential bias such as residual confounding for smoking and pregnancy symptoms. The current guideline by the WHO is to limit caffeine intake during pregnancy to a maximum of 300 mg/day (2–3 cups/day) and even more stringent are the Nordic Nutrition and American College of Obstetricians and Gynecologists recommendations to limit caffeine exposure to 200 mg/day (2010).

A review article on the evaluation of the reproductive and developmental risks of caffeine (Brent et al. 2011) has examined a number of epidemiological studies during the period of 2000–2010 on growth retardation in animals fed with caffeine (coffee). They concluded that the level of exposure necessary to produce fetal growth retardation to a pregnant women had to be significantly higher and far above the highest possible caffeine level of exposure which a pregnant woman would be exposed. They also concluded that there is a need for controlled trials to isolate cause and effect (Chen et al. 2014, Martin 2013; Pacheco et al. 2007).

11.2.3 Maternal Coffee Intake and Potential Complications

11.2.3.1 Gestational Diabetes Mellitus (GDM)

The Epidemiology Branch at the Eunice Kennedy Shriver National Institute of Child Health and Human Development, National Institute of Health, Bethesda, MD examined a population of non-diabetic women with singleton pregnancies in the Danish National Birth Cohort (n = 71,239) and estimated the relative risks for the association between first trimester coffee and tea intake or estimated total caffeine intake and GDM. Their results suggest that moderate first trimester coffee and tea consumption were not associated with GDM increased risk; on the contrary it could have a protective effect (Hinkle et al. 2014).

11.2.3.2 Anemia and Iron Deficiency

There are reports of findings from the 1980s in rats (Munoz et al. 1986) and among pregnant women and their infants (Munoz et al. 1988), which indicated that maternal coffee intake may contribute to maternal and infant anemia due to iron deficiency. Animal studies also investigated the effect of coffee consumption and the absorption of zinc and iron (Morck et al. 1983; Pecoud et al. 1975). The results suggested a significant reduction in the zinc and iron absorption due to coffee intake. A review article on the consequences for the newborn of chronic coffee consumption during gestation and lactation concluded (Nehlig and Debry 1994) that maternal caffeine consumption in moderate amounts ($\leq$300 mg of caffeine equivalent to 2–3 cups a day) has no measurable consequences in the fetus and newborn infant. No recent studies were found to confirm those previous results. It is possible that the increased addition of important minerals to cereals and the required use of vitamins and minerals supplementation during pregnancy might have adjusted those increased needs during gestation and in the infants.

11.2.3.3 Febrile Seizures

Febrile seizures are quite common during childhood, caused by genetic or environmental factors in early life. The association of maternal coffee intake and exposure to cigarettes and alcohol was evaluated by the Aarhus Birth Cohort consisted of 25,196 children of mothers scheduled to deliver between 1989 and 1996 (Vestergaard et al. 2005). No association was found between maternal coffee intake during pregnancy and risk of febrile seizures.

11.2.3.4 Primary Sclerosing Cholangitis (PSC)

A recent study from Norway on maternal coffee consumption, smoking and hormones and the risk of PSC was published (Andersen et al. 2014). They observed 240 patients with PSC from Oslo University Hospital and they concluded that coffee consumption and smoking can actually protect against development of PSC.

11.2.3.5 Coffee Effects on Adult Son's Semen Quality

Association between prenatal coffee, current caffeine exposure and semen quality and levels of reproductive hormones was evaluated in 347 sons out of 5109 selected for a follow-up study from The Danish Pregnancy Cohort. Semen and blood samples were analyzed (Ramlau-Hansen et al. 2008). Caffeine intake was shown to be associated with increased levels of testosterone, but no clear association was found with semen quality. Another study with a subset of The Child Health and Development Studies involving the participation of 338 adult sons from a follow-up study where 196 participants donated semen for analysis (Cirillo et al. 2011). Semen samples were analyzed for sperm concentration, motility and morphology. It was found a proportionate 25 % reduction on sperm count, 13 % decrease in motility and 25 % decline in the normal morphology. Those studies provide suggestive evidence that maternal coffee use during pregnancy may impair the reproductive development of the male fetus. However, those results were from a very high prenatal exposure, corresponding to 5 cups of coffee use per day, much higher than the recommended dose of no more than 2 cups per day during pregnancy.

11.2.3.6 Strabismus

A review of medical records for children in the Danish National Birth Cohort identified 1321 cases of strabismus from a total of 96,842 children born between 1996 and 2003 (Torp-Pedersen et al. 2010). Maternal smoking was associated with a significantly elevated risk of strabismus, light maternal alcohol consumption was associated with decreased risk and no association was found with maternal coffee or tea drinking.

11.2.4 Coffee and Risk of Cancers

11.2.4.1 Childhood Acute Leukemia

Epidemiological studies on the association between coffee consumption during pregnancy and childhood acute lymphoblastic leukemia (ALL), the most common subtype among acute leukemia in US, have been inconsistent (Cheng et al. 2014). Two French population-based case control studies from 2005 (Menegaux et al. 2005) and 2007 (Menegaux et al. 2007) concluded that over 280 incident cases and 288 controls that maternal coffee consumption during pregnancy was associated with childhood acute leukemia, ORs increasing in ALL with coffee consumption (OR = 1.1 [0.7–1.8], OR = 2.4 [1.3–4.7] and OR = 3.1 [1.0–9.5]), respectively, for < or = 3, 4–8 and >8 cups/day. The latter, otherwise, concluded that maternal coffee consumption was not significantly related to AL. Only the highest intake of coffee ($\geq$3 cups/day) had a significant correlation with mothers that were also nonsmokers, over a population of 472 of AL cases and 567 case controls. A recent meta-analysis study published on American Journal of Obstetrics and Gynecology (Chen et al. 2014) on this particular issue has concluded that it is suggestive of an increase in the risk of ALL with maternal coffee consumption. They found a linear dose-response relationship between coffee consumption and childhood AL.

Caffeine may act as a topoisomerase II inhibitor, a DNA repair inhibitor or a carcinogen metabolism inhibitor (Ferguson and Philpott 2008; Ross et al. 1996). These actions could induce chromosomal translocations and aberrations, such as on chromosome 11q23, which was taken as a cause for the pathogenesis of infant leukemia (Ross et al. 1996). However, it is our understanding that the inconsistency found in the studies might be attributable to the complex chemistry of coffee and to other compounds present in much higher concentration in coffee, such as the chlorogenic acids (CGA's) (Santos 2010).

11.2.4.2 Coffee Constituent Chlorogenic Acid and Cancer Prevention

Coffee is the major dietary source of chlorogenic acids (CGA's), a 200 mL cup is reported to contain a range from 70 to 350 mg, much higher than the amount of caffeine (Burgos-Moron et al. 2012). CGA's have been known for their antioxidant properties (Kono et al. 1997; Sato et al. 2011). Burgos-Moron's group of researchers reported for the first time that CGA induces high levels of topoisomerase I and topoisomerase II-DNA complexes in cells. Topo I and topo II are nuclear enzymes that introduce single- or double-strand breaks in the DNA to solve topological problems associated with DNA replication, transcription, recombination and chromatin remodeling. Their study showed that lung cancer cells were more sensitive than normal lung fibroblasts to the cytotoxic activity of CGA, suggesting that CGA may induce selective killing of cancer cells and consequently a possible cancer preventative activity.

11.2.4.3 Coffee and Risk of Testicular Cancer

It has been suggested that increased risk for testicular cancer in the world may be due to exposures during fetal development (Bray et al. 2006; Maffezzini 2007; Huyghe et al. 2003). The Child Health and Development Studies, a 40-year follow-up of more than 20,000 pregnancies between 1959 and 1967 found only 20 cases of testicular cancer diagnosed through 2003 among sons with a maternal interview during pregnancy. Compared with controls, mothers of testicular cancer cases were more likely to drink alcohol and less likely to drink coffee (Mongraw-Chaffin et al. 2009).

11.3 Coffee Effects in Infancy and Early Childhood

Caffeine is a widely used psychoactive substance by both adults and children. It is classified as a stimulant drug that is typically used for its ability to arouse the central nervous system. Children and adolescents are the fastest growing population of caffeine users with an increase of 70 % in the past 30 years (Harnack et al. 1999). This section will review the effects of the use of caffeine-containing beverages (coffee, tea, chocolate, soft drinks) on the behavior and development of children.

It is important to recall that although coffee is the main source of caffeine in the adult diet, it seems that soft drinks is the preferred route of caffeine administration by children and adolescents (Frary et al. 2005). Much of the data from the studies that this section will cover will be a reflection of their drinking habits and not directly related to coffee consumption.

A review of effects of caffeine on development and behavior in infancy and childhood, published in 2002 by researchers from of the National Institute of Mental Health, NIH, Bethesda (Castellanos and Rapoport 2002), examined studies found in the literature in the previous decade. They found that the number of papers indexed were small; in consequence they have attributed this scarce literature to the continued practice to recommend women to moderate coffee consumption. They reported a few positive associations such as gestational coffee drinking linked to iron deficiency in children (Engle et al. 1999) and heavy caffeine intake (≥400 mg/day) associated with increased risk of sudden infant death syndrome (SIDS) (Ford et al. 1998). Overall the results were beneficial, such as that children reported fewer adverse effects while exhibiting greater objective changes in activity and task performance (Rapoport et al. 1981). Additionally a meta-analysis of 9 studies with 96 children with Attention Deficit/Hyperactivity Disorder (ADHD) showed a beneficial effect on parental ratings of aggressive or disruptive behavior (Stein et al. 1996). Finally the report concluded that there was little evidence that would warrant grave concern about the use of moderate doses of caffeine in most situations and that the effects of caffeine in children seem- to be modest and generally innocuous.

Animal studies looking at chronic caffeine treatment during the pre-pubertal period in adult spontaneously hypertensive rats (SHR), an animal model for the study of ADHD, demonstrated for the first time that caffeine or methylphenidate (Ritalin™) improved cognitive deficits in adulthood (Pires et al. 2010). Methylphenidate is the most accepted pharmacological treatment for ADHD. However, its use during adolescence may cause long-lasting neurobiological developmental consequences in rodents. In this case, the search of an alternative/complimentary treatment such as caffeine could be of help.

The effect of caffeine and technology on sleep duration and body mass index was the focus of a study with 625 children aged between 6 and 20 years old, from the National Sleep Foundation's Sleep in America Poll (Calamaro et al. 2012). The study found that on a typical day around 30 % of the children consumed a cup or can of caffeinated beverage, almost half of the sample had a television in their bedroom and 10 % or fewer had a computer and a phone. Children averaged 9.5 h of sleep each night, even though children aged between 6 and 10 years need 10–11 h of sleep per night. A complex relationship between caffeine intake and the use of technology such as television and computers in the bedroom was found. Shortened sleep (15 min less) was associated with drinking a cup of caffeinated beverage, but greater loss of sleep (45 min less, on average) was found when 2 other items were present: television and computer. However, no individual technology item influenced the hours of sleep the child obtained.

Another review on caffeine use in children published in 2009 (Temple 2009) discussed probable mechanisms for the caffeine effects seen in adults as well as in children, mainly through its binding to adenosine receptors in the brain. The main concern is the fact that children and adolescents are still developing some parts of the brain. Some of these areas such as orbitofrontal and temporal lobes (Giedd 1999; Sowell et al. 1999) contain adenosine receptors that can be potentially affected by caffeine (Svenningsson et al. 1997). However, the data available regarding caffeine use in children is very scarce and do not allow drawing any solid conclusions. More research needs to be conducted in this area.

11.4 Coffee Effects in Adolescence

11.4.1 Effects of Acute Doses of Caffeine in Adolescents

Acute administration of doses of 50, 100 and 200 mg of caffeine (equivalent to 1–2 cups of coffee, 100 mg of caffeine per cup on average) to a group of adolescents followed by monitoring cardiovascular responses (blood pressure) and food intake has been performed (Temple et al. 2010). The main effects of an acute caffeine dose was on heart rate (HR) and diastolic blood pressure (DBP), with HR decreasing but DBP increasing with increasing caffeine dose. High caffeine consumers (>50 mg/day) reported to use caffeine to stay awake in the form of coffee, tea, sodas and energy drinks. Boys were more likely than girls to report use of caffeine for a "rush" or simply for more energy or exercise performance. They concluded that the acute effects of caffeine in adolescents are moderated by gender and frequency of caffeine use.

11.4.2 Effects of Acute and Chronic Dosing of Caffeine in Animals

Rats undergo developmental changes in the brain during adolescence with many parallels to human brain development during the comparable stage (Crews et al. 2007, Spear 2000). Therefore, adolescent rats are a valuable model in exploring interactions of drugs with the development of the brain. A study used rats on the 28th postnatal day-of-age (P28) to test as adolescents and in the age-range between P65 to P95 to test as young adults (Rhoads et al. 2011). The purpose of the study was to examine the effects of acute and chronic dosing of caffeine in adolescent rats and compare with adult rats. Their results showed that adolescent rats present similar responses to the initial acute dose of caffeine but showed greater signs of tolerance and dependence than adults after regular coffee consumption. They concluded that adaptive changes such as tolerance and withdrawal symptoms may be occurring faster or to a greater extent in the still-developing adolescent brain.

11.4.3 Coffee and Weight Gain

A subset of 5502 girls from the Growing up Today Study from all over U.S. aged 14–21 year., returned surveys in 2001 reporting the past-year recreational Internet time, sleep, coffee (with caffeine) and alcohol consumption. Their height and weight were also reported in 2000. The investigators examined whether excessive recreational Internet time, insufficient sleep, regular coffee and/or alcoholic beverages consumption promote weight gain (Berkey et al. 2008). They found that females aged 18+ years presented a high correlation between more Internet time, more alcohol and less sleep with same year BMI increase. The study also found no evidence that drinking coffee promotes weight gain.

Some studies were looking at coffee and other caffeine containing beverages consumption concerned with the establishment of future habits in adult life that could lead to disorders such as diabetes and obesity (Koivusilta et al. 2001; O'Dea and Wilson 2006). Koivusilta et al. (2001) concluded that poor dietary choices (high consumption of soft drinks, sweets, snack foods, take away and large food portions) were associated with a high BMI in children and adolescents and in adult life. The findings were most notable in those who either skipped breakfast and/or prepared a breakfast of poor nutritional quality (Ortega et al. 1998). O'Dea in her study also looked at educational background and socioeconomic status (SES) in a large (n=4441) national study of school children aged 6–18 years. (O'Dea and Wilson 2006). In this study, non-nutritious fluids were considered as soft drink, water, coffee and tea; and that students with low SES were more likely to skip breakfast or to consume a non-nutritious breakfast instead. Boys and girls of low SES as compared with middle/high SES had significantly greater BMI and consequently greater risk of overweight and obesity.

11.4.4 Coffee Consumption and Vitamin D

The effects of coffee (caffeine) consumption and levels of vitamin D are controversial. It has been demonstrated that caffeine negatively influences calcium balance by reducing renal reabsorption of calcium (Massey and Whiting 1993). Another study found that methylxanthine, theophylline and caffeine inhibit the conversion of 25-hydroxyvitamin D3 to 1,25- dihydroxyvitamin D3 in the renal tubules of vitamin D deficient chicks, leading to an increase in the vitamin D circulating levels (Taft et al. 1984). Moderate coffee consumption has no effect on bone health (Massey and Whiting 1993). A recent study with 330 randomly selected Saudi adolescents aged 11–14 years. from the existing Biomarkers Screening in Riyadh Program found that serum vitamin D levels increases as coffee and tea consumption increases; and that increase is independent of physical activity, sun exposure, gender, age and BMI (Al-Othman et al. 2012).

11.4.5 Effects of Caffeine on Sleep Disorders in Animals

Caffeine is a CNS stimulant known to affect sleep-wake regulation (Fredholm et al. 1999; Schwierin et al. 1996). Adolescence is a critical period for brain maturation in which a huge reorganization of cortical connectivity takes place. A study with the purpose of establishing a rat model to assess the relationship between cortical maturation and sleep utilized caffeine as a short-term stimulant (Olini et al. 2013). The amount of caffeine administered to the rats was equivalent to 3–4 cups of coffee a day (100 mg caffeine/cup of coffee on average). The study showed that caffeine interferes with cortical maturation during adolescence in rats, a critical period of brain development. Also these results might be of clinical importance since the critical period of synapse elimination during adolescence is associated with increased incidence of psychiatric and mood disorders such as schizophrenia, anxiety, substance of abuse and personality disorders.

11.5 Conclusion

The majority of the literature reviewed in this section was related to the effects of caffeine in the early stages of human life, during pregnancy, neonates and early childhood; and the major concern being the possible damage to the various phases of child development.

The studies that involved those periods of human development referred to number of cups of coffee as the major source of caffeine in the adult diet. Some controversial results are still present, due to the lack of data in terms of type of caffeinated beverage (coffee/tea/sodas), type of coffee (caffeinated/decaffeinated) and correlation with in vivo levels of caffeine in the body. Nevertheless, the majority of the studies presented here concluded that coffee was not related with any of the adverse effects or development of diseases focused on the respective research papers. On the contrary, there were some studies that concluded that coffee in moderation is good for health.

As we move into late childhood and adolescence, the major source of caffeine is not coffee, but sodas and energy drinks. The number of papers are considerably smaller and the results cannot be attributed to coffee consumption, reason why they were not addressed in this chapter (Bernstein et al. 1994, 2002; Oberstar et al. 2002; Orbeta et al. 2006; Reissig et al. 2009). Many more studies are still necessary in order to better clarify those questions that continue unanswered. In the meantime, the best advice remains to use coffee/caffeine in moderation and enjoy your early morning beverage without guilt (Santos and Lima 2009).

References

Al-Othman A, Al-Musharaf S, Al-Daghri NM, Yakout S, Alkharfy KM, Al-Saleh Y, Al-Attas OS, Alokail MS, Moharram O, Sabico S, Kumar S, Chrousos GP (2012) Tea and coffee consumption in relation to vitamin D and calcium levels in Saudi adolescents. Nutr J 11:56

Andersen IM, Tengesdal G, Lie BA, Boberg KM, Karlsen TH, Hov JR (2014) Effects of coffee consumption, smoking, and hormones on risk for primary sclerosing cholangitis. Clin Gastroenterol Hepatol 12:1019–1028

Berkey CS, Rockett HR, Colditz GA (2008) Weight gain in older adolescent females: the internet, sleep, coffee, and alcohol. J Pediatr 153:635–639, 639 e1

Bernstein GA, Carroll ME, Crosby RD, Perwien AR, Go FS, Benowitz NL (1994) Caffeine effects on learning, performance, and anxiety in normal school-age children. J Am Acad Child Adolesc Psychiatry 33:407–415

Bernstein GA, Carroll ME, Thuras PD, Cosgrove KP, Roth ME (2002) Caffeine dependence in teenagers. Drug Alcohol Depend 66:1–6

Bray F, Richiardi L, Ekbom A, Pukkala E, Cuninkova M, Moller H (2006) Trends in testicular cancer incidence and mortality in 22 European countries: continuing increases in incidence and declines in mortality. Int J Cancer 118:3099–3111

Brent RL, Christian MS, Diener RM (2011) Evaluation of the reproductive and developmental risks of caffeine. Birth Defects Res B Dev Reprod Toxicol 92:152–187

Browne ML (2006) Maternal exposure to caffeine and risk of congenital anomalies: a systematic review. Epidemiology 17:324–331

Browne ML, Bell EM, Druschel CM, Gensburg LJ, Mitchell AA, Lin AE, Romitti PA, Correa A (2007) Maternal caffeine consumption and risk of cardiovascular malformations. Birth Defects Res A Clin Mol Teratol 79:533–543

Burgos-Moron E, Calderon-Montano JM, Orta ML, Pastor N, Perez-Guerrero C, Austin C, Mateos S, Lopez-Lazaro M (2012) The coffee constituent chlorogenic acid induces cellular DNA damage and formation of topoisomerase I- and II-DNA complexes in cells. J Agric Food Chem 60:7384–7391

Buscariollo DL, Fang X, Greenwood V, Xue H, Rivkees SA, Wendler CC (2014) Embryonic caffeine exposure acts via A1 adenosine receptors to alter adult cardiac function and DNA methylation in mice. PLoS One 9:e87547

Butler MA, Iwasaki M, Guengerich FP, Kadlubar FF (1989) Human cytochrome P-450PA (P-450IA2), the phenacetin O-deethylase, is primarily responsible for the hepatic 3-demethylation of caffeine and N-oxidation of carcinogenic arylamines. Proc Natl Acad Sci U S A 86:7696–7700

Calamaro CJ, Yang K, Ratcliffe S, Chasens ER (2012) Wired at a young age: the effect of caffeine and technology on sleep duration and body mass index in school-aged children. J Pediatr Health Care 26:276–282

Castellanos FX, Rapoport JL (2002) Effects of caffeine on development and behavior in infancy and childhood: a review of the published literature. Food Chem Toxicol 40:1235–1242

Chen LW, Wu Y, Neelakantan N, Chong M, Pan A, Van Dam RM (2014) Maternal caffeine intake during pregnancy is associated with risk of low birth weight: a systematic review and dose inverted question markresponse meta-analysis. BMC Med 12:174

Cheng J, Su H, Zhu R, Wang X, Peng M, Song J, Fan D (2014) Maternal coffee consumption during pregnancy and risk of childhood acute leukemia: a metaanalysis. Am J Obstet Gynecol 210:151 e1–151 e10

Chiaffarino F, Parazzini F, Chatenoud L, Ricci E, Tozzi L, Chiantera V, Maffioletti C, Fedele L (2006) Coffee drinking and risk of preterm birth. Eur J Clin Nutr 60:610–613

Cirillo PM, Cohn BA, Krigbaum NY, Lee M, Brazil C, Factor-Litvak P (2011) Effect of maternal coffee, smoking and drinking behavior on adult son's semen quality: prospective evidence from the Child Health and Development Studies. J Dev Orig Health Dis 2:375–386

Collier SA, Browne ML, Rasmussen SA, Honein MA (2009) Maternal caffeine intake during pregnancy and orofacial clefts. Birth Defects Res A Clin Mol Teratol 85:842–849

Crews F, He J, Hodge C (2007) Adolescent cortical development: a critical period of vulnerability for addiction. Pharmacol Biochem Behav 86:189–199

Dlugosz L, Belanger K, Hellenbrand K, Holford TR, Leaderer B, Bracken MB (1996) Maternal caffeine consumption and spontaneous abortion: a prospective cohort study. Epidemiology 7:250–255

Engle PL, Vasdias T, Howard I, Romero-Abal ME, Quan De Serrano J, Bulux J, Solomons NW, Dewey KG (1999) Effects of discontinuing coffee intake on iron deficient Guatemalan toddlers' cognitive development and sleep. Early Hum Dev 53:251–269

Fenster L, Hubbard AE, Swan SH, Windham GC, Waller K, Hiatt RA, Benowitz N (1997) Caffeinated beverages, decaffeinated coffee, and spontaneous abortion. Epidemiology 8:515–523

Ferguson LR, Philpott M (2008) Nutrition and mutagenesis. Annu Rev Nutr 28:313–329

Ford RP, Schluter PJ, Mitchell EA, Taylor BJ, Scragg R, Stewart AW (1998) Heavy caffeine intake in pregnancy and sudden infant death syndrome. New Zealand Cot Death Study Group. Arch Dis Child 78:9–13

Frary CD, Johnson RK, Wang MQ (2005) Food sources and intakes of caffeine in the diets of persons in the United States. J Am Diet Assoc 105:110–113

Fredholm BB, Battig K, Holmen J, Nehlig A, Zvartau EE (1999) Actions of caffeine in the brain with special reference to factors that contribute to its widespread use. Pharmacol Rev 51:83–133

Giedd J (1999) Brain development, IX: human brain growth. Am J Psychiatry 156:4

Grubben MJ, Boers GH, Blom HJ, Broekhuizen R, De Jong R, Van Rijt L, DE Ruijter E, Swinkels DW, Nagengast FM, Katan MB (2000) Unfiltered coffee increases plasma homocysteine concentrations in healthy volunteers: a randomized trial. Am J Clin Nutr 71:480–484

Gu L, Gonzalez FJ, Kalow W, Tang BK (1992) Biotransformation of caffeine, paraxanthine, theobromine and theophylline by cDNA-expressed human CYP1A2 and CYP2E1. Pharmacogenetics 2:73–77

Han XM, Ou-Yang DS, Lu PX, Jiang CH, Shu Y, Chen XP, Tan ZR, Zhou HH (2001) Plasma caffeine metabolite ratio (17X/137X) in vivo associated with G-2964A and C734A polymorphisms of human CYP1A2. Pharmacogenetics 11:429–435

Harnack L, Stang J, Story M (1999) Soft drink consumption among US children and adolescents: nutritional consequences. J Am Diet Assoc 99:436–441

Hinkle SN, Laughon SK, Catov JM, Olsen J, Bech BH (2015) First trimester coffee and tea intake and risk of gestational diabetes mellitus: a study within a national birth cohort. BJOG 122(3):420–8

Huyghe E, Matsuda T, Thonneau P (2003) Increasing incidence of testicular cancer worldwide: a review. J Urol 170:5–11

Johansen AM, Wilcox AJ, Lie RT, Andersen LF, Drevon CA (2009) Maternal consumption of coffee and caffeine-containing beverages and oral clefts: a population-based case-control study in Norway. Am J Epidemiol 169:1216–1222

Koivusilta LK, Rimpela AH, Rimpela M, Vikat A (2001) Health behavior-based selection into educational tracks starts in early adolescence. Health Educ Res 16:201–214

Kono Y, Kobayashi K, Tagawa S, Adachi K, Ueda A, Sawa Y, Shibata H (1997) Antioxidant activity of polyphenolics in diets. Rate constants of reactions of chlorogenic acid and caffeic acid with reactive species of oxygen and nitrogen. Biochim Biophys Acta 1335:335–342

Kumar A, Kumar V, Singh J, Hooda A, Dutta S (2012) Drug-induced discoloration of teeth: an updated review. Clin Pediatr (Phila) 51:181–185

Lean ME, Crozier A (2012) Coffee, caffeine and health: what's in your cup? Maturitas 72:171–172

Little J, Cardy A, Munger RG (2004) Tobacco smoking and oral clefts: a meta-analysis. Bull World Health Organ 82:213–218

Maffezzini M (2007) TC incidence increasing: spread the word. Eur Urol 51:596–597

Martin CH (2013) Higher coffee intake in pregnancy linked to prolonged gestation, and caffeine intake linked with babies being small for gestational age. Evid Based Nurs 0: 1

Massey LK, Whiting SJ (1993) Caffeine, urinary calcium, calcium metabolism and bone. J Nutr 123:1611–1614

Medical SH (2004) Coffee health risks: for the moderate drinker, coffee is safe says Harvard Women's Health Watch [Online]. Harvard Medical Publications, Boston. Available: https://www.health.harvard.edu/press_releases/coffee_health_risk. Accessed 20 Jan 2015

Menegaux F, Steffen C, Bellec S, Baruchel A, Lescoeur B, Leverger G, Nelken B, Philippe N, Sommelet D, Hemon D, Clavel J (2005) Maternal coffee and alcohol consumption during pregnancy, parental smoking and risk of childhood acute leukaemia. Cancer Detect Prev 29:487–493

Menegaux F, Ripert M, Hemon D, Clavel J (2007) Maternal alcohol and coffee drinking, parental smoking and childhood leukaemia: a French population-based case-control study. Paediatr Perinat Epidemiol 21:293–299

Mongraw-Chaffin ML, Cohn BA, Anglemyer AT, Cohen RD, Christianson RE (2009) Maternal smoking, alcohol, and coffee use during pregnancy and son's risk of testicular cancer. Alcohol 43:241–245

Morck TA, Lynch SR, Cook JD (1983) Inhibition of food iron absorption by coffee. Am J Clin Nutr 37:416–420

Munoz L, Keen CL, Lonnerdal B, Dewey KG (1986) Coffee intake during pregnancy and lactation in rats: maternal and pup hematological parameters and liver iron, zinc and copper concentration. J Nutr 116:1326–1333

Munoz LM, Lonnerdal B, Keen CL, Dewey KG (1988) Coffee consumption as a factor in iron deficiency anemia among pregnant women and their infants in Costa Rica. Am J Clin Nutr 48:645–651

Nehlig A, Debry G (1994) Consequences on the newborn of chronic maternal consumption of coffee during gestation and lactation: a review. J Am Coll Nutr 13:6–21

Oberstar JV, Bernstein GA, Thuras PD (2002) Caffeine use and dependence in adolescents: one-year follow-up. J Child Adolesc Psychopharmacol 12:127–135

O'dea JA, Wilson R (2006) Socio-cognitive and nutritional factors associated with body mass index in children and adolescents: possibilities for childhood obesity prevention. Health Educ Res 21:796–805

Olini N, Kurth S, Huber R (2013) The effects of caffeine on sleep and maturational markers in the rat. PLoS One 8:e72539

Orbeta RL, Overpeck MD, Ramcharran D, Kogan MD, Ledsky R (2006) High caffeine intake in adolescents: associations with difficulty sleeping and feeling tired in the morning. J Adolesc Health 38:451–453

Ortega RM, Requejo AM, Lopez-Sobaler AM, Quintas ME, Andres P, Redondo MR, Navia B, Lopez-Bonilla MD, Rivas T (1998) Difference in the breakfast habits of overweight/obese and normal weight schoolchildren. Int J Vitam Nutr Res 68:125–132

Pacheco AH, Barreiros NS, Santos IS, Kac G (2007) Caffeine Consumption during pregnancy and prevalence of low birth weight and prematurity: a systematic review. Cad Saude Publica 23:2807–2819

Pecoud A, Donzel P, Schelling JL (1975) Effect of foodstuffs on the absorption of zinc sulfate. Clin Pharmacol Ther 17:469–474

Pires VA, Pamplona FA, Pandolfo P, Prediger RD, Takahashi RN (2010) Chronic caffeine treatment during prepubertal period confers long-term cognitive benefits in adult spontaneously hypertensive rats (SHR), an animal model of attention deficit hyperactivity disorder (ADHD). Behav Brain Res 215:39–44

Ramlau-Hansen CH, Thulstrup AM, Bonde JP, Olsen J, Bech BH (2008) Semen quality according to prenatal coffee and present caffeine exposure: two decades of follow-up of a pregnancy cohort. Hum Reprod 23:2799–2805

Rapoport JL, Jensvold M, Elkins R, Buchsbaum MS, Weingartner H, Ludlow C, Zahn TP, Berg CJ, Neims AH (1981) Behavioral and cognitive effects of caffeine in boys and adult males. J Nerv Ment Dis 169:726–732

Reissig CJ, Strain EC, Griffiths RR (2009) Caffeinated energy drinks–a growing problem. Drug Alcohol Depend 99:1–10

Rhoads DE, Huggler AL, Rhoads LJ (2011) Acute and adaptive motor responses to caffeine in adolescent and adult rats. Pharmacol Biochem Behav 99:81–86

Ross JA, Potter JD, Reaman GH, Pendergrass TW, Robison LL (1996) Maternal exposure to potential inhibitors of DNA topoisomerase II and infant leukemia (United States): a report from the Children's Cancer Group. Cancer Causes Control 7:581–590

Sachse C, Brockmoller J, Bauer S, Roots I (1999) Functional significance of a C → A polymorphism in intron 1 of the cytochrome P450 CYP1A2 gene tested with caffeine. Br J Clin Pharmacol 47:445–449

Santos RM (2010) Our 'black box' cup of coffee: what is inside? Res Pharmaceutica 1:60–63

Santos RM, Lima DR (1989) Coffee as a medicinal plant and vitamin source for smokers. Italian J Chest Dis 43:56–59

Santos RM, Lima DR (2007) Coffee, the revolutionary drink for pleasure and health. Xlibris, Pittsburgh

Santos RM, Lima DR (2009) An unashamed defense of coffee: 101 reasons to drink coffee without guilt. Xlibris, Pittsburgh

Sato Y, Itagaki S, Kurokawa T, Ogura J, Kobayashi M, Hirano T, Sugawara M, Iseki K (2011) In vitro and in vivo antioxidant properties of chlorogenic acid and caffeic acid. Int J Pharm 403:136–138

Schmidt RJ, Romitti PA, Burns TL, Murray JC, Browne ML, Druschel CM, Olney RS (2010) Caffeine, selected metabolic gene variants, and risk for neural tube defects. Birth Defects Res A Clin Mol Teratol 88:560–569

Schwierin B, Borbely AA, Tobler I (1996) Effects of N6-cyclopentyladenosine and caffeine on sleep regulation in the rat. Eur J Pharmacol 300:163–171

Sowell ER, Thompson PM, Holmes CJ, Jernigan TL, Toga AW (1999) In vivo evidence for post-adolescent brain maturation in frontal and striatal regions. Nat Neurosci 2:859–861

Spear LP (2000) The adolescent brain and age-related behavioral manifestations. Neurosci Biobehav Rev 24:417–463

Stein MA, Krasowski M, Leventhal BL, Phillips W, Bender BG (1996) Behavioral and cognitive effects of methylxanthines. A meta-analysis of theophylline and caffeine. Arch Pediatr Adolesc Med 150:284–288

Svenningsson P, Hall H, Sedvall G, Fredholm BB (1997) Distribution of adenosine receptors in the postmortem human brain: an extended autoradiographic study. Synapse 27:322–335

Taft JL, French M, Danks JA, Larkins RG (1984) Opposing actions of methylxanthines and dibutyryl cyclic AMP on 1,25 dihydroxyvitamin D3 production and calcium fluxes in isolated chick renal tubules. Biochem Biophys Res Commun 121:355–363

Temple JL (2009) Caffeine use in children: what we know, what we have left to learn, and why we should worry. Neurosci Biobehav Rev 33:793–806

Temple JL, Dewey AM, Briatico LN (2010) Effects of acute caffeine administration on adolescents. Exp Clin Psychopharmacol 18:510–520

Torfs CP, Christianson RE (2000) Effect of maternal smoking and coffee consumption on the risk of having a recognized Down syndrome pregnancy. Am J Epidemiol 152:1185–1191

Torp-Pedersen T, Boyd HA, Poulsen G, Haargaard B, Wohlfahrt J, Holmes JM, Melbye M (2010) In-utero exposure to smoking, alcohol, coffee, and tea and risk of strabismus. Am J Epidemiol 171:868–875

Urgert R, Van Vliet T, Zock PL, Katan MB (2000) Heavy coffee consumption and plasma homocysteine: a randomized controlled trial in healthy volunteers. Am J Clin Nutr 72:1107–1110

Vestergaard M, Wisborg K, Henriksen TB, Secher NJ, Ostergaard JR, Olsen J (2005) Prenatal exposure to cigarettes, alcohol, and coffee and the risk for febrile seizures. Pediatrics 116:1089–1094

Wendler CC, Busovsky-Mcneal M, Ghatpande S, Kalinowski A, Russell KS, Rivkees SA (2009) Embryonic caffeine exposure induces adverse effects in adulthood. FASEB J 23:1272–1278

Yoon PW, Rasmussen SA, Lynberg MC, Moore CA, Anderka M, Carmichael SL, Costa P, Druschel C, Hobbs CA, Romitti PA, Langlois PH, Edmonds LD (2001) The national birth defects prevention study. Public Health Rep 116(Suppl 1):32–40

Chapter 12
Ethical Considerations in Development of Future Therapies for Women and Children

Toby Schonfeld

Abstract Translational toxicology has the potential to equip healthcare providers with new strategies to address health effects from exposure to toxic agents, especially for women. Yet in many cases, the existence of developmental milestones is coextensive with vulnerability, such that these populations merit special protections when it comes to their participation in the very research that would yield these strategies. This chapter reviews the ethical considerations and regulatory limitations that obtain to these groups of research participants and then applies these considerations to the fundamental concepts in translational toxicology. The focus of this chapter is the development of future therapies. First, the chapter reviews the criteria for what makes research ethical, and then describes the ethical and regulatory considerations that attach to the kinds of projects necessary for the development of future therapies in translational toxicology. Following this, the chapter details considerations unique to each experimental strategy (prevention, mitigation, and reversal), and finally includes several general ethical considerations for the discipline as a whole.

Keywords Ethics • Regulation • Vulnerable populations • Toxicology and pregnancy • Research ethics

Translational toxicology has the potential to equip healthcare providers with new strategies to address health effects from exposure to toxic agents. Rather than simply advising patients to avoid exposures – advice often difficult to follow when the exposures are outside of the patient's control – this new field may provide strategies for protecting, mitigating, or reversing adverse effects of environmental exposures. Such strategies are particularly desirable in populations where developmental milestones may provide opportune windows for intervention, and therefore (pregnant) women, fetuses, and children are the targets for therapy. Yet in many cases, the

T. Schonfeld (✉)
Program in Human Research Ethics and Oversight, Office of the Science Advisor, U.S. Environmental Protection Agency, 1200 Pennsylvania Ave., NW Mailcode: 8105R, Washington, DC 20460, USA
e-mail: schonfeld.toby@epa.gov

C.L. Hughes, M.D. Waters (eds.), *Translational Toxicology*, Molecular and Integrative Toxicology, DOI 10.1007/978-3-319-27449-2_12

existence of developmental milestones is coextensive with vulnerability, such that these populations merit special protections when it comes to their participation in the very research that would yield these strategies.

In this chapter, I will review the ethical considerations and regulatory limitations that obtain to these groups of research participants and then apply these considerations to the fundamental concepts in translational toxicology. While there are certainly a host of clinical ethics issues that will be related to the initiation of any proposed therapy, the focus of this chapter is the *development* of future therapies. Therefore, I will restrict my comments to those related to the research participation of these groups. And as a side note: while animal studies are necessarily prior to trials with humans, the ethics of animal experimentation is also beyond the scope of this work.

I will begin the discussion of research by reviewing the criteria for what makes research ethical, and then describe the ethical and regulatory considerations that attach to the kinds of projects necessary for the development of future therapies in translational toxicology. Following this, I will describe considerations unique to each experimental strategy (prevention, mitigation, and reversal), and finally identify several general ethical considerations for the discipline as a whole. This work will set out the framework for those considering the development of one of the proposed strategies to develop therapies for women or children in a way that is accessible, thought-provoking, and practically applicable to study design.

12.1 What Makes Clinical Research Ethical?

In an influential article in the *Journal of the American Medical Association*, Emanuel and colleagues identified seven features that constitute ethical biomedical research (Emanuel et al. 2000). Ten years later, Emanuel and colleagues updated their thinking to reflect the contemporary nature of clinical research (Emanuel et al. 2011). They argue that none of the regulatory guidelines are sufficiently broad or specific enough to include both the ethical considerations for the context of research, nor are they sufficiently action-guiding for researchers who endeavor to involve humans as participants in their studies. As a response, their (now) eight-faceted approach to clinical research creates a framework which, when considered in its entirety during both the planning and implementation stages of the research, will enable researchers to have a solid ethical foundation for their research project.

While the standard approach to biomedical research is the randomized clinical trial (RCT), Emanuel et al.'s framework is geared toward any research that aims "to improve health and healthcare" (Emanuel et al. 2011, p. 125). Trials of chemopreventive agents or other pharmaceuticals (even those "generally-recognized-as-safe") often take the same form as an RCT, and therefore this framework is directly applicable. But even for those observational studies or social and/or behavioral modification efforts that differ in format from an RCT, to the extent that the goal is to improve health (either of the participants directly or of future patients), the framework will still serve as an important foundational reference point.

In what follows, I discuss each of the eight constitutive elements in turn, applying examples from translational toxicology to demonstrate how the ethical considerations would influence study design and conduct.

12.1.1 Collaborative Partnership

Fundamental to the ethical considerations in research is the notion that research is done *with* people, not to them (Weijer and Emanuel 2000). As a result, it is helpful to think of research participants as partners in the enterprise (and why some have moved away from the terminology of "subjects," which may suggest a lower position in the research hierarchy than investigators). Partnering with participants not only helps to guard against exploitation by having participants help design fair and just study practices, but it also helps to ensure that the proposed research meets the needs of the community (Emanuel et al. 2011). Consider a study that attempts to reduce nicotine exposure to women and fetuses by getting pregnant women to quit smoking. Without partnering with the targeted audience, it will be impossible to know the context in which the pregnant women are making the choice to smoke and therefore know whether or not the study design is optimized. For example, for pregnant women in high stress environments, smoking may provide the only "escape" or the only feature of their lives over which they have control. Mitigating exposure in these contexts, then, must address the underlying rationale for the smoking, rather than merely the smoking behavior itself in order to be successful. Partnering with members of this community demonstrates an attitude of mutual respect and helps to ensure a fair sharing of the benefits and burdens of research participation (Emanuel et al. 2011).

12.1.2 Social Value

In order for research to be beneficial, it must have social value: it must lead to improvements in health or healthcare or sufficiently advance knowledge so that such improvements are possible in the future (Emanuel et al. 2011). Without this value, there is no ethical justification for enrolling participants in a protocol because there will be no possibility for benefit to offset the risks of participation. Note that this is true even for observational research: asking participants questions or having them participate in a focus group may, at least, waste their time and, depending on the questions, expose them to psychological or social harm, for no benefit. This does not mean that every research project must confer direct benefit on the participants; rather, the possibility of generalizable knowledge on a societal scale can also justify the conduct of research involving human participants. With respect to research attempting to mitigate or reverse exposure, investigators should be cautious to ensure that their studies have sufficient statistical power for the results to be meaningful to a wider audience and that the strategy proposed can be practically implemented by others in the community (Emanuel et al. 2011). For example, a strategy that involves physically moving participants away from the environment where the exposure is occurring (e.g. to a new school or a new house) is likely to be impractical to be implemented on a large scale. Instead, consider approaches to research that are adaptable to communities who may not have the same resources as the research team.

12.1.3 Scientific Validity

Every research project should begin with a clear hypothesis (or null hypothesis), an approach that is designed to answer the scientific question, and a data analysis plan that is appropriate to the methods selected (Emanuel et al. 2011). As with social value, research projects that lack scientific validity will yield no generalizable results and therefore will result in the exploitation of participants (because there is no possibility of benefit to offset the risks of participation). As translational toxicology begins to mature, there are at least two significant challenges that researchers will face when they move into the health arena. The first is ensuring that participants retain access to whatever healthcare services they are routinely entitled, regardless of whether or not accessing those healthcare services cohere with the goals of the study (Emanuel et al. 2011). So, for example, investigators may study over-the-counter or prescription drugs used by pregnant women in an observational study design, but may not restrict a woman's access to pharmaceuticals generally available or prescribed by her healthcare provider. Secondly, estimating sufficient statistical power for studies that aim to improve health often require different considerations from the types of studies that environmental scientists conduct. Those engaging in these new strategies to develop therapeutics to exposure must consider these alternative approaches to study design and participant recruitment.

12.1.4 Fair Participant Selection

Science should dictate which individuals are targeted for participation, not convenience to the investigator or predictions about which kinds of people a recruitment scheme will be easier to attract. Rather, in order to minimize the possibility of exploitation, participants should be chosen because they meet scientific goals and therefore enhance the social utility of the research. For researchers who are developing new therapeutic strategies, it is also important to remember the responsibility to minimize risk in both designing the study and selecting the participants. Because the target for much translational toxicology research will be individuals at or near developmental milestones, many of them will fall into the "vulnerable" category (see next section). This means that extra research protections will need to be in place in order to ensure risks are minimized for participants from these groups. In some cases, their vulnerability is precisely what makes them appropriate participants for the research, which can cause extra complications (Schonfeld 2013). Regardless, choosing participants fairly is essential to the ethical conduct of research.

12.1.5 Favorable Risk-Benefit Ratio

All research carries risk, even if the risk is simply time or inconvenience spent on activities the participant would not otherwise choose. Yet for research to be ethical, on balance the research must favor benefits over risks (Emanuel et al. 2011). One way to do this is to ensure that risks are minimized to the greatest extent possible consistent with sound scientific design. Capitalizing on procedures already happening as part of clinical care (e.g. a routine blood draw where an extra vial can be drawn) minimizes risks. But enhancing the benefits to the participants and the community in which they reside is another way to ensure a favorable risk-benefit ratio (Emanuel et al. 2011). Suppose you are concerned about the effect of maternal diet on the development of Autism Spectrum Disorder (ASD) in children, but you have reason to believe that folic acid intake at the appropriate stage of development may be protective against ASD (Lyall et al. 2014). Providing folic acid to all participants in the study, free of charge, is one way to maximize benefits to participants since we know that folic acid is beneficial for many other aspects of development (Kim et al. 2014).

Risks and benefits can be categorized by type, magnitude, and frequency, and it is important to carefully articulate these in the research design phase (Emanuel et al. 2011). Otherwise, a comparison of risks and benefits may fail to accurately capture the trade-offs involved in research participation. Regardless, benefits and risks conferred on research participants are limited to the risks of the research interventions only. So if, as suggested earlier, investigators are going to capitalize on a routine blood draw and simply take an extra vial of blood, then the risks of the blood draw itself are <u>not</u> risks of the research. Rather, the risks conferred on the participant are the risks of taking the *extra* blood. Finally, when there are no direct benefits to participants in the study, it is important to consider the societal benefits carefully in comparison to the individual risks to participants (Emanuel et al. 2011). This is a common situation for early Phase drug trials, where the safety of the pharmaceutical is part of what investigators are trying to establish. Any study, however, that does not offer individual-level benefits must be extra careful to minimize risks to the greatest extent possible.

12.1.6 Independent Review

In order to ensure regulatory compliance with the Common Rule (see below), all[1] research studies that involve human subjects must be reviewed by an independent body (known as an Institutional Review Board in the US and a Research Ethics

[1] Some studies are in fact exempt from IRB review; see 45 CFR 46.101 (b).

Board or Research Review Board in other parts of the world). But there are important ethical reasons for this, too. Third-party review of proposed research guards against conflict of interest among the investigators. In addition, by submitting the protocol to an independent, diversely-constituted body, broader considerations about the research can be brought to bear. Sometimes it is easier for a third party to identify and address issues with study design, subject recruitment, and informed consent precisely because it has fresh eyes to devote to the issue. The diverse expertise on something like an IRB can be very useful in helping to refine a study design to ensure compliance with the previously-mentioned concepts. Consider a research proposal that suggests a rigorous exercise regimen for a particular group of post-menopausal women as a strategy to reverse the toxic effects of a series of environmental exposures. It might be that a geriatrician on a review board knows of data that would help to bolster the study's hypothesis, or she might have information about a particular risk that could be conferred by this strategy that needs to be addressed before the research can go forward. In either case, the review board serves to facilitate the conduct of ethical research by helping to ensure that the risks to subjects are minimized. In this way, independent review of research protects research subjects, investigators, and the institution/organization that sponsors the research.

12.1.7 Informed Consent

Obtaining informed consent from research participants respects the autonomy of participants by ensuring that they can make a decision about whether or not the research activity coheres well with their values, goals, and priorities. To accomplish this, investigators must (1) provide information about the study in a cognitively-appropriate, non-jargoned fashion; (2) ensure that potential subjects understand the risks and benefits of participating in the trial; and (3) describe to participants any alternatives to participation, including the right not to participate, to ensure that individuals are freely choosing participation (Emanuel et al. 2011). Designing a consent process that respects subjects and their context, capacities, and community is not easy; investigators must be sensitive to the cognitive capacity, social and economic status, and specific contexts of their participants in order to ensure that consent will be truly voluntary (Emanuel et al. 2011). There are a whole host of vulnerabilities that may influence one's ability to give truly informed consent (Kipnis 2003), and investigators must consider these features ahead of time and plan accordingly. For example, suppose a researcher is interested in mitigating the role of endocrine disruptors in pre-teens. Adolescents are particularly sensitive to confidentiality concerns in healthcare, and have reported instances in which they either withheld information or failed to seek help in the first place because of concerns about their confidentiality not being respected (Sankar et al. 2003). Researchers who want to involve pre-teens in a study, then, should consider carefully what added protections they can reasonably offer to this group who is particularly sensitive to information sharing.[2]

12.1.8 Respect for Participants

Informed consent does not end when the research participant signs the informed consent document. Rather, informed consent is a process that continues throughout the duration of the study. As new information becomes available to the research team, it is essential that team members communicate with active participants in a way that is appropriate to the individuals. Similarly, investigators have the responsibility to monitor the well-being of their subjects and to act accordingly (e.g. remove participants from the trial if they are experience significant adverse events from the study agent). As part of the voluntary nature of consent, participants must always be free to withdraw from the study without penalty; however it is the research team's responsibility to inform the participant if he or she needs to take certain precautions when leaving a study for safety reasons (e.g. titrate down a pharmaceutical rather than stop "cold turkey"). Finally, part of respecting participants as equal partners in the research process includes returning research results to them after the research has concluded and the data have been analyzed. This demonstrates to participants the value of their time in the study, even if the null hypothesis has not been disproven.

12.2 Ethical and Regulatory Considerations with Research involving Women and Children as Participants

Because of the focus on developmental milestones as an ideal opportunity for intervention regarding exposure, the majority of research that will be conducted in translational toxicology involves pregnant women or children as the primary participants. In many respects, this is quite laudable since these two groups have historically been excluded from participation in potentially beneficial research (Shields and Lyerly 2013; Diekema 2006), giving rise to the term "therapeutic orphans". There are several reasons for these exclusions, most of them having to do with risk aversion. Researchers and sponsors have been loathe to do anything that exposes children or fetuses to risk, for both legal and moral reasons: no one wants the legal liability of a birth defect, nor do they want to be responsible for harming children. Yet the consequence of this reluctance is a clinical situation where the vast majority of treatments for childhood illnesses are still "off label" – that is, lacking the appropriate scientific

[2]This is true even though permission to participate must be obtained from an adolescent's parents since they are not at the legal age of consent. Regardless, getting teens to assent to research participation is essential, and just as context-specific as consent in other populations.

data to demonstrate efficacy (and, relatedly, treatment toxicities) and where "reducing adult dosing" simply will not work (Palmaro et al. 2014; Frattarelli et al. 2014).

Similarly, pregnant women – and, in fact, those "potentially pregnant" (Merton 1993) – have been excluded from clinical research because of the risks to a fetus from investigational interventions. Ironically, both of these situations have led to the same consequence: a dearth of information about how to care for pregnant women (and, by extension, their fetuses) in the context of illness, disease, and discomfort: a situation that may, in fact, place both pregnant women and children at GREATER harm than if data were collected. Consider the historical example of Thalidomide, where babies were born with severe birth defects as a result of a medication commonly provided to pregnant women as an anti-emetic. No one wants to create or be responsible for the effects of the next Thalidomide. Yet the irony is that the widespread harm to fetuses resulted from *excluding* women from clinical trials; in fact, had rigorous studies been done of this and similar drugs, the magnitude of the harm to children could likely have been attenuated (Lyerly et al. 2009; Friedman 2012). Instead, the response has been to exclude pregnant women – and most women of childbearing potential (Schonfeld 2013) – from clinical trials. As Ruth Macklin argues, "the most compelling reason [for including pregnant women in a greater number of clinical trials] is the need for scientific evidence gathered under rigorous scientific conditions, in which fewer women and their fetuses would be placed at risk than the much larger number who are exposed to medications once they come to market" (Macklin 2010, p. 632).

Some argue that it is restrictive regulations that prohibit the advancement of research involving these populations, while others claim that these regulations offer fundamental protections for those who want to involve these groups to participate in research. Regardless, it is important to understand the regulatory context prior to designing studies involving pregnant women or children as participants.

12.2.1 International Research Regulations

There are several international guidelines that offer assistance to investigators when designing trials, although to be maximally applicable for research they include only general statements about "vulnerability" when referring to pregnant women and children. For example, the Declaration of Helsinki (Appendix I) includes a section on "Vulnerable Groups and Individuals," but there simply states that "[s]ome groups and individuals are particularly vulnerable and may have an increased likelihood of being wronged or incurring additional harm," and as a result deserve "specifically considered protection" (World Medical Association 2013). An example of such protections specifically listed in the Declaration is the investigator's assurance that the proposed research could not be conducted adequately with a non-vulnerable population.

The International Ethical Guidelines for Biomedical Research Involving Human Subjects (2002) from the Council for International Organizations of Medical

Sciences (CIOMS) also incorporates general language about vulnerability and consent, but has additional language specific to children and pregnant women. Guideline 14, "Research Involving Children," includes considerations such as first conducting the research with adults, when possible, and obtaining both consent from parents and assent from children to participate in the study (CIOMS 2002). Guideline 17, "Pregnant Women as Research Subjects," describes the default status for pregnant women as able to be included in research, which differs greatly from the US regulations (see below). The guideline reminds investigators about their responsibility to clearly and accurately describe the risks and benefits of research participation but makes no specific reference to the level of risk to the woman or her fetus that is acceptable; rather, there is a recognition in these guidelines that the well-being of one is inextricably linked to the well-being of the other. Still, the general principles hold that the research should be important to be carried out in this population, and the evidence of pre-clinical and clinical studies should be provided whenever possible (CIOMS 2002).

Additionally, many nations have their own national research ethics committees and associated guidance, and some of those organizations have joined international consortia or offer conferences to share best practices and establish common processes and approaches (e.g. the European Network of Research Ethics Committees [EURECNET]; Asia-Pacific Research Ethics Conference [APREC]; the National Health and Medical Research Council of Australia, etc.). Researchers planning translational toxicological research in those areas should consult the relevant guidance documents.

12.2.2 Research Regulations in the USA

The U.S. regulatory context for conducting research with pregnant women and children as participants is somewhat complicated. The regulations from the Department of Health and Human Services (DHHS) can be found at Part 46 of title 45 of the Code of Federal Regulations (CFR). The primary set of protections for human subjects, Subpart A, is known as the "Common Rule" on account of the fact that 18 federal agencies in addition to DHHS have agreed to adopt those provisions for federally-funded research that involves human participants. Included in this subpart are the requirements for informed consent and for independent review of the research, conducted by Institutional Review Boards (IRBs). Moreover, the Common Rule also includes guidance for conducting research with groups identified as requiring "additional protections." The regulations refer to these groups as "vulnerable populations": "When some or all of the subjects are likely to be vulnerable to coercion or undue influence, such as children, prisoners, pregnant women, mentally disabled persons, or economically or educationally disadvantaged persons, additional safeguards have been included in the study to protect the rights and welfare of these subjects" (45 CFR 46.111b). Note that both pregnant women and children are included in the list of those likely to be vulnerable to coercion and undue influence. As a way of specifying additional safeguards, both groups have special regulatory sections dedicated to their "protections;" Subpart B for "Pregnant Women, Human Fetuses, and Neonates involved in Research" and Subpart D for children (see Appendix II).[3]

There are several features of each of the subparts that deserve special mention, and so we will pay particular attention to them here. Note that these are not the only considerations; for the full text of the regulations, please see Appendix II. Here I simply highlight some of the key features that should be considered by researchers in the design phase of studies.

12.2.2.1 Special Considerations involving Pregnant Women as Research Participants

First, Subpart B makes clear that agents should not be tried in pregnant women prior to having been studied in pregnant animals and non-pregnant human participants. Even if this is an intervention that is specifically designed for pregnant women and/or their fetuses, the regulations require that information on reproductive and general toxicities be available before designing studies with pregnant participants. This is true even though (a) animal studies do not always translate well to human studies (Rhrissorrakrai et al. 2014), and (b) pharmacokinetics are different in pregnancy, and as a result agents may operate very differently in pregnant women than in other participants (Lyerly et al. 2008).

Secondly, note that the regulations require separate risk and benefit considerations for the pregnant woman and the fetus. Essentially, the regulations ask investigators to consider to whom the possibility of direct benefit obtains. If there is no possibility of direct benefit to the woman or to the fetus, then the research is not approvable if the risk posed by the interaction is greater than minimal to the fetus.[4]

The third consideration relates to the risk/benefit calculus described above. The regulations tie consent requirements directly to fetal risk: if the risk of the intervention is not greater than minimal to the fetus, then the pregnant woman's consent is sufficient for the research to proceed. This is true even if there is no prospect of

[3] Note that while the DHHS regulations apply to research funded by DHHS (NIH, etc.), the subparts may not apply to projects funded by other federal agencies if they have not adopted those parts of the regulation. For example, the EPA has not adopted Subpart C (regulations involving prisoners), so researchers using EPA funds exclusively are not bound by those requirements. In addition, some institutions apply the subparts to all research, regardless of funding – known in common parlance as "checking the box." For those institutions that do not check the box, then research that does not receive funding from DHHS is not subject to those regulatory requirements.

[4] DHHS does reserve the right to approve research that does not meet these requirements if they agree there is "an opportunity to understand, prevent, or alleviate a serious problem affecting the health or welfare of pregnant women, fetuses, or neonates" and the requisite approval criteria are met [45 CFR 46.207].

direct benefit for the woman or the fetus, as mentioned above. However, in situations where there is the possibility for the intervention to provide direct benefit to the fetus alone, consent of both the father and the pregnant woman are required. This "two-parent consent" holds regardless of the fact that some interventions that hold out the prospect of direct benefit to the fetus may create significant risk for the pregnant woman herself (e.g. surgical correction of fetal myelomeningocele (Adzick 2010; Cohen et al. 2014)). It is the one place in the federal regulations where someone with decisional capacity is unable to give her own consent (alone) for a procedure that will happen to her body (Schonfeld 2013).

12.2.2.2 Special Considerations involving Children as Research Participants

For children, there are two risk classifications listed (but not defined) within the regulations: minimal risk and a minor increase over minimal risk. These categories are important for determining (a) what kind of research is approvable, and (b) whether consent from one parent or two parents is necessary in order for the research to proceed. Additionally, investigators must identify whether or not there is the possibility of direct benefit to the children participating in the study. It is these considerations (the possibility of direct benefit as compensatory for risks incurred on study) that make some research approvable that otherwise would not be.

Finally, researchers must get assent from the children who are participating in the study in addition to consent from the parents (often termed "permission" in this context since the parents are authorizing the participation of others). Certainly, assent will not be possible for the very young or for those who are unable to understand the intricacies of a research protocol. However, explaining in a very general way what the research is and why the child is being asked to participate in it (e.g. "we are trying to understand what you breathe into your lungs during recess", what the procedures may entail (e.g., "we will ask you to blow into a tube before and after recess"), and what alternatives there are (e.g. "you can still play with your friends at recess if you do not want to be part of this research") gives the child an opportunity to have some control over what happens to her body. Many institutions go by the rule of "7 s": up to age 7, no assent is required. From age 7-14, assent should be obtained by describing the basic study design and procedures in an age-appropriate way and assessing the child's willingness to participate. From age 14 forward, children have meaningful veto power so that researchers will not enroll them in a study to which they do not assent, regardless of whether or not their parents provided permission for them to participate. The idea here is that children at this stage are capable of enough understanding and self-determination to weigh the benefits and burdens of participation and have a deciding hand in determining the course of their own future.

12.3 Experimental Strategies in Translational Toxicology

Translational toxicology aims at three experimental strategies for addressing environmental exposures that produce adverse health outcomes: preventing the exposure (or preventing the exposure from having negative health effects), mitigating the adverse health effects of the exposure, or reversing the effects of the exposure. While laudable in their innovative approaches to address clinical outcomes of environmental exposures, each of these strategies must be coupled with careful sensitivity to the ethical issues that obtain to the proposed research projects. In each of these cases, I will identify examples of possible research projects and highlight the ethical and regulatory challenges related to each one. This is not to say that this research cannot be conducted; quite the contrary, my purpose here is to facilitate the design of ethical research by proactively identifying the issues investigators must consider.

12.3.1 Prevention Research

Consider an experimental strategy that attempts to restrict the caloric intake of pregnant women in order to prevent the development of negative metabolic outcomes (like obesity) in the fetus (Hughes et al. 2013a). As part of this study, women of normal weight are asked to eat no more than 35 kcal/kg each day, divided roughly into roughly 40–50 % complex carbohydrates, 20 % lean protein, and 30–40 % good fats.

To the extent that the goal is to prevent the negative health outcomes associated with pediatric and adolescent obesity, ascertaining the effect of restriction of maternal caloric intake is intriguing. But investigators must be careful to design the trial so that it takes into account the risks and benefits conferred not just on the fetus, but also on the pregnant women whose calories are restricted. As with any trial, risks should be described in relationship to likelihood and severity in a way that a woman can assess the risk-benefit relationship for herself and determine whether or not participation coheres with her goals, values, and priorities. For example, suppose that a woman is interested in participating in the research precisely because her two other children are obese. However, the requirements of the protocol may be such that, as a busy mother of two with two part-time jobs, she lacks the time necessary to prepare the healthy meals that are the central feature of the research. She usually just runs through the drive-through at a local restaurant, or else makes easy things that please her kids (like spaghetti) because she knows she always has the ingredients in the house – she has little time to shop for fresh fruits and vegetables. In such a case, while she is committed to the outcomes of the research and values its goals, the study design simply does not work with her life. She wants the best for her baby, but isn't willing or able to sacrifice the time she spends with her kids to do the shopping and cooking that the protocol requires. However, an alternative research design, where subjects are given the pre-prepared meals (e.g. weekly deliveries cataloged by meal and date) that she simply has to put in a bowl or heat in the microwave might, in fact, work for her. This is a way that partnering with the research community during the study design phase could create a research context that is sensitive to the needs of the target population.

As a second example, consider an experimental strategy where women with no known underlying disease conditions (and who therefore take no prescription medications) planning to become pregnant enroll in a trial where they agree to refrain from taking any over the counter (OTC) medications, including herbal supplements. Given the number of potential influences intrauterine chemical exposures can have on fetuses, elimination of one source of chemical exposures could be postulated as a preventive strategy. However, it is difficult to adequately communicate the risks of this participation, as every pregnancy, and every woman, is different. She might develop symptoms that cause her significant discomfort, which could otherwise be relieved with OTC medications. And since it is difficult to predict how one will feel in that situation, some would call into question her ability to give truly informed consent here. Certainly, there would need to be provisions for attrition on such a study, both for women who change their minds about refraining from OTC use, or for those who develop a condition that requires prescription treatment – both situations that would likely result in the participant's withdrawal from the study. And if the attrition rate is too great, then the research may be in danger of not being completed in a timely fashion or in a way that facilitates the statistical analysis described in the research plan. In that case, then, the risk/benefit ratio of the study has changed negatively for <u>all</u> participants as the potential for societal benefit has significantly decreased. This outcome may happen even if the investigators try to recruit those who are least likely to withdraw (say, women with a history of previous pregnancies during which they took no or few OTC medications); in addition, that recruitment strategy would call into question the generalizability of their results given that two-thirds of pregnant women are prescribed a drug during their pregnancy (Andrade et al. 2004; Daw et al. 2012; Yang et al. 2008). Ironically, this protocol might also fail to get through some IRBs, since the "no medication" rule while on study may in fact put women and fetuses at risk if they, for example, delay necessary medical care because of a desire to stay on the study, or if they refuse to take a standard anti-emetic that would enable them to consume the nutrition required for successful fetal development because of study restrictions.

12.3.2 Mitigation Research

For those environmental exposures that are unavoidable or inevitable, mitigating the negative effects of those exposures may lead to better overall health outcomes. One strategy proposed for this is to use "Generally-Recognized-as-Safe" (GRAS) agents that can be tested rigorously in a population exposed to a particular hazard (Hughes et al. 2013b). In theory, using GRAS agents lessens the risk of the study, which is particularly important to already vulnerable populations like pregnant women and children. However, it is not entirely clear that this strategy is substantially safer than other investigational agents that do not carry the GRAS label or that regulatory bodies would see them in this way. Consider, for example, the use of green tea, long hailed as, among other things, an antioxidant stemming from the polyphenols and catechins in the tea (Hughes et al. 2013a). It is conceivable that researchers might design a trial of concentrated green tea catechins as a food additive for peri- and post-menopausal women as an anti-aging strategy to preserve structure and function despite a long history of exposure to environmental toxins. Yet preliminary data are not conclusive about the benefits of green tea extracts, and in fact suggest that adverse effects disproportionately affect women (Abdel-Rahman et al. 2011). Given the regulatory requirements described earlier, it is plausible to think that in the face of these data, IRBs would insist that these risks be calculated in the risk/benefit relationship and described in the consent process. That is to say, IRBs may view this through the same lens as any other investigational agent.

This brings up the question about GRAS agents in general: what do we mean, exactly, when we say an agent is "safe" or term it an "ethical pharmaceutical"? GRAS agents are not tested through the same phased drug trial system that applies to drugs that are looking to be approved by the U.S. Food and Drug Administration (FDA) (21 CFR 170.20). Rather, manufacturers themselves make the determination that their substance is safe, with safe being defined as "a reasonable certainty in the minds of competent scientists that the substance is not harmful under the intended conditions of use" (21 CFR 170.3(i)). But careful analyses of the agents often produce complicated results. Many agents are safe in one preparation but may be toxic in another or with a different population (Abdel-Rahman et al. 2011). Plus, there is no standard mechanism for evaluating safety; rather, each substance is evaluated individually. Given that researchers would be proposing *novel* uses for these GRAS agents in the kind of research proposed here, one wonders whether or not the assurances of safety would persuade IRBs that the agent in question is appropriate for testing. Given the pharmacokinetic challenges to pediatric (Sage et al. 2014) and obstetric (Lyerly et al. 2008) drug development, one wonders about the challenges posed by even something like a GRAS agent.

There are occasions when manufacturers submit their information to the FDA regarding their GRAS substance. These notifications are reviewed by a panel convened by the FDA, who then responds to manufacturers only if they do not agree with the determination – no news is good news in this case (Neltner et al. 2013). However, a recent study calls into question the objectivity of these panels by pointing out that "between 1997 and 2012…financial conflicts of interest were ubiquitous in determination that an additive food was GRAS"(Neltner et al. 2013); p. E4). Additionally, at least 1 of 10 individuals served on more than 75 % of all panels convened for these food additive GRAS determinations. It is possible that this speaks to the need for expertise in these areas, but it may also speak to problems with objectivity and integrity. Regardless, to the extent that the IRB is charged with guarding against conflicts of interest, these are issues to which they would likely attend.

The upshot of all of this is that it is not at all clear that even GRAS agents would gain special privileges in research involving children and pregnant women as participants. This is partly because of the risk-averse nature of society, but also is because of the current regulatory structure. Regardless, researchers should not assume that their studies will receive "easier" handling because they involve GRAS agents rather than standard pharmaceuticals.

12.3.3 Reversal Research

The final strategy possible in translational toxicology is to attempt to undo the harm caused by toxins after the exposure has occurred. An example of this kind of research is the introduction of N-acetylcysteine (NAC) to reverse the negative effects of smoking by changing some biomarkers that may reduce the cancer-causing effects of the toxins (Hughes et al. 2013b, p. 3). The idea, then, is that NAC could be introduced into pregnant women to both reverse the damage to themselves done by smoking (or by passively being exposed to secondhand smoke) and to simultaneously mitigate the risks of smoking to the fetus by mitigating the DNA damage done by the toxins (Hughes et al. 2013b).

Such interventions look like they have the potential to promote direct benefit to both the pregnant woman and the fetus. The question then becomes what the level of risk is to both the woman and her fetus of the NAC. It is certainly true that smoking is a risky activity – but smoking is not part of the research. Rather, that is a background condition that sets the stage for the intervention, and therefore the risks of smoking do not factor into the risk/benefit calculus of the research. Instead, investigators must consider the risks and benefits of the intervention on its own merits in order for it to be approvable under the regulations – including, in this case, any differences that obtain to the dangers of first-hand compared with second-hand smoke (Kalkbrenner et al. 2014). Certainly, investigators would have to provide data about pregnant animal studies as well as additional information about clinical studies involving non-pregnant adults for the research to move forward. But the challenge really comes with the risk assessment. Assuring IRBs and other oversight groups that the intervention is "safe enough" to use in pregnancy will be an uphill battle for investigators. Consider that as of 2007, there were only 12 drugs approved for use in pregnant women, and 10 of them involved how to get the baby out (Lyerly et al. 2008)! The current risk-averse research climate rests the burden of proof on the investigators regarding safety, and the burden comes to a suspicious public. Most research involving pregnant women currently is observational in nature, with very few intervention studies being approved. And while there are both good scientific and ethical reasons to change this (Lyerly et al. 2008, 2009), there would have to be a sea change in the way that pregnant women and fetuses are viewed before this kind of research is likely to be able to move forward.

12.4 General Ethical Issues Related to the Development of Future Therapies

There are a few remaining ethical issues to discuss related to the development of future therapies. These are points for investigators to consider as they begin designing their trials.

12.4.1 Clinical vs. Biological Significance of Results

Many times in toxicological research, end points rest on biologically significant markers. Yet it is the case that not all biologically significant results will also be clinically significant. To the extent that biological significance must be established prior to clinical significance, then this makes sense. But it does create a particular challenge related to informed consent of participants. Suppose that an endpoint of a particular intervention is reduction of airway inflammation. Airway inflammation may not translate into anything that participants would notice (depending on the severity, etc.). Investigators, then, must be very careful to explain this distinction to participants in a way that they understand it. Otherwise, the risk of therapeutic misconception is great: participants may expect to receive clinical benefit from participation in the trial. One way to address this is to add surrogate endpoints like biomarkers onto other studies that are looking at clinical significance. This will give researchers access to data they may not otherwise have, while at the same time minimizing additional risk and burden to participants.

12.4.2 Social/Behavioral Interventions vs. Pharmaceutical/ Chemical Interventions

Some argue that social and behavioral interventions are preferable over chemical interventions because the "risk" is lower, as is the possibility for adverse effects. This is not always the case. Questionnaires are one thing; behavioral modification is a different beast entirely. Consider the woman who is asked to curtail or change her activities in a fundamental way during her pregnancy in order to reduce exposures to her fetus (see, for example, (Lyall et al. 2014)). Such behavioral changes can have significant costs financially, socially, and emotionally. Given that we as a society have not been particularly successful at getting the population to modify behavior to reduce the most common killer of Americans – cardiovascular disease – there is reason to suspect that there are burdens to behavior change not broadly considered by those groups who recommend such behavioral changes. Even for those populations who are particularly motivated to make a change, desire does not always equate to success.

12.4.3 Developmental Milestones in Context

There is good scientific reason to intervene at important development milestones to optimize the ability to address adverse health effects caused by toxins. However, it is important to remember that all prenatal fetal exposures entail that the pregnant woman is also exposed to an agent. In some cases, as with NAC for reversing the effects of smoking exposure, there may be benefit to both parties. But in cases where the benefit is solely or largely conferred on the fetus, investigators must consider the context in which this research will occur: through the woman's body. Her welfare is just as important as that of the fetus and must be treated as such. Therefore, "development milestone opportunities" must be considered as part of the overall strategy of research.

12.5 Conclusion

Translational toxicology holds promise for addressing the adverse health effects of environmental exposures. Indeed, investigating options directly with the groups of participants who stand to gain the most from interventions is both scientifically sound and morally laudable. Yet there are ethical and regulatory considerations that attach to research involving participants at several important developmental milestones. Attending to these issues in the early design stages of a research project can help to ensure that the research proceeds according to best practices in ethics and passes regulatory muster.

Appendix I: WMA Declaration of Helsinki – Ethical Principles for Medical Research Involving Human Subjects

Adopted by the 18th WMA General Assembly, Helsinki, Finland, June 1964 and amended by the:

29th WMA General Assembly, Tokyo, Japan, October 1975
35th WMA General Assembly, Venice, Italy, October 1983
41st WMA General Assembly, Hong Kong, September 1989
48th WMA General Assembly, Somerset West, Republic of South Africa, October 1996
52nd WMA General Assembly, Edinburgh, Scotland, October 2000
53rd WMA General Assembly, Washington DC, USA, October 2002 (Note of Clarification added)
55th WMA General Assembly, Tokyo, Japan, October 2004 (Note of Clarification added)

59th WMA General Assembly, Seoul, Republic of Korea, October 2008
64th WMA General Assembly, Fortaleza, Brazil, October 2013

Preamble

1. The World Medical Association (WMA) has developed the Declaration of Helsinki as a statement of ethical principles for medical research involving human subjects, including research on identifiable human material and data.

   The Declaration is intended to be read as a whole and each of its constituent paragraphs should be applied with consideration of all other relevant paragraphs.
2. Consistent with the mandate of the WMA, the Declaration is addressed primarily to physicians. The WMA encourages others who are involved in medical research involving human subjects to adopt these principles.

General Principles

3. The Declaration of Geneva of the WMA binds the physician with the words, "The health of my patient will be my first consideration," and the International Code of Medical Ethics declares that, "A physician shall act in the patient's best interest when providing medical care."
4. It is the duty of the physician to promote and safeguard the health, well-being and rights of patients, including those who are involved in medical research. The physician's knowledge and conscience are dedicated to the fulfilment of this duty.
5. Medical progress is based on research that ultimately must include studies involving human subjects.
6. The primary purpose of medical research involving human subjects is to understand the causes, development and effects of diseases and improve preventive, diagnostic and therapeutic interventions (methods, procedures and treatments). Even the best proven interventions must be evaluated continually through research for their safety, effectiveness, efficiency, accessibility and quality.
7. Medical research is subject to ethical standards that promote and ensure respect for all human subjects and protect their health and rights.
8. While the primary purpose of medical research is to generate new knowledge, this goal can never take precedence over the rights and interests of individual research subjects.
9. It is the duty of physicians who are involved in medical research to protect the life, health, dignity, integrity, right to self-determination, privacy, and confidentiality of personal information of research subjects. The responsibility for the protection of research subjects must always rest with the physician or other

health care professionals and never with the research subjects, even though they have given consent.

10. Physicians must consider the ethical, legal and regulatory norms and standards for research involving human subjects in their own countries as well as applicable international norms and standards. No national or international ethical, legal or regulatory requirement should reduce or eliminate any of the protections for research subjects set forth in this Declaration.
11. Medical research should be conducted in a manner that minimises possible harm to the environment.
12. Medical research involving human subjects must be conducted only by individuals with the appropriate ethics and scientific education, training and qualifications. Research on patients or healthy volunteers requires the supervision of a competent and appropriately qualified physician or other health care professional.
13. Groups that are underrepresented in medical research should be provided appropriate access to participation in research.
14. Physicians who combine medical research with medical care should involve their patients in research only to the extent that this is justified by its potential preventive, diagnostic or therapeutic value and if the physician has good reason to believe that participation in the research study will not adversely affect the health of the patients who serve as research subjects.
15. Appropriate compensation and treatment for subjects who are harmed as a result of participating in research must be ensured.

Risks, Burdens and Benefits

16. In medical practice and in medical research, most interventions involve risks and burdens.

    Medical research involving human subjects may only be conducted if the importance of the objective outweighs the risks and burdens to the research subjects.
17. All medical research involving human subjects must be preceded by careful assessment of predictable risks and burdens to the individuals and groups involved in the research in comparison with foreseeable benefits to them and to other individuals or groups affected by the condition under investigation.

    Measures to minimise the risks must be implemented. The risks must be continuously monitored, assessed and documented by the researcher.
18. Physicians may not be involved in a research study involving human subjects unless they are confident that the risks have been adequately assessed and can be satisfactorily managed.

When the risks are found to outweigh the potential benefits or when there is conclusive proof of definitive outcomes, physicians must assess whether to continue, modify or immediately stop the study.

Vulnerable Groups and Individuals

19. Some groups and individuals are particularly vulnerable and may have an increased likelihood of being wronged or of incurring additional harm.

    All vulnerable groups and individuals should receive specifically considered protection.
20. Medical research with a vulnerable group is only justified if the research is responsive to the health needs or priorities of this group and the research cannot be carried out in a non-vulnerable group. In addition, this group should stand to benefit from the knowledge, practices or interventions that result from the research.

Scientific Requirements and Research Protocols

21. Medical research involving human subjects must conform to generally accepted scientific principles, be based on a thorough knowledge of the scientific literature, other relevant sources of information, and adequate laboratory and, as appropriate, animal experimentation. The welfare of animals used for research must be respected.
22. The design and performance of each research study involving human subjects must be clearly described and justified in a research protocol.

The protocol should contain a statement of the ethical considerations involved and should indicate how the principles in this Declaration have been addressed. The protocol should include information regarding funding, sponsors, institutional affiliations, potential conflicts of interest, incentives for subjects and information regarding provisions for treating and/or compensating subjects who are harmed as a consequence of participation in the research study.

In clinical trials, the protocol must also describe appropriate arrangements for post-trial provisions.

Research Ethics Committees

23. The research protocol must be submitted for consideration, comment, guidance and approval to the concerned research ethics committee before the study begins. This committee must be transparent in its functioning, must be independent of the researcher, the sponsor and any other undue influence and must be duly qualified. It must take into consideration the laws and regulations of the country or countries in which the research is to be performed as well as applicable international norms and standards but these must not be allowed to reduce or eliminate any of the protections for research subjects set forth in this Declaration.

The committee must have the right to monitor ongoing studies. The researcher must provide monitoring information to the committee, especially information about any serious adverse events. No amendment to the protocol may be made without consideration and approval by the committee. After the end of the study, the researchers must submit a final report to the committee containing a summary of the study's findings and conclusions.

Privacy and Confidentiality

24. Every precaution must be taken to protect the privacy of research subjects and the confidentiality of their personal information.

Informed Consent

25. Participation by individuals capable of giving informed consent as subjects in medical research must be voluntary. Although it may be appropriate to consult family members or community leaders, no individual capable of giving informed consent may be enrolled in a research study unless he or she freely agrees.
26. In medical research involving human subjects capable of giving informed consent, each potential subject must be adequately informed of the aims, methods, sources of funding, any possible conflicts of interest, institutional affiliations of the researcher, the anticipated benefits and potential risks of the study and the discomfort it may entail, post-study provisions and any other relevant aspects of the study. The potential subject must be informed of the right to refuse to participate in the study or to withdraw consent to participate at any time without reprisal. Special attention should be given to the specific information needs of individual potential subjects as well as to the methods used to deliver the information.

After ensuring that the potential subject has understood the information, the physician or another appropriately qualified individual must then seek the potential subject's freely-given informed consent, preferably in writing. If the consent cannot be expressed in writing, the non-written consent must be formally documented and witnessed.

All medical research subjects should be given the option of being informed about the general outcome and results of the study.

27. When seeking informed consent for participation in a research study the physician must be particularly cautious if the potential subject is in a dependent relationship with the physician or may consent under duress. In such situations the informed consent must be sought by an appropriately qualified individual who is completely independent of this relationship.

28. For a potential research subject who is incapable of giving informed consent, the physician must seek informed consent from the legally authorised representative. These individuals must not be included in a research study that has no likelihood of benefit for them unless it is intended to promote the health of the group represented by the potential subject, the research cannot instead be performed with persons capable of providing informed consent, and the research entails only minimal risk and minimal burden.
29. When a potential research subject who is deemed incapable of giving informed consent is able to give assent to decisions about participation in research, the physician must seek that assent in addition to the consent of the legally authorised representative. The potential subject's dissent should be respected.
30. Research involving subjects who are physically or mentally incapable of giving consent, for example, unconscious patients, may be done only if the physical or mental condition that prevents giving informed consent is a necessary characteristic of the research group. In such circumstances the physician must seek informed consent from the legally authorised representative. If no such representative is available and if the research cannot be delayed, the study may proceed without informed consent provided that the specific reasons for involving subjects with a condition that renders them unable to give informed consent have been stated in the research protocol and the study has been approved by a research ethics committee. Consent to remain in the research must be obtained as soon as possible from the subject or a legally authorised representative.
31. The physician must fully inform the patient which aspects of their care are related to the research. The refusal of a patient to participate in a study or the patient's decision to withdraw from the study must never adversely affect the patient-physician relationship.
32. For medical research using identifiable human material or data, such as research on material or data contained in biobanks or similar repositories, physicians must seek informed consent for its collection, storage and/or reuse. There may be exceptional situations where consent would be impossible or impracticable to obtain for such research. In such situations the research may be done only after consideration and approval of a research ethics committee.

Use of Placebo

33. The benefits, risks, burdens and effectiveness of a new intervention must be tested against those of the best proven intervention(s), except in the following circumstances:

Where no proven intervention exists, the use of placebo, or no intervention, is acceptable; or

Where for compelling and scientifically sound methodological reasons the use of any intervention less effective than the best proven one, the use of placebo, or no intervention is necessary to determine the efficacy or safety of an intervention

and the patients who receive any intervention less effective than the best proven one, placebo, or no intervention will not be subject to additional risks of serious or irreversible harm as a result of not receiving the best proven intervention.

Extreme care must be taken to avoid abuse of this option.

Post-Trial Provisions

34. In advance of a clinical trial, sponsors, researchers and host country governments should make provisions for post-trial access for all participants who still need an intervention identified as beneficial in the trial. This information must also be disclosed to participants during the informed consent process.

Research Registration and Publication and Dissemination of Results

35. Every research study involving human subjects must be registered in a publicly accessible database before recruitment of the first subject.
36. Researchers, authors, sponsors, editors and publishers all have ethical obligations with regard to the publication and dissemination of the results of research. Researchers have a duty to make publicly available the results of their research on human subjects and are accountable for the completeness and accuracy of their reports. All parties should adhere to accepted guidelines for ethical reporting. Negative and inconclusive as well as positive results must be published or otherwise made publicly available. Sources of funding, institutional affiliations and conflicts of interest must be declared in the publication. Reports of research not in accordance with the principles of this Declaration should not be accepted for publication.

Unproven Interventions in Clinical Practice

37. In the treatment of an individual patient, where proven interventions do not exist or other known interventions have been ineffective, the physician, after seeking expert advice, with informed consent from the patient or a legally authorised representative, may use an unproven intervention if in the physician's judgement it offers hope of saving life, re-establishing health or alleviating suffering. This

intervention should subsequently be made the object of research, designed to evaluate its safety and efficacy. In all cases, new information must be recorded and, where appropriate, made publicly available.

Appendix II: Code of Federal Regulations

Subpart B	Additional Protections for Pregnant Women, Human Fetuses and Neonates Involved in Research
	Source:66 FR 56778, Nov. 13, 2001, unless otherwise noted.

§46.201 To what do these regulations apply?

(a) Except as provided in paragraph (b) of this section, this subpart applies to all research involving pregnant women, human fetuses, neonates of uncertain viability, or nonviable neonates conducted or supported by the Department of Health and Human Services (DHHS). This includes all research conducted in DHHS facilities by any person and all research conducted in any facility by DHHS employees.
(b) The exemptions at §46.101(b)(1) through (6) are applicable to this subpart.
(c) The provisions of §46.101(c) through (i) are applicable to this subpart. Reference to State or local laws in this subpart and in §46.101(f) is intended to include the laws of federally recognized American Indian and Alaska Native Tribal Governments.
(d) The requirements of this subpart are in addition to those imposed under the other subparts of this part.

§46.202 Definitions

The definitions in §46.102 shall be applicable to this subpart as well. In addition, as used in this subpart:

(a) Dead fetus means a fetus that exhibits neither heartbeat, spontaneous respiratory activity, spontaneous movement of voluntary muscles, nor pulsation of the umbilical cord.
(b) Delivery means complete separation of the fetus from the woman by expulsion or extraction or any other means.
(c) Fetus means the product of conception from implantation until delivery.
(d) Neonate means a newborn.
(e) Nonviable neonate means a neonate after delivery that, although living, is not viable.
(f) Pregnancy encompasses the period of time from implantation until delivery. A woman shall be assumed to be pregnant if she exhibits any of the pertinent presumptive signs of pregnancy, such as missed menses, until the results of a pregnancy test are negative or until delivery.

(g) Secretary means the Secretary of Health and Human Services and any other officer or employee of the Department of Health and Human Services to whom authority has been delegated.
(h) Viable, as it pertains to the neonate, means being able, after delivery, to survive (given the benefit of available medical therapy) to the point of independently maintaining heartbeat and respiration. The Secretary may from time to time, taking into account medical advances, publish in the FEDERAL REGISTER guidelines to assist in determining whether a neonate is viable for purposes of this subpart. If a neonate is viable then it may be included in research only to the extent permitted and in accordance with the requirements of subparts A and D of this part.

§46.203 Duties of IRBs in connection with research involving pregnant women, fetuses, and neonates.

In addition to other responsibilities assigned to IRBs under this part, each IRB shall review research covered by this subpart and approve only research which satisfies the conditions of all applicable sections of this subpart and the other subparts of this part.

§46.204 Research involving pregnant women or fetuses.

Pregnant women or fetuses may be involved in research if all of the following conditions are met:

(a) Where scientifically appropriate, preclinical studies, including studies on pregnant animals, and clinical studies, including studies on nonpregnant women, have been conducted and provide data for assessing potential risks to pregnant women and fetuses;
(b) The risk to the fetus is caused solely by interventions or procedures that hold out the prospect of direct benefit for the woman or the fetus; or, if there is no such prospect of benefit, the risk to the fetus is not greater than minimal and the purpose of the research is the development of important biomedical knowledge which cannot be obtained by any other means;
(c) Any risk is the least possible for achieving the objectives of the research;
(d) If the research holds out the prospect of direct benefit to the pregnant woman, the prospect of a direct benefit both to the pregnant woman and the fetus, or no prospect of benefit for the woman nor the fetus when risk to the fetus is not greater than minimal and the purpose of the research is the development of important biomedical knowledge that cannot be obtained by any other means, her consent is obtained in accord with the informed consent provisions of subpart A of this part;
(e) If the research holds out the prospect of direct benefit solely to the fetus then the consent of the pregnant woman and the father is obtained in accord with the informed consent provisions of subpart A of this part, except that the father's consent need not be obtained if he is unable to consent because of unavailability, incompetence, or temporary incapacity or the pregnancy resulted from rape or incest.

(f) Each individual providing consent under paragraph (d) or (e) of this section is fully informed regarding the reasonably foreseeable impact of the research on the fetus or neonate;
(g) For children as defined in §46.402(a) who are pregnant, assent and permission are obtained in accord with the provisions of subpart D of this part;
(h) No inducements, monetary or otherwise, will be offered to terminate a pregnancy;
(i) Individuals engaged in the research will have no part in any decisions as to the timing, method, or procedures used to terminate a pregnancy; and
(j) Individuals engaged in the research will have no part in determining the viability of a neonate.

§46.205 Research involving neonates.

(a) Neonates of uncertain viability and nonviable neonates may be involved in research if all of the following conditions are met:

  a. Where scientifically appropriate, preclinical and clinical studies have been conducted and provide data for assessing potential risks to neonates.
  b. Each individual providing consent under paragraph (b)(2) or (c)(5) of this section is fully informed regarding the reasonably foreseeable impact of the research on the neonate.
  c. Individuals engaged in the research will have no part in determining the viability of a neonate.
  d. The requirements of paragraph (b) or (c) of this section have been met as applicable.

(b) Neonates of uncertain viability. Until it has been ascertained whether or not a neonate is viable, a neonate may not be involved in research covered by this subpart unless the following additional conditions have been met:

  a. The IRB determines that:

    i. The research holds out the prospect of enhancing the probability of survival of the neonate to the point of viability, and any risk is the least possible for achieving that objective, or
    ii. The purpose of the research is the development of important biomedical knowledge which cannot be obtained by other means and there will be no added risk to the neonate resulting from the research; and
    iii. The legally effective informed consent of either parent of the neonate or, if neither parent is able to consent because of unavailability, incompetence, or temporary incapacity, the legally effective informed consent of either parent's legally authorized representative is obtained in accord with subpart A of this part, except that the consent of the father or his legally authorized representative need not be obtained if the pregnancy resulted from rape or incest.

(c) Nonviable neonates. After delivery nonviable neonate may not be involved in research covered by this subpart unless all of the following additional conditions are met:

a. Vital functions of the neonate will not be artificially maintained;
b. The research will not terminate the heartbeat or respiration of the neonate;
c. There will be no added risk to the neonate resulting from the research;
d. The purpose of the research is the development of important biomedical knowledge that cannot be obtained by other means; and
e. The legally effective informed consent of both parents of the neonate is obtained in accord with subpart A of this part, except that the waiver and alteration provisions of §46.116(c) and (d) do not apply. However, if either parent is unable to consent because of unavailability, incompetence, or temporary incapacity, the informed consent of one parent of a nonviable neonate will suffice to meet the requirements of this paragraph (c)(5), except that the consent of the father need not be obtained if the pregnancy resulted from rape or incest. The consent of a legally authorized representative of either or both of the parents of a nonviable neonate will not suffice to meet the requirements of this paragraph (c)(5).

(d) Viable neonates. A neonate, after delivery, that has been determined to be viable may be included in research only to the extent permitted by and in accord with the requirements of subparts A and D of this part.

§46.206 Research involving, after delivery, the placenta, the dead fetus or fetal material.

(a) Research involving, after delivery, the placenta; the dead fetus; macerated fetal material; or cells, tissue, or organs excised from a dead fetus, shall be conducted only in accord with any applicable federal, state, or local laws and regulations regarding such activities.
(b) If information associated with material described in paragraph (a) of this section is recorded for research purposes in a manner that living individuals can be identified, directly or through identifiers linked to those individuals, those individuals are research subjects and all pertinent subparts of this part are applicable.

§46.207 Research not otherwise approvable which presents an opportunity to understand, prevent, or alleviate a serious problem affecting the health or welfare of pregnant women, fetuses, or neonates.

(a) The Secretary will conduct or fund research that the IRB does not believe meets the requirements of §46.204 or §46.205 only if:
(b) The IRB finds that the research presents a reasonable opportunity to further the understanding, prevention, or alleviation of a serious problem affecting the health or welfare of pregnant women, fetuses or neonates; and

(c) The Secretary, after consultation with a panel of experts in pertinent disciplines (for example: science, medicine, ethics, law) and following opportunity for public review and comment, including a public meeting announced in the FEDERAL REGISTER, has determined either:

   a. That the research in fact satisfies the conditions of §46.204, as applicable; or
   b. The following:

      i. The research presents a reasonable opportunity to further the understanding, prevention, or alleviation of a serious problem affecting the health or welfare of pregnant women, fetuses or neonates;
      ii. The research will be conducted in accord with sound ethical principles; and
      iii. Informed consent will be obtained in accord with the informed consent provisions of subpart A and other applicable subparts of this part.

Subpart D	Additional Protections for Children Involved as Subjects in Research
	Source:48 FR 9818, March 8, 1983, unless otherwise noted.

§46.401 To what do these regulations apply?

(a) This subpart applies to all research involving children as subjects, conducted or supported by the Department of Health and Human Services.

   a. This includes research conducted by Department employees, except that each head of an Operating Division of the Department may adopt such nonsubstantive, procedural modifications as may be appropriate from an administrative standpoint.
   b. It also includes research conducted or supported by the Department of Health and Human Services outside the United States, but in appropriate circumstances, the Secretary may, under paragraph (e) of §46.101 of subpart A, waive the applicability of some or all of the requirements of these regulations for research of this type.

(b) Exemptions at §46.101(b)(1) and (b)(3) through (b)(6) are applicable to this subpart. The exemption at §46.101(b)(2) regarding educational tests is also applicable to this subpart. However, the exemption at §46.101(b)(2) for research involving survey or interview procedures or observations of public behavior does not apply to research covered by this subpart, except for research involving observation of public behavior when the investigator(s) do not participate in the activities being observed.
(c) The exceptions, additions, and provisions for waiver as they appear in paragraphs (c) through (i) of §46.101of subpart A are applicable to this subpart.

[48 FR 9818, Mar.8, 1983; 56 FR 28032, June 18, 1991; 56 FR 29757, June 28, 1991.]

§46.402 Definitions.

The definitions in §46.102 of subpart A shall be applicable to this subpart as well. In addition, as used in this subpart:

(a) *Children* are persons who have not attained the legal age for consent to treatments or procedures involved in the research, under the applicable law of the jurisdiction in which the research will be conducted.
(b) *Assent* means a child's affirmative agreement to participate in research. Mere failure to object should not, absent affirmative agreement, be construed as assent.
(c) *Permission* means the agreement of parent(s) or guardian to the participation of their child or ward in research.
(d) *Parent* means a child's biological or adoptive parent.
(e) *Guardian* means an individual who is authorized under applicable State or local law to consent on behalf of a child to general medical care.

§46.403 IRB duties.

In addition to other responsibilities assigned to IRBs under this part, each IRB shall review research covered by this subpart and approve only research which satisfies the conditions of all applicable sections of this subpart.

§46.404 Research not involving greater than minimal risk.

HHS will conduct or fund research in which the IRB finds that no greater than minimal risk to children is presented, only if the IRB finds that adequate provisions are made for soliciting the assent of the children and the permission of their parents or guardians, as set forth in §46.408.

§46.405 Research involving greater than minimal risk but presenting the prospect of direct benefit to the individual subjects.

HHS will conduct or fund research in which the IRB finds that more than minimal risk to children is presented by an intervention or procedure that holds out the prospect of direct benefit for the individual subject, or by a monitoring procedure that is likely to contribute to the subject's well-being, only if the IRB finds that:

(a) The risk is justified by the anticipated benefit to the subjects;
(b) The relation of the anticipated benefit to the risk is at least as favorable to the subjects as that presented by available alternative approaches; and
(c) Adequate provisions are made for soliciting the assent of the children and permission of their parents or guardians, as set forth in §46.408.

§46.406 Research involving greater than minimal risk and no prospect of direct benefit to individual subjects, but likely to yield generalizable knowledge about the subject's disorder or condition.

HHS will conduct or fund research in which the IRB finds that more than minimal risk to children is presented by an intervention or procedure that does not hold out the prospect of direct benefit for the individual subject, or by a monitoring procedure which is not likely to contribute to the well-being of the subject, only if the IRB finds that:

(a) The risk represents a minor increase over minimal risk;
(b) The intervention or procedure presents experiences to subjects that are reasonably commensurate with those inherent in their actual or expected medical, dental, psychological, social, or educational situations;
(c) The intervention or procedure is likely to yield generalizable knowledge about the subjects' disorder or condition which is of vital importance for the understanding or amelioration of the subjects' disorder or condition; and
(d) Adequate provisions are made for soliciting assent of the children and permission of their parents or guardians, as set forth in §46.408.

§46.407 Research not otherwise approvable which presents an opportunity to understand, prevent, or alleviate a serious problem affecting the health or welfare of children.

HHS will conduct or fund research that the IRB does not believe meets the requirements of §46.404, §46.405, or §46.406 only if:

(a) the IRB finds that the research presents a reasonable opportunity to further the understanding, prevention, or alleviation of a serious problem affecting the health or welfare of children; and
(b) the Secretary, after consultation with a panel of experts in pertinent disciplines (for example: science, medicine, education, ethics, law) and following opportunity for public review and comment, has determined either:

    a. that the research in fact satisfies the conditions of §46.404, §46.405, or §46.406, as applicable, or (2) the following:

        i. the research presents a reasonable opportunity to further the understanding, prevention, or alleviation of a serious problem affecting the health or welfare of children;
        ii. the research will be conducted in accordance with sound ethical principles;
        iii. adequate provisions are made for soliciting the assent of children and the permission of their parents or guardians, as set forth in §46.408.

§46.408 Requirements for permission by parents or guardians and for assent by children.

(a) In addition to the determinations required under other applicable sections of this subpart, the IRB shall determine that adequate provisions are made for soliciting the assent of the children, when in the judgment of the IRB the children are capable of providing assent. In determining whether children are capable of assenting, the IRB shall take into account the ages, maturity, and psychological state of the children involved. This judgment may be made for all children to be involved in research under a particular protocol, or for each child, as the IRB deems appropriate. If the IRB determines that the capability of some or all of the children is so limited that they cannot reasonably be consulted or that the intervention or procedure involved in the research holds out a prospect of direct benefit that is important to the health or well-being of the children and is available only in the context of the research, the assent of the children is not a necessary condition for proceeding with the research. Even where the IRB determines that the subjects are capable of assenting, the IRB may still waive the assent requirement under circumstances in which consent may be waived in accord with §46.116 of Subpart A.

(b) In addition to the determinations required under other applicable sections of this subpart, the IRB shall determine, in accordance with and to the extent that consent is required by §46.116 of Subpart A, that adequate provisions are made for soliciting the permission of each child's parents or guardian. Where parental permission is to be obtained, the IRB may find that the permission of one parent is sufficient for research to be conducted under §46.404 or §46.405. Where research is covered by §§46.406 and 46.407 and permission is to be obtained from parents, both parents must give their permission unless one parent is deceased, unknown, incompetent, or not reasonably available, or when only one parent has legal responsibility for the care and custody of the child.

(c) In addition to the provisions for waiver contained in §46.116 of subpart A, if the IRB determines that a research protocol is designed for conditions or for a subject population for which parental or guardian permission is not a reasonable requirement to protect the subjects (for example, neglected or abused children), it may waive the consent requirements in Subpart A of this part and paragraph (b) of this section, provided an appropriate mechanism for protecting the children who will participate as subjects in the research is substituted, and provided further that the waiver is not inconsistent with federal, state, or local law. The choice of an appropriate mechanism would depend upon the nature and purpose of the activities described in the protocol, the risk and anticipated benefit to the research subjects, and their age, maturity, status, and condition.

(d) Permission by parents or guardians shall be documented in accordance with and to the extent required by §46.117 of subpart A.

(e) When the IRB determines that assent is required, it shall also determine whether and how assent must be documented.

§46.409 Wards.

(a) Children who are wards of the state or any other agency, institution, or entity can be included in research approved under §46.406 or §46.407 only if such research is:

   a. Related to their status as wards; or
   b. Conducted in schools, camps, hospitals, institutions, or similar settings in which the majority of children involved as subjects are not wards.

(b) If the research is approved under paragraph (a) of this section, the IRB shall require appointment of an advocate for each child who is a ward, in addition to any other individual acting on behalf of the child as guardian or in loco parentis. One individual may serve as advocate for more than one child. The advocate shall be an individual who has the background and experience to act in, and agrees to act in, the best interests of the child for the duration of the child's participation in the research and who is not associated in any way (except in the role as advocate or member of the IRB) with the research, the investigator(s), or the guardian organization.

References

Abdel-Rahman A, Anyangwe N, Carlacci L, Casper S, Danam RP, Enongene E, Erives G, Fabricant D, Gudi R, Hilmas CJ, Hines F, Howard P, Levy D, Lin Y, Moore RJ, Pfeiler E, Thurmond TS, Turujman S, Walker NJ (2011) The safety and regulation of natural products used as foods and food ingredients. Toxicol Sci 123:333–348

Adzick NS (2010) Fetal myelomeningocele: natural history, pathophysiology, and in-utero intervention. Semin Fetal Neonatal Med 15:9–14

Andrade SE, Gurwitz JH, Davis RL, Chan KA, Finkelstein JA, Fortman K, Mcphillips H, Raebel MA, Roblin D, Smith DH, Yood MU, Morse AN, Platt R (2004) Prescription drug use in pregnancy. Am J Obstet Gynecol 191:398–407

Cohen AR, Couto J, Cummings JJ, Johnson A, Joseph G, Kaufman BA, Litman RS, Menard MK, Moldenhauer JS, Pringle KC, Schwartz MZ, Walker WO, Jr. Warf BC, Wax JR (2014) Position statement on fetal myelomeningocele repair. Am J Obstet Gynecol 210:107–111

Council, for International Organizations of Medical Sciences(CIOMS), C. F. I. O. O. M. S (2002) International ethical guidelines for biomedical research involving human subjects. World Medical Association, Geneva

Daw JR, Mintzes B, Law MR, Hanley GE, Morgan SG (2012) Prescription drug use in pregnancy: a retrospective, population-based study in British Columbia, Canada (2001–2006). Clin Ther 34:239–249.e2

Diekema DS (2006) Conducting ethical research in pediatrics: a brief historical overview and review of pediatric regulations. J Pediatr 149:S3–S11

Emanuel EJ, Wendler D, Grady C (2000) What makes clinical research ethical? JAMA 283:2701–2711

Emanuel E, Wendler D, Grady C (2011) An ethical framework for biomedical research. In: Emanuel EJ, Grady C, Crouch RA, Lie RK, Miller FG, Wendler D (eds) The oxford textbook of clinical research ethics. Oxford University Press, New York

Frattarelli DA, Galinkin JL, Green TP, Johnson TD, Neville KA, Paul IM, van den Anker JN (2014) Off-label use of drugs in children. Pediatrics 133:563–567

Friedman JM (2012) ABCDXXX: the obscenity of postmarketing surveillance for teratogenic effects. Birth Defects Res A Clin Mol Teratol 94:670–676

Hughes C, Waters M, Allen D, Obasanjo I (2013a) Translational toxicology: a developmental focus for integrated research strategies. BMC Pharmacol Toxicol 14:51

Hughes C, Waters M, Obasanjo I, Allen D (2013b) Translational developmental toxicology: prospects for protective therapeutic obstetrical and neonatal interventions. J Neonatal Biol 2:8

Kalkbrenner AE, Schmidt RJ, Penlesky AC (2014) Environmental chemical exposures and autism spectrum disorders: a review of the epidemiological evidence. Curr Probl Pediatr Adolesc Health Care 44:277–318

Kim MW, Ahn KH, Ryu KJ, Hong SC, Lee JS, Nava-Ocampo AA, Oh MJ, Kim HJ (2014) Preventive effects of folic acid supplementation on adverse maternal and fetal outcomes. PLoS One 9:e97273

Kipnis K (2003) Seven vulnerabilities in the pediatric research subject. Theor Med Bioeth 24:107–120
Lyall K, Schmidt RJ, Hertz-Picciotto I (2014) Maternal lifestyle and environmental risk factors for autism spectrum disorders. Int J Epidemiol 43:443–464
Lyerly AD, Little MO, Faden R (2008) The second wave: toward responsible inclusion of pregnant women in research. Int J Fem Approaches Bioeth 1:5–22
Lyerly AD, Mitchell LM, Armstrong EM, Harris LH, Kukla R, Kuppermann M, Little MO (2009) Risk and the pregnant body. Hastings Cent Rep 39:34–42
Macklin R (2010) Enrolling pregnant women in biomedical research. Lancet 375:632–633
Merton V (1993) The exclusion of pregnant, pregnable, and once-pregnable people (a.k.a. women) from biomedical research. Am J Law Med 19:369–451
Neltner TG, Alger HM, O'Reilly JT, Krimsky S, Bero LA, Maffini MV (2013) Conflicts of interest in approvals of additives to food determined to be generally recognized as safe: out of balance. JAMA Int Med 173:2032–2036
Palmaro A, Bissuel R, Renaud N, Durrieu G, Escourrou B, Oustric S, Montastruc JL, Lapeyre-Mestre M (2014) Off-label prescribing in pediatric outpatients. Pediatrics 135:49–58
Rhrissorrakrai K, Belcastro V, Bilal E, Norel R, Poussin C, Mathis C, Dulize RH, Ivanov NV, Alexopoulos L, Rice JJ, Peitsch M, Stolovitzky G, Meyer P, Hoeng J (2014) Understanding the limits of animal models as predictors of human biology: lessons learned from the sbv IMPROVER Species Translation Challenge. Bioinformatics 31:471–483
Sage DP, Kulczar C, Roth W, Liu W, Knipp GT (2014) Persistent pharmacokinetic challenges to pediatric drug development. Front Genet 5:281
Sankar P, Mora S, Merz JF, Jones NL (2003) Patient perspectives of medical confidentiality: a review of the literature. J Gen Intern Med 18:659–669
Schonfeld T (2013) The perils of protection: vulnerability and women in clinical research. Theor Med Bioeth 34:189–206
Shields KE, Lyerly AD (2013) Exclusion of pregnant women from industry-sponsored clinical trials. Obstet Gynecol 122:1077–1081
Weijer C, Emanuel EJ (2000) Ethics. Protecting communities in biomedical research. Science 289:1142–1144
World Medical Association (2013) Declaration of Helsinki: ethical principles for medical research involving human subjects. Available at: http://www.wma.net/en/30publications/10policies/b3/. Accessed 2/3/2015
Yang T, Walker MC, Krewski D, Yang Q, Nimrod C, Garner P, Fraser W, Olatunbosun O, Wen SW (2008) Maternal characteristics associated with pregnancy exposure to FDA category C, D, and X drugs in a Canadian population. Pharmacoepidemiol Drug Saf 17:270–277

Index

C.L. Hughes, M.D. Waters (eds.), *Translational Toxicology*, Molecular and Integrative Toxicology, DOI 10.1007/978-3-319-27449-2

MIX
Papier aus verantwortungsvollen Quellen
Paper from responsible sources
FSC® C105338

If you have any concerns about our products,
you can contact us on
ProductSafety@springernature.com

In case Publisher is established outside the EU,
the EU authorized representative is:
Springer Nature Customer Service Center GmbH
Europaplatz 3, 69115 Heidelberg, Germany

Printed by Libri Plureos GmbH
in Hamburg, Germany